5 Modern Mathematics for Schools

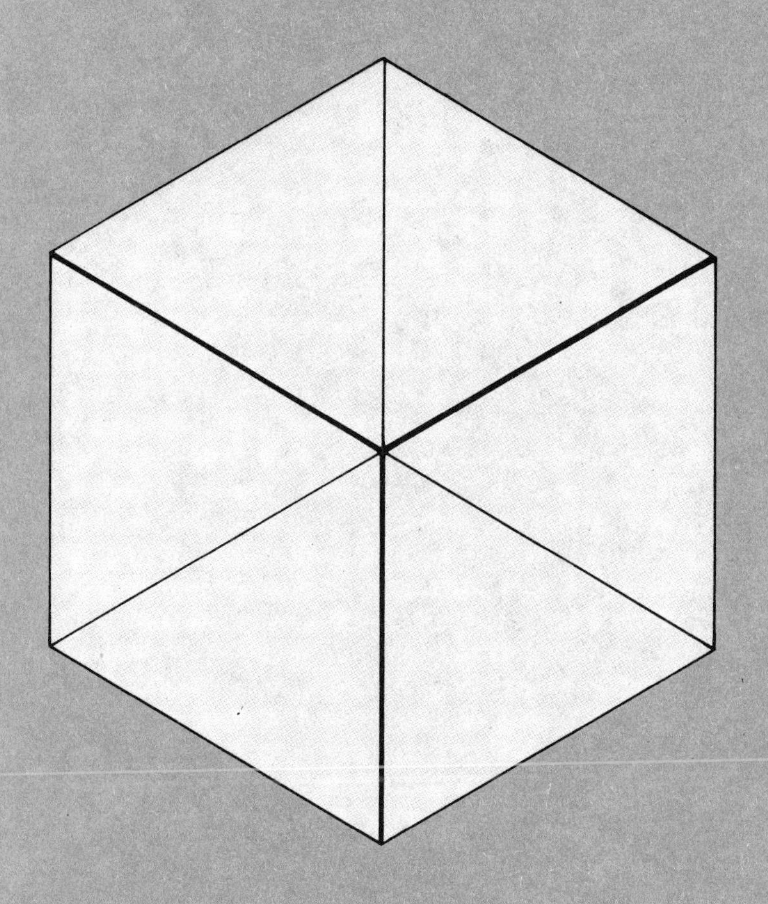

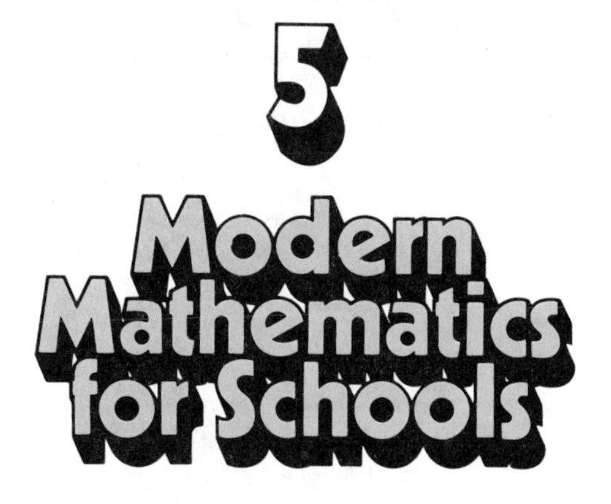

Second Edition

Scottish Mathematics Group

Blackie

Chambers

Blackie & Son Limited
Bishopbriggs · Glasgow G64 3NZ
Furnival House · 14-18 High Holborn
London WC1V 6BX

W & R Chambers Limited
11 Thistle Street · Edinburgh EH2 1DG

First published 1973
Reprinted 1982

Designed by James W. Murray

International Standard Book Numbers
Pupils' Book
Blackie 0 216 89416 6
Chambers 0 550 75915 8
Teachers' Book
Blackie 0 216 89417 4
Chambers 0 550 75925 5

Printed in Great Britain by
Thomson Litho Ltd, East Kilbride, Scotland
Set in 10pt Monotype Times Roman

Members associated with this book

W. T. Blackburn
Dundee College of Education

W. Brodie
Trinity Academy

C. Clark
Formerly of Lenzie Academy

D. Donald
Formerly of Robert Gordon's College

R. A. Finlayson
Allan Glen's School

Elizabeth K. Henderson
Westbourne School for Girls

J. L. Hodge
Dundee College of Education

J. Hunter
University of Glasgow

R. McKendrick
Langside College

W. More
Formerly of High School of Dundee

Helen C. Murdoch
Hutchesons' Girls' Grammar School

A. G. Robertson
John Neilson High School

A. G. Sillitto
Formerly of Jordanhill College of Education

A. A. Sturrock
Grove Academy

Rev. J. Taylor
St. Aloysius' College

E. B. C. Thornton
Bishop Otter College

J. A. Walker
Dollar Academy

P. Whyte
Hutchesons' Boys' Grammar School

H. S. Wylie
Govan High School

Contributor of the Computer Studies

A. W. McMeeken
Dundee College of Education

Preface

Book 1 of the original series *Modern Mathematics for Schools* was first published in July 1965. This revised series has been produced in order to take advantage of the experience gained in the classroom with the original textbooks and to reflect the changing mathematical needs in recent years, particularly as a result of the general move towards some form of comprehensive education.

Throughout the whole series, the text and exercises have been cut or augmented wherever this was considered to be necessary, and nearly every chapter has been completely rewritten. In order to cater more adequately for the wider range of pupils now taking certificate-oriented courses, the pace has been slowed down in the earlier books in particular, and parallel sets of A and B exercises have been introduced where appropriate. The A sets are easier than the B sets, and provide straightforward but comprehensive practice; the B sets have been designed for the more able pupils, and may be taken in addition to, or instead of, the A sets. From Book 4 onwards a basic exercise, which should be taken by all pupils, is sometimes followed by a harder one on the same work in order to give abler pupils an extra challenge, or further practice; in such a case the numbering is, for example, Exercise 2 followed by Exercise 2B. It is hoped that this arrangement, along with the *Graph Workbook for Modern Mathematics,* will allow considerable flexibility of use, so that while all the pupils in a class may be studying the same topic, each pupil may be working examples which are appropriate to his or her aptitude and ability.

In 1974 the first of a series of pads of expendable *Mathsheets* was published in order to provide simpler and more practical material;

each pad contains a worksheet for every exercise in the corresponding textbook. This increases still further the flexibility of the *Modern Mathematics for Schools* 'package' by providing three closely related levels of work—mathsheets, A exercises and B exercises. *Mathsheets* to accompany Books 1, 2 and 3 are now available.

Each chapter is backed up by a summary, and by revision exercises; in addition, cumulative revision exercises have been introduced at the end of alternate books. A new feature is the series of Computer Topics from Book 4 to Book 7. These form an elementary introduction to computer studies, and are primarily intended to give pupils some appreciation of the applications and influence of computers in modern society.

Books 1 to 7 provide a suitable course for many modern Ordinary Level and Ordinary Grade syllabuses in mathematics, including the University of London GCE Syllabus C, the Associated Examining Board Syllabus Mathematics 105, the Cambridge Local Syndicate Syllabus C, and the Scottish Certificate of Education. Books 8 and 9 complete the work for the Scottish Higher Grade Syllabus, and provide a good preparation for all Advanced Level and Sixth Year Syllabuses, both new and traditional.

Related to this revised series of textbooks are the *Modern Mathematics Newsletters* (No. 5, February 1975), the *Teacher's Editions* of the textbooks, the *Graph Workbook for Modern Mathematics,* the *Three-Figure Tables for Modern Mathematics,* and the booklets of *Progress Papers for Modern Mathematics.* These Progress Papers consist of short, quickly marked objective tests closely connected with the textbooks. There is one booklet for each textbook, containing A and B tests on each chapter, so that teachers can readily assess their pupils' attainments, and pupils can be encouraged in their progress through the course.

The separate headings of Algebra, Geometry, Arithmetic, and later Trigonometry and Calculus, have been retained in order to allow teachers to develop the course in the way they consider best. Throughout, however, ideas, material and methods are integrated *within* each branch of mathematics and *across* the branches; the opportunity to do this is indeed one of the more obvious reasons for teaching this kind of mathematics in the schools—for it is *mathematics* as a whole that is presented.

Pupils are encouraged to find out facts and discover results for themselves, to observe and study the themes and patterns that pervade mathematics today. As a course based on this series of books progresses, a certain amount of equipment will be helpful, particularly in the development of geometry. The use of calculating machines, slide rules, and computers is advocated where appropriate, but these instruments are not an essential feature of the work.

While fundamental principles are emphasized, and reasonable attention is paid to the matter of structure, the width of the course should be sufficient to provide a useful experience of mathematics for those pupils who do not pursue the study of the subject beyond school level. An effort has been made throughout to arouse the interest of all pupils and at the same time to keep in mind the needs of the future mathematician.

The introduction of mathematics in the Primary School and recent changes in courses at Colleges and Universities have been taken into account. In addition, the aims, methods, and writing of these books have been influenced by national and international discussions about the purpose and content of courses in mathematics, held under the auspices of the Organization for Economic Co-operation and Development and other organizations.

The authors wish to express their gratitude to the many teachers who have offered suggestions and criticisms concerning the original series of textbooks; they are confident that as a result of these contacts the new series will be more useful than it would otherwise have been.

Contents

Algebra

Geometry

Arithmetic

Trigonometry

Computer Studies

Notation

Sets of numbers

Different countries and different authors give different notations and definitions for the various sets of numbers. In this series the following are used:

E The universal set

ϕ The empty set

N The set of natural numbers {1, 2, 3, ...}

W The set of whole numbers {0, 1, 2, 3, ...}

Z The set of integers {..., – 2, – 1, 0, 1, 2, ...}

Q The set of rational numbers

R The set of real numbers

The set of prime numbers {2, 3, 5, 7, 11, ...}

Algebra

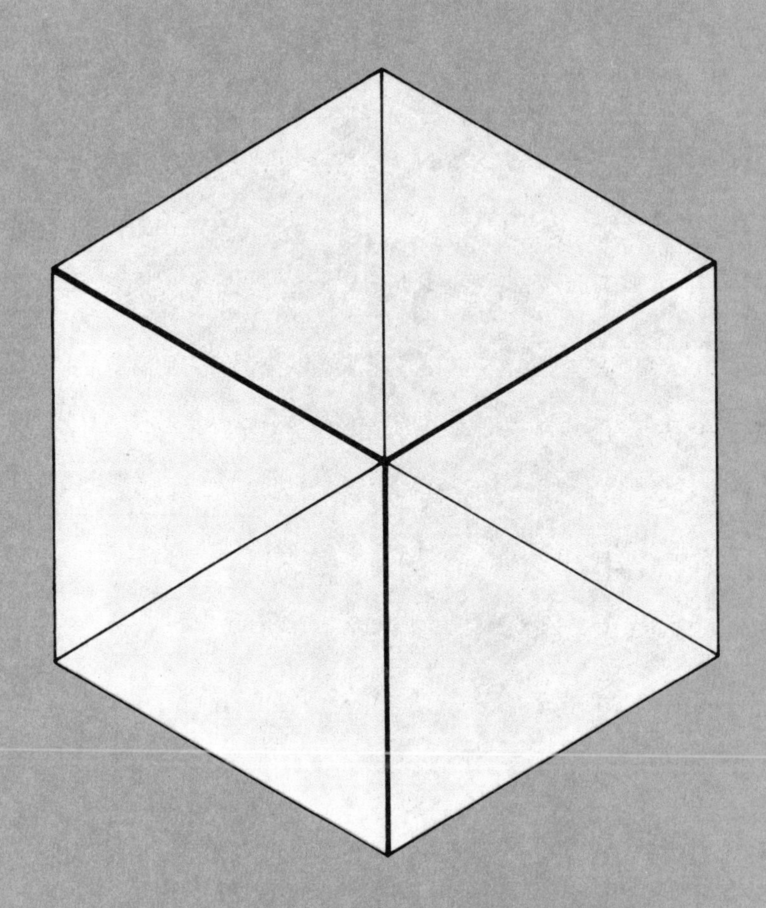

Algebra

Sums, Products and Squares

1 Addition of like terms and expressions

We have already met expressions like $7a-12a$ and $15y-4-5y+7$; we now study more elaborate expressions and see how they can be simplified. Simplification consists of finding an expression that is equal to the given expression but simpler than it.

Example 1. Simplify $3x^3+4x^2+x^3-2x^2$.

$$\begin{aligned} &3x^3+4x^2+x^3-2x^2 \\ =\ &3x^3+x^3+4x^2-2x^2 \qquad \text{(commutative law)} \\ =\ &(3+1)x^3+(4-2)x^2 \qquad \text{(distributive law)} \\ =\ &4x^3+2x^2 \end{aligned}$$

With practice, some of the steps may be omitted.

Example 2. Find the sum of $4x^2-3xy-2y^2$ and $-7x^2+5xy-8y^2$.

$$\begin{aligned} &4x^2-3xy-2y^2+(-7x^2+5xy-8y^2) \\ =\ &4x^2-3xy-2y^2-7x^2+5xy-8y^2 \\ =\ &4x^2-7x^2-3xy+5xy-2y^2-8y^2 \\ =\ &-3x^2+2xy-10y^2 \end{aligned}$$

Note how *like* terms have been grouped together in the second-last line so that the distributive law can be more easily applied. With practice, however, this line may be omitted, the work being done mentally.

Exercise 1

Simplify the following:

1 $6a+4a$ 2 $6a-4a$ 3 $-6a+4a$ 4 $-6a-4a$

5 $5x^2+4x^2$ 6 $4x^2-x^2$ 7 $3x^2+3x^2$ 8 x^2-x^2

9 $3ab-2ab$ 10 $8a^3+3a^3$ 11 $xy+yx$ 12 $3pq-qp$

13 $4x+3y-2x+y$ 14 $2a-2b-2a+2b$

15 $5x^2+x+2x^2-x$ 16 $3y^2-4y^2+7y-y$

17 $k^3+2k^3+5k-8k$

18 $5p^3-4p-3p^3+6p$

19 $x^2-xy+3x^2-2xy$

20 $4a^2-4ab-a^2-ab$

21 $2(x+y)+3(x-y)$

22 $3(a-b)+2(a+b)$

23 $5(x-2)+2(x+1)$

24 $4(y+3)+3(y-4)$

25 $2(x^2+1)+4(x^2-1)$

26 $5(y^2+y)+3(y^2-y)$

Find the sum of:

27 $5x+7$ and $3x+4$

28 $3a+7$ and $4a-7$

29 x^2+10 and $-x^2+10$

30 y^2+2y and y^2+y

31 $2a+3b$ and $4a-3b$

32 $6x-6y$ and $6x-6y$

33 $a+b+2$ and $a-b-1$

34 $2p+3q+4$ and $p-q-4$

35 x^2+2x+1 and x^2+x+3

36 $2x^2-x+3$ and x^2-2x-3

37 a^2+a+1 and $-a^2-2a-3$

38 b^2+2b+4 and $-3b^2-2b+4$

39 $a+3b+5c$ and $a-2b-2c$

40 $7p-4q-r$ and $-p-q-r$

2 *Subtraction of expressions*

In this Section we use the distributive law again:

$$a(b+c) = ab+ac\text{; also } a(b-c) = ab-ac$$

And we must remember that:
the product of a positive number and a negative number is negative;
the product of two negative numbers is positive.

Example 1

(i) $2(b+c)$
$= 2b+2c$

(ii) $3(b-c)$
$= 3b-3c$

(iii) $-4(x+y)$
$= -4x-4y$

(iv) $-(x-y)$
$= -1(x-y)$
$= -x+y$

Example 2. Subtract $3x-4$ from $2x+5$.

$$\begin{aligned} & 2x+5-(3x-4) \\ &= 2x+5-1(3x-4) \\ &= 2x+5-3x+4 \\ &= -x+9 \end{aligned}$$

Note. We have seen that subtracting a number b from a number a is the same as *adding the inverse* of b to a, i.e. $a-b = a+(-b)$.

If we apply this principle in *Example 2* we obtain:

$$\begin{aligned} &2x+5-(3x-4) \\ =\ &2x+5+(-3x+4) \\ =\ &2x+5-3x+4 \\ =\ &-x+9 \end{aligned}$$

Exercise 2

Simplify:

1 $-(2x-3)$
2 $-(3x+1)$
3 $-(x-1)$
4 $-(2-3a)$
5 $-(4+b)$
6 $-(-2c+3)$
7 $-(q-p)$
8 $-(-a-b+c)$
9 $-(2a-3b+4c)$
10 $5x-(x-2)$
11 $4x-(x+1)$
12 $3x-(x-3)$
13 $6-(2x-3)$
14 $4-(3x+2)$
15 $2-(4x+2)$

Subtract:

16 $2x+4$ from $3x+7$
17 $5y+3$ from $10y-9$
18 k^2-3 from $2k^2+1$
19 $5b^2-1$ from $4b^2+1$
20 $-x^2+x$ from $3x^2+5x$
21 $3a^2-4a$ from $6a^2-7a$
22 $a+b+3$ from $3a+4b-3$
23 $3x-3y-1$ from $x-y+5$
24 x^2+x+1 from x^2-x+1
25 a^2+2a-3 from $3a^2-2a+4$
26 $2(x+3)$ from $3(x-1)$
27 $5(y-2)$ from $4(y+1)$
28 $6(x^2-2)$ from $4(2x^2-3)$
29 $8(1-2y^2)$ from $5(2-3y^2)$

Exercise 2B

Simplify:

1 $4a^2-5ab-2ab-a^2$
2 a^2+a+a^2-a
3 $1{\cdot}3p-2{\cdot}7q+3{\cdot}7p+1{\cdot}7q$
4 $\frac{1}{2}t^2-2t+\frac{1}{3}-\frac{1}{2}t^2-2t+\frac{2}{3}$
5 $(2x^3+5x^2-7)+(3x^3-8x^2+5)$
6 $(\frac{1}{3}a^2+\frac{1}{4}a-2)+(\frac{5}{3}a^2+\frac{3}{4}a+2)$
7 $(-2u-3v+4w)+(-4w-5v-2u)$

8 $7i-2j+3k-(2i+k-2j)$

9 $a^2-2a-5-(2a^2+a-3)$

10 $2(2x+3y)-3(x-4y)+2(x+y+3)$

11 $a(b+c)-b(c+a)-c(a+b)$

12 $a^2+\frac{1}{3}a-2-(\frac{1}{3}a^2+\frac{1}{3}a+2)$

13 $p(q-r)+q(r-p)+r(p-q)$

14 $3x(x-3)-2x(x-1)+x(2x-2)$

Note. In Exercises 1, 2 and 2B operations have been carried out on expressions like $3x^2-4x+5$ or $3x^2y+2xy^2+4y-7$. Expressions like these are called *polynomials* in one and two variables respectively. Polynomials with two terms, as for example, $3x-5$, are sometimes called *binomials* and those with three terms, such as $3x^2-4x+1$, are called *trinomials.*

3 Equations and inequations

Examples. Solve the following, where x is a variable on the set of real numbers.

1

$$\begin{aligned} & 5(x-4)+3(1-x) = 10 \\ \Leftrightarrow\ & 5x-20+3-3x = 10 \\ \Leftrightarrow\ & 5x-3x = 10+17 \\ \Leftrightarrow\ & 2x = 27 \\ \Leftrightarrow\ & x = \tfrac{27}{2} \end{aligned}$$

The solution set is $\{\frac{27}{2}\}$

2

$$\begin{aligned} & 3x-2(3x-1) > 7 \\ \Leftrightarrow\ & 3x-6x+2 > 7 \\ \Leftrightarrow\ & -3x > 5 \\ \Leftrightarrow\ & (-\tfrac{1}{3})\times -3x < (-\tfrac{1}{3})\times 5 \\ \Leftrightarrow\ & x < -\tfrac{5}{3} \end{aligned}$$

The solution set is $\{x : x < -\frac{5}{3}, x \in R\}$

Exercise 3

Find the solution sets of the following equations and inequations, where x is a variable on the set of real numbers:

1 $2(x+3)-1 = 7$

2 $5(x-1)+2 > 12$

3 $4x-3(x+1) = 0$

4 $6x-2(x-1) < 10$

5 $6-(x-2) = 3x$

6 $10-(2x+2) \geqslant -4x$

7 $3(x+2)+2(x-1) = 8$

8 $4(x-3)+5(x+1) > 0$

9 $6(x+2)-3(x-4) = -6$

10 $8(x-1)-2(x+5) \leqslant 12$

11 $4x-2(1-3x)=3$

12 $x-4(2x+1)>10$

13 $5(2x-3)-2(6x-5)=3(3x-4)$

14 $2(5x-1)-3(2x-1)\leqslant 4(1+3x)$

15 $10(1-x)=5(2x-1)+3(3-4x)$

4 *Expressing a product of factors as a sum of terms*

As explained in the note opposite, expressions with two terms, such as $x+2$ or $x+3$ are called *binomials*. We can find the product of two binomials by using the distributive law as follows:

$$\begin{aligned} &(x+2)(x+3)\\ &= x(x+3)+2(x+3)\\ &= x^2+3x+2x+6\\ &= x^2+5x+6 \end{aligned}$$

Also, by drawing a rectangle $(x+3)$ units long and $(x+2)$ units broad, and by dividing up the rectangle as shown in Figure 1, the result can be illustrated diagrammatically.

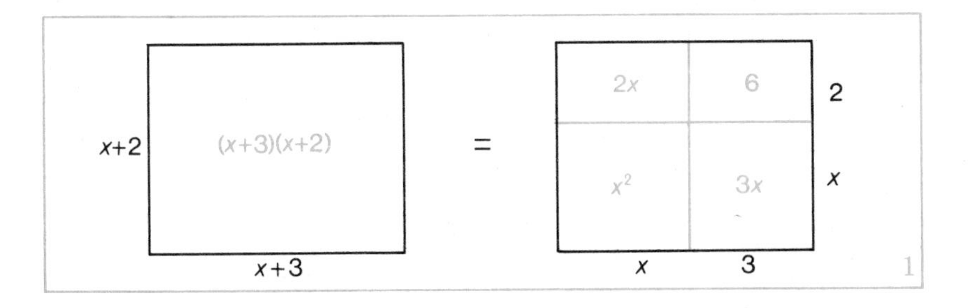

1

Example 1. $(x-2)(x+5)$
$$\begin{aligned} &= x(x+5)-2(x+5)\\ &= x^2+5x-2x-10\\ &= x^2+3x-10 \end{aligned}$$

Example 2. $(2x-4)(3x-7)$
$$\begin{aligned} &= 2x(3x-7)-4(3x-7)\\ &= 6x^2-14x-12x+28\\ &= 6x^2-26x+28 \end{aligned}$$

Exercise 4

For revision, simplify the products in questions ***1–24***:

1	$5\times(-3)$	*2*	-3×2	*3*	$-2\times(-3)$	*4*	$-5\times(-4)$
5	$10\times(-10)$	*6*	$-3\times(-1)$	*7*	-7×4	*8*	$-1\times(-1)$
9	$-2x\times 3$	*10*	$-4\times(-5y)$	*11*	$-3x\times(-4)$	*12*	$-3m\times 5$
13	$-4a\times 2b$	*14*	$-2x\times(-5y)$	*15*	$-4x\times 0$	*16*	$3p\times(-5p)$
17	$2(x+3)$	*18*	$3(x-4)$	*19*	$-4(3x+1)$	*20*	$-7(2x-5)$
21	$x(2x+1)$	*22*	$x(3x-4)$	*23*	$x(1-x)$	*24*	$-x(x-2)$

Draw a diagram, as in Figure 1, to illustrate each of the following products, and hence write down the product as a sum of terms:

25 $(x+1)(x+3)$ *26* $(x+3)(x+4)$ *27* $(a+1)(a+1)$

Use the distributive law to express each of the following products as a sum of terms:

28	$(x+2)(x+5)$	*29*	$(y+3)(y+7)$	*30*	$(z+1)(z+1)$
31	$(a+3)(2a+1)$	*32*	$(2b+3)(2b+3)$	*33*	$(3c+2)(4c+5)$
34	$(x+2)(x-1)$	*35*	$(x+3)(x-4)$	*36*	$(x-2)(x+5)$
37	$(a-2)(a-3)$	*38*	$(b-4)(b-4)$	*39*	$(c-1)(c-5)$
40	$(d-2)(3d+1)$	*41*	$(e+4)(2e-5)$	*42*	$(2f+3)(4f-1)$

No doubt you will have discovered that the product of two binomials gives a polynomial with four terms, two of which can often be combined. The following scheme shows how the result can be obtained mentally:

$$(x+2)(x-5)$$

(1) (4) (2) (3)

$$= x^2-3x-10 \quad (2x-5x=-3x)$$

Notice that the 'inner' and 'outer' products (2) and (3) combine to give the middle term of the answer.

Exercise 5

Express the following products as sums of terms:

1	$(x+1)(x+3)$	2	$(x+3)(x+5)$	3	$(x+4)(x+2)$
4	$(a+5)(a+5)$	5	$(a+1)(a+6)$	6	$(a+2)(a+3)$
7	$(x+1)(2x+3)$	8	$(x+2)(3x+1)$	9	$(x+3)(2x+5)$
10	$(2a+3)(2a+3)$	11	$(3b+1)(3b+1)$	12	$(2c+3)(4c+1)$

* * *

13	$(x-2)(x-3)$	14	$(x-1)(x-2)$	15	$(x-3)(x-4)$
16	$(a-6)(a-6)$	17	$(a-5)(a-1)$	18	$(a-1)(a-1)$
19	$(x-1)(2x-5)$	20	$(x-2)(4x-1)$	21	$(x-3)(2x-3)$
22	$(2a-1)(2a-1)$	23	$(2b-3)(2b-1)$	24	$(2c-5)(3c-2)$

* * *

25	$(x+5)(x-2)$	26	$(y+4)(y-1)$	27	$(z+3)(z-2)$
28	$(p-1)(p+2)$	29	$(q-2)(q+3)$	30	$(r+1)(r-1)$
31	$(s+2)(2s-1)$	32	$(t-3)(4t+1)$	33	$(u-1)(3u+4)$
34	$(2v+3)(2v-3)$	35	$(3w+1)(2w-1)$	36	$(4x-3)(5x+2)$

* * *

37	$(a+6)(a+2)$	38	$(b-4)(b-3)$	39	$(c+5)(c-2)$
40	$(d-2)(d-7)$	41	$(x+4)(2x-1)$	42	$(2y-3)(y+2)$
43	$(2z+4)(3z+2)$	44	$(3k-5)(2k-3)$	45	$(4m-2)(5m+3)$
46	$(x+y)(x+2y)$	47	$(x-y)(x-3y)$	48	$(x+2y)(x-5y)$

Note. When a product of two polynomials is expressed as a sum of terms, we sometimes say that the product has been *expanded* and that this sum is the *expansion* of the product.

Example. Expand and simplify $(3x-2)(2x^2+5x-4)$.

$$\begin{aligned}&(3x-2)(2x^2+5x-4)\\ &= 3x(2x^2+5x-4)-2(2x^2+5x-4)\\ &= 6x^3+15x^2-12x-4x^2-10x+8\\ &= 6x^3+11x^2-22x+8\end{aligned}$$

Exercise 5B

Expand, and simplify where possible:

1 $(a+b)(2a+3b)$ 2 $(x-y)(2x-y)$ 3 $(2p+q)(p-2q)$

4 $(3m-2n)(3m+2n)$ 5 $(s+3t)(3s-2t)$ 6 $(2u-4v)(3u-2v)$

7 $(5+2x)(3+x)$ 8 $(1-y)(1+3y)$ 9 $(4-3x)(5-4x)$

10 $(x+1)(y+2)$ 11 $(a+3)(b-1)$ 12 $(c-4)(d-5)$

13 $(x+2)(x^2+3x+1)$ 14 $(3x-2)(2x^2-x-5)$

15 $(4t+5)(2t^2-7t+3)$ 16 $(a+b)(a^2-ab+b^2)$

17 $(a-b)(a^2+ab+b^2)$ 18 $(p^2+q^2)(p^2-q^2)$

19 $(x^2-4)(2x^2+1)$ 20 $(x^2-1)(3x^2-4x-5)$

5 Two important squares

(i)
$$\begin{aligned}(a+b)^2 &= (a+b)(a+b)\\ &= a(a+b)+b(a+b)\\ &= a^2+ab+ab+b^2\\ &= a^2+2ab+b^2\end{aligned}$$

(ii)
$$\begin{aligned}(a-b)^2 &= (a-b)(a-b)\\ &= a(a-b)-b(a-b)\\ &= a^2-ab-ab+b^2\\ &= a^2-2ab+b^2\end{aligned}$$

These two results are very useful and you should become familiar with them. Note their form carefully:

$$(a+b)^2 = a^2+2ab+b^2$$

$$(a-b)^2 = a^2-2ab+b^2$$

Example 1. $(x+5)^2$
$$= x^2+2(x)(5)+(5)^2$$
$$= x^2+10x+25$$

Example 2. $(2x-3y)^2$
$$= (2x)^2-2(2x)(3y)+(3y)^2$$
$$= 4x^2-12xy+9y^2$$

Exercise 6

Expand each of the following squares:

1 $(x+y)^2$ 2 $(p+q)^2$ 3 $(m-n)^2$ 4 $(u-v)^2$

5 $(x+3)^2$ 6 $(x+5)^2$ 7 $(x+1)^2$ 8 $(x+10)^2$

9 $(a-2)^2$ 10 $(a-3)^2$ 11 $(a-4)^2$ 12 $(a-8)^2$

13 $(2x+1)^2$ 14 $(3x+2)^2$ 15 $(5x+3)^2$ 16 $(10x+1)^2$

17 $(2a-3)^2$ 18 $(3a-1)^2$ 19 $(2a-5)^2$ 20 $(5a-4)^2$

21 $(a+2b)^2$ 22 $(3x+y)^2$ 23 $(4c-d)^2$ 24 $(2m-3n)^2$

Copy the following, and fill in the blank spaces:

25 $(x+\dots)^2 = \dots+\dots+9$ 26 $(y-\dots)^2 = \dots-\dots+25$

27 $(a+\dots)^2 = \dots+2a+\dots$ 28 $(b-\dots)^2 = \dots-2b+\dots$

29 $(2x+\dots)^2 = \dots+\dots+9$ 30 $(3y-\dots)^2 = \dots-30y+\dots$

Simplify:

31 $(x+4)^2+(x+2)^2$ 32 $(a-5)^2+(a-1)^2$

33 $(y+3)^2-(y-3)^2$ 34 $(b-6)^2-(b+4)^2$

Expand:

35 $(x^2+3)^2$ 36 $(y^2-2)^2$ 37 $(a^2+b^2)^2$ 38 $(p^2-q^2)^2$

39 $\left(a+\frac{1}{a}\right)^2$ 40 $\left(a-\frac{1}{a}\right)^2$ 41 $\left(2x+\frac{1}{2x}\right)^2$ 42 $\left(3x-\frac{1}{x}\right)^2$

6 *Some applications of the products and squares*

(i) Solution of equations

Example. Solve the equation $2x^2-(x-3)^2 = (x+2)(x-1)$, where x is a variable on the set of real numbers.

$$\begin{aligned}
& 2x^2-(x-3)^2 = (x+2)(x-1) \\
\Leftrightarrow\ & 2x^2-(x^2-6x+9) = x^2+x-2 \\
\Leftrightarrow\ & 2x^2-x^2+6x-9 = x^2+x-2 \\
\Leftrightarrow\ & 2x^2-x^2-x^2+6x-x = -2+9 \\
\Leftrightarrow\ & 5x = 7 \\
\Leftrightarrow\ & x = \tfrac{7}{5}
\end{aligned}$$

Exercise 7

Solve the equations in questions ***1–14***, where x is a variable on the set of real numbers:

1 $(x+2)^2 = x^2+12$

2 $(x-3)^2 = x^2+15$

3 $(x+5)(x+3) = x^2+11$

4 $(x-4)(x-2) = x^2+2$

5 $(x+10)^2 = x^2+100$

6 $(x-1)^2 = x^2+3$

7 $(x+4)^2 = (x+3)(x+6)$

8 $(x+1)(x+2) = (x-1)^2$

9 $(x-12)^2 = x^2+144$

10 $(2x+4)^2 = 4x(x+1)$

11 $(2x-1)(x-1) = 2x(x+1)$

12 $x(2x+1) = (x-1)(2x-3)$

13 $(2x-3)^2 = (2x+1)^2$

14 $(x+7)(x-7) = (x+9)(x-5)$

15 From Figure 2, use the theorem of Pythagoras to form an equation in x. Solve this equation, and write down the lengths of the sides of the triangle.

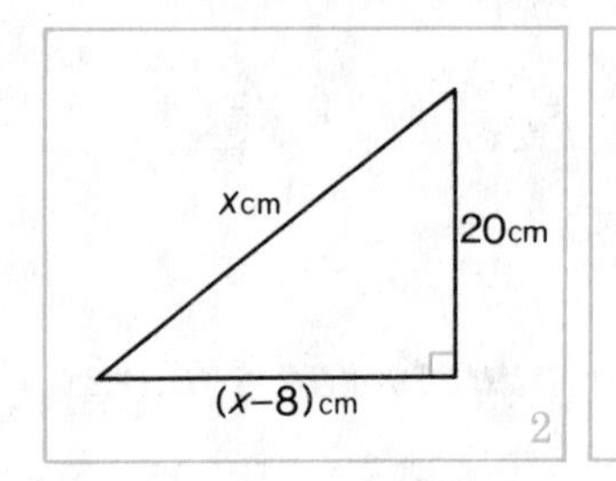

2

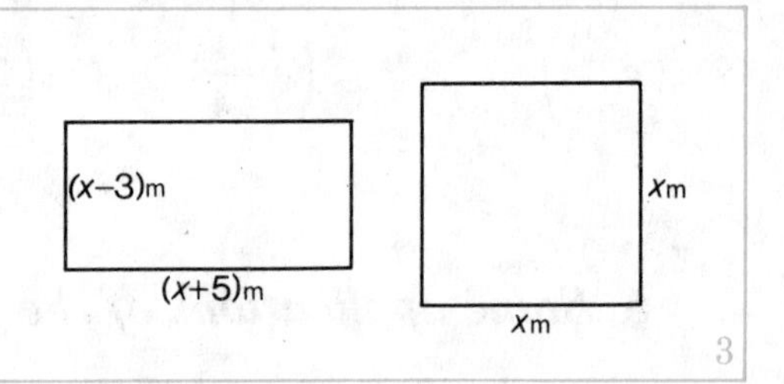

3

16 The rectangle and square in Figure 3 have the same area. Form an equation in x, and solve it. Write down the dimensions of the rectangle and square.

Exercise 7B

Solve the equations in questions ***1–6***, where x is a variable on the set of real numbers:

1 $x^2-(x-5)^2 = 50$

2 $3-(x+2)^2 = 7-x^2$

3 $x^2-(x-9)^2 = 9$

4 $76-(12-x)^2 = 64-x^2$

5 $2(x-1)^2-(x+2)^2 = x^2+2$

6 $7-(x-2)(x-3) = -(x+1)^2$

7 The hypotenuse of a right-angled triangle is $(5x+5)$ cm long, and the lengths of the other two sides are $(4x+8)$ cm and $(3x-5)$ cm.

Form an equation, and solve it. Hence write down the lengths of the three sides.

8 A square metal plate of side 4 cm is heated so that each side increases in length by t cm. Show this in a diagram, and write down expressions for the original area, the new area and the increase in area of the plate. If $t = 0{\cdot}1$, calculate the increase in area.

9 A metal washer has an internal radius of r mm and a width of w mm, as shown in Figure 4. Write down an expression for the external radius, and show that the shaded area of the washer, A mm^2, is given by $A = \pi w(2r+w)$.

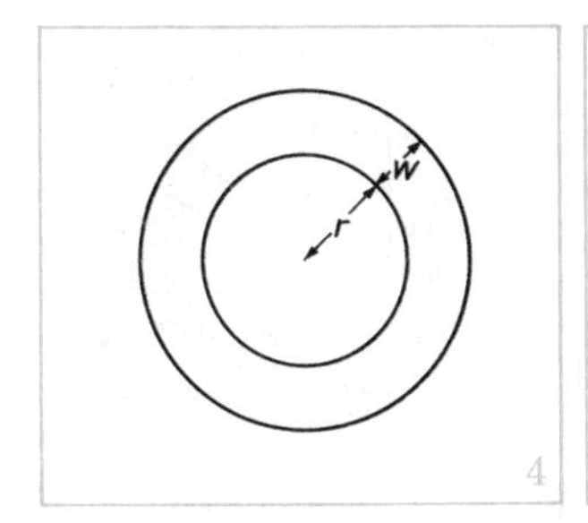

4

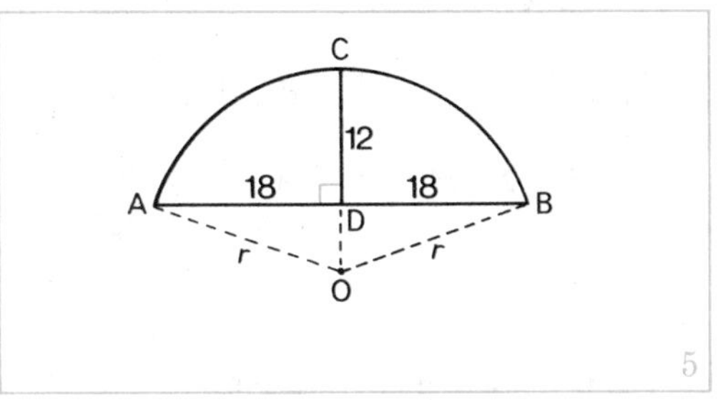

5

10 In Figure 5, ACB is an arc of a circle centre O. The length of chord AB is 36 metres and the greatest height of the arc above the chord AB is 12 metres. Denote the radius of the circle by r metres.

a Express the length of OD in terms of r.

b By applying the theorem of Pythagoras to triangle OAD, form an equation in r.

c Solve this equation for r.

(ii) Identities

$2x = 6$ is an *open sentence*; it becomes a *true* sentence when x is replaced by 3, and a *false* sentence for any other replacement for x. $y^2 \times y = y^3$ is a *true* sentence for *all* real number replacements for y, and is called an *identity*.
$(a+b)^2 = a^2+2ab+b^2$ and $(a-b)^2 = a^2-2ab+b^2$ are also *identities*.

To prove that an equation is an identity we need to show that its left side is identical to its right side.

Example 1. Prove that $(p+q)^2-4pq = (p-q)^2$ is an identity.

$$\text{Left side} = (p+q)^2-4pq = p^2+2pq+q^2-4pq = p^2-2pq+q^2$$

$$\text{Right side} = (p-q)^2 = p^2-2pq+q^2$$

So the left side is the same as the right side, and we have an identity.

Example 2. A set of three whole numbers which are the measures of length of sides of a right-angled triangle is called a *Pythagorean triple.* Show that x^2+y^2, x^2-y^2 and $2xy$, where $x > y$, always give Pythagorean triples. (See Figure 6).

To prove this, we must verify that $(x^2+y^2)^2 = (x^2-y^2)^2+(2xy)^2$.

Left side

$$= (x^2+y^2)^2$$
$$= x^4+2x^2y^2+y^4$$

Right side

$$= (x^2-y^2)^2+(2xy)^2$$
$$= x^4-2x^2y^2+y^4+4x^2y^2$$
$$= x^4+2x^2y^2+y^4$$

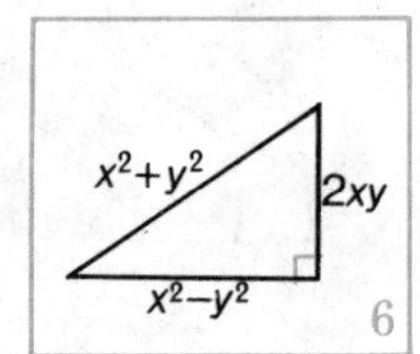

6

It follows that $(x^2+y^2)^2 = (x^2-y^2)^2+(2xy)^2$ is an identity.

If x and y are replaced by whole numbers such that $x > y$, a Pythagorean triple will be obtained. For example, if $x = 2$ and $y = 1$,

$$x^2+y^2 = 2^2+1^2 = 5$$
$$x^2-y^2 = 2^2-1^2 = 3$$
$$2xy = 2\times 2\times 1 = 4$$

giving the measures of length of sides 3, 4 and 5.

Exercise 8

Prove that each of the following equations is an identity:

1 $x(x-1) = x^2-x$

2 $4x(1-x) = 4(x-x^2)$

3 $(a-b)^2 = (b-a)^2$

4 $(3m-n)^2 = (n-3m)^2$

5 $(a+2)^2-4a = a^2+4$

6 $(x-y)^2+2xy = x^2+y^2$

7 $(a+b)(a-b) = a^2-b^2$

8 $(x-3)^2+12x = (x+3)^2$

9 $(x+1)^2 = (x+3)(x-1)+4$

10 $(y-3)^2-(3-y)^2 = 0$

11 $(1+a^2)(1+b^2) = (a+b)^2+(ab-1)^2$

12 $(ax+by)^2+(bx-ay)^2 = (a^2+b^2)(x^2+y^2)$

13 Using x^2+y^2, x^2-y^2 and $2xy$, where $x > y$, as measures of the lengths of the sides of a right-angled triangle (as shown in Worked Example 2 opposite), replace x and y by various pairs of numbers to obtain sets of Pythagorean triples.

Summary

1 *Addition of like terms and expressions*

$$2x^2 - x + x^2 - 3x$$
$$= 3x^2 - 4x$$

$$3(a-b) + 2(a+b)$$
$$= 3a - 3b + 2a + 2b$$
$$= 5a - b$$

2 *Subtraction of expressions*

$$-(2x-1)$$
$$= -1(2x-1)$$
$$= -2x+1$$

$$2x+5-(3x-4)$$
$$= 2x+5-3x+4$$
$$= -x+9$$

3 *Expressing sums as products*

(i) *Using the distributive law*

$$(2x-4)(3x-7)$$
$$= 2x(3x-7) - 4(3x-7)$$
$$= 6x^2 - 14x - 12x + 28$$
$$= 6x^2 - 26x + 28$$

(ii) *Using the scheme*:

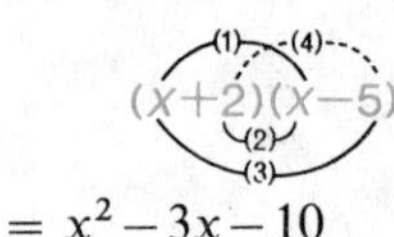

$$= x^2 - 3x - 10$$

4 *Two important squares*

$$(a+b)^2 = a^2 + 2ab + b^2$$
$$(a-b)^2 = a^2 - 2ab + b^2$$

These are *identities*, i.e. they are true for *all* replacements for a and b from the set of real numbers.

$$(x+5)^2$$
$$= x^2 + 10x + 25$$

$$(y-1)^2$$
$$= y^2 - 2y + 1$$

$$(2m-3n)^2$$
$$= 4m^2 - 12mn + 9n^2$$

5 *Applications to equations* $(x \in R)$

$$(x+1)(x+2) = (x-1)^2$$
$$\Leftrightarrow \quad x^2 + 3x + 2 = x^2 - 2x + 1$$
$$\Leftrightarrow \quad 3x + 2x = 1 - 2$$
$$\Leftrightarrow \quad 5x = -1$$
$$\Leftrightarrow \quad x = -\tfrac{1}{5}$$

$$x^2 - (x-5)^2 = 50$$
$$\Leftrightarrow x^2 - (x^2 - 10x + 25) = 50$$
$$\Leftrightarrow \quad x^2 - x^2 + 10x - 25 = 50$$
$$\Leftrightarrow \quad 10x = 75$$
$$\Leftrightarrow \quad x = 7{\cdot}5$$

An Introduction to Matrices

1 *Matrix notation*

A *matrix* is a rectangular array of numbers arranged in rows and columns.

Information is often presented in matrix form in everyday life and in mathematics and science. Here are some examples.

Example 1. Bus fares, in pence.

	Number of stages 1–3	4–6	over 6
Adult	4	6	8
Juvenile	2	3	4

Example 2. Top of English Football League Division I Table, season 1971–72.

	P	W	D	L	F	A	Points
Derby County	42	24	10	8	69	33	58
Leeds United	42	24	9	9	73	31	57
Liverpool	42	24	9	9	64	30	57
Manchester City	42	23	11	8	77	45	57
Arsenal	42	22	8	12	58	40	52
Tottenham Hotspur	42	19	13	10	63	42	51

Example 3. The coefficients of the variables in a system of equations.

(i) *Equations*	(ii) *Coefficient of x*	*Coefficient of y*
$3x + 4y = 5$	3	4
$2x - 6y = 7$	2	-6

When the row and column headings are omitted, as often happens, the array of numbers is enclosed in round (or square) brackets, and called a *matrix*. For example, here are the matrices from *Examples 1, 2* and *3*.

$$\begin{pmatrix} 4 & 6 & 8 \\ 2 & 3 & 4 \end{pmatrix} \quad \begin{pmatrix} 42 & 24 & 10 & 8 & 69 & 33 & 58 \\ 42 & 24 & 9 & 9 & 73 & 31 & 57 \\ 42 & 24 & 9 & 9 & 64 & 30 & 57 \\ 42 & 23 & 11 & 8 & 77 & 45 & 57 \\ 42 & 22 & 8 & 12 & 58 & 40 & 52 \\ 42 & 19 & 13 & 10 & 63 & 42 & 51 \end{pmatrix} \quad \begin{pmatrix} 3 & 4 \\ 2 & -6 \end{pmatrix}$$

Each number in the array is called an *entry* or an *element* of the matrix, and is identified by stating the row and column in which it appears. In *Example 1*, 6 is the entry in the first row and second column.

Exercise 1

1 Answer questions ***a–e*** for the matrix $\begin{pmatrix} 1 & 2 & 3 & 4 \\ 5 & 6 & 7 & 8 \\ 9 & 10 & 11 & 12 \end{pmatrix}$

a State (*1*) the number of rows (*2*) the number of columns.

b List the elements in the second row.

c List the elements in the third column.

d Write down the entry in (*1*) the first row and first column (*2*) the third row and third column (*3*) the second row and fourth column.

e State the rows and columns which describe the positions of these entries:

(*1*) 4 (*2*) 9 (*3*) 8 (*4*) 11 (*5*) 7 (*6*) 5

2 Write down matrices from the following, and in each case state the number of rows and columns:

a Biscuit recipes (masses in units of 25 g, milk in spoonsful)

	Flour	*Butter*	*Sugar*	*Milk*	*Eggs*
Butter biscuits	12	4	4	0	0
Plain biscuits	8	0	1	2	1

b Cricket bowling analysis

	Overs	*Maidens*	*Runs*	*Wickets*
Brown	20	8	72	4
Ford	19	9	65	4
Gray	12	4	28	2

3 Write down examples of matrices with numerical elements arranged in:

a 2 rows and 2 columns *b* 3 rows and 2 columns

c 1 row and 4 columns *d* 5 rows and 1 column.

4 For each of the following matrices, state:

a the number of rows and columns

b the entry in the first row and first column.

(*1*) $\begin{pmatrix} 2 & 4 \\ 6 & 8 \end{pmatrix}$ (*2*) $\begin{pmatrix} a & b & c \\ p & q & r \end{pmatrix}$ (*3*) (5)

(*4*) $\begin{pmatrix} 1 \\ 0 \\ -1 \end{pmatrix}$ (*5*) $\begin{pmatrix} -2 & -1 & 0 \\ 0 & 1 & 2 \\ 1 & 0 & -1 \end{pmatrix}$ (*6*) $(d \ e \ f \ g \ h)$

5 The top of the Scottish Football League Division I Table for season 1971–72 is shown below.

	P	W	D	L	F	A	Points
Celtic	34	28	4	2	96	28	60
Aberdeen	34	21	8	5	80	26	50
Rangers	34	21	2	11	71	38	44
Hibernian	34	19	6	9	62	34	44
Dundee	34	14	13	7	59	38	41
Heart of Midlothian	34	13	13	8	53	49	39

In the corresponding matrix of numbers:

a how many rows and how many columns are there?

b in which rows or columns are the entries
(*1*) all the same (*2*) all greater than 50 (*3*) all even numbers?

6 In each of the following systems of equations write down the matrix of coefficients of the variables.

a $\begin{aligned} 2x+3y &= 0 \\ 6x+4y &= 1 \end{aligned}$ *b* $\begin{aligned} 3x-\ y &= 4 \\ x+2y &= 3 \end{aligned}$ *c* $\begin{aligned} -x+\ y &= 7 \\ 5x-4y &= 1 \end{aligned}$

7 Write down the addition and multiplication tables for numbers given in binary form, then show the content of each table as a matrix.

8 Find examples of information presented in matrix form in newspapers or magazines.

2 The order of a matrix; equal matrices

A matrix is often denoted by a capital letter. For example,

$$A = \begin{pmatrix} 4 & 6 & 8 \\ 2 & 3 & 4 \end{pmatrix} \qquad B = \begin{pmatrix} 3 & 4 \\ 2 & -6 \end{pmatrix}$$

The *order* of a matrix is given by the number of rows, followed by the number of columns. A has 2 rows and 3 columns, and so is said to be of order 2×3 (read '2 by 3'). When the number of rows in a matrix is the same as the number of columns, the matrix is called a *square matrix*; B is a square matrix of order 2.

Exercise 2

1 State the order of each of the following matrices:

a $\begin{pmatrix} 3 & 1 & 4 \\ 2 & 3 & 5 \end{pmatrix}$ *b* $\begin{pmatrix} 3 & 2 \\ 1 & 3 \\ 4 & 5 \end{pmatrix}$ *c* $\begin{pmatrix} a & h & g \\ h & b & f \\ g & f & c \end{pmatrix}$

d $\begin{pmatrix} 3 & 2 & 1 & 4 \\ 1 & 2 & 3 & 0 \\ 2 & 0 & 1 & 3 \end{pmatrix}$ *e* $(-1 \quad -2 \quad -3)$ *f* $\begin{pmatrix} u \\ v \\ w \end{pmatrix}$

Note. **e** is a *row matrix*, and **f** is a *column matrix*.

2 Write down the total number of entries in each matrix in question *1*. Do you see a quick way to find the answers?

3 How many entries are there in:

a a 3×3 matrix *b* a 4×3 matrix *c* a 1×1 matrix

d a $1 \times n$ matrix *e* an $m \times n$ matrix *f* a square $n \times n$ matrix?

4 Write down an example of:

a a 2×4 matrix *b* a 3×3 matrix *c* a 1×1 matrix

d a 3×1 matrix.

Two matrices A and B are said to be *equal* when

(i) they are of the same order, and

(ii) their corresponding entries are equal.

For example, if $A = \begin{pmatrix} p & q \\ r & s \end{pmatrix}$ and $B = \begin{pmatrix} p & q \\ r & s \end{pmatrix}$, then $A = B$.

5 Which of the following matrices are equal?

$A = (1 \quad 2 \quad 3)$ $\qquad B = (3 \quad 2 \quad 1)$ $\qquad C = (1 \quad 2 \quad 3)$

$D = \begin{pmatrix} 2 \\ -1 \end{pmatrix}$ $\qquad E = \begin{pmatrix} 2 \\ 1 \end{pmatrix}$ $\qquad F = \begin{pmatrix} 1 \\ 2 \end{pmatrix}$

$G = \begin{pmatrix} 2 \\ 1 \end{pmatrix}$ $\qquad H = \begin{pmatrix} 1 & 2 \\ 3 & 4 \end{pmatrix}$ $\qquad J = \begin{pmatrix} -1 & -2 \\ -3 & -4 \end{pmatrix}$

$K = \begin{pmatrix} 1 & 3 \\ 2 & 4 \end{pmatrix}$ $\qquad L = \begin{pmatrix} 1 & 2 \\ 3 & 4 \end{pmatrix}$

6 What is the order of each matrix in question *5*?

7 Find x and y in each of the following:

a $\begin{pmatrix} x & 2y \\ 0 & 3 \end{pmatrix} = \begin{pmatrix} 1 & 8 \\ 0 & 3 \end{pmatrix}$ $\qquad$ *b* $\begin{pmatrix} x+3 \\ 2-y \end{pmatrix} = \begin{pmatrix} 5 \\ 1 \end{pmatrix}$

c $\begin{pmatrix} 2x & 0 \\ 0 & 2y \end{pmatrix} = \begin{pmatrix} 6 & 0 \\ 0 & -8 \end{pmatrix}$ $\qquad$ *d* $(3x \quad -y) = (12 \quad 2)$

e $\begin{pmatrix} x+y \\ x-y \end{pmatrix} = \begin{pmatrix} 4 \\ 6 \end{pmatrix}$ $\qquad$ *f* $\begin{pmatrix} x+y \\ x-y \end{pmatrix} = \begin{pmatrix} 7 \\ 7 \end{pmatrix}$

From a given matrix A, a new matrix can be formed by writing row 1 as column 1, row 2 as column 2, and so on. This new matrix is called the *transpose* of A, and is denoted by A' (read as 'A transpose'). The rows of A are the columns of A', and the columns of A are the rows of A'.

For example, if $A = \begin{pmatrix} 1 & 2 \\ 3 & 4 \\ 5 & 6 \end{pmatrix}$ then $A' = \begin{pmatrix} 1 & 3 & 5 \\ 2 & 4 & 6 \end{pmatrix}$

8 Write down the transpose of each matrix in question *1*, and state the order of each transpose.

9 $P = \begin{pmatrix} x & 5 \\ 3 & y \end{pmatrix}$ and $Q = \begin{pmatrix} 4 & 3 \\ 5 & -2 \end{pmatrix}$. Find x and y, given that $P' = Q$.

3 Addition of matrices

Now that we know what a matrix is, can we usefully define *addition* on a set of matrices?

Example. Bob and Jim, who are close rivals in the mathematics class, compare their marks in mathematics and statistics at the end of the second term.

	First term		*Second term*		*Total*	
	Bob	Jim	Bob	Jim	Bob	Jim
Mathematics	82	78	75	80	157	158
Statistics	68	72	70	78	138	150

Setting out this information in matrix form, it is reasonable to write:

$$\begin{pmatrix} 82 & 78 \\ 68 & 72 \end{pmatrix} + \begin{pmatrix} 75 & 80 \\ 70 & 78 \end{pmatrix} = \begin{pmatrix} 157 & 158 \\ 138 & 150 \end{pmatrix}$$

This method of combining matrices is called *addition* of matrices.

If A and B are two matrices of the same order, the sum of A and B, denoted by $A+B$, is the matrix obtained by adding the entries of A and the corresponding entries of B.

The matrix $A+B$ will be of the same order as each of A and B. It is not possible to add two matrices of different orders.

Example 1. $\begin{pmatrix} 3 \\ -1 \end{pmatrix} + \begin{pmatrix} 5 \\ 1 \end{pmatrix} = \begin{pmatrix} 3+5 \\ -1+1 \end{pmatrix} = \begin{pmatrix} 8 \\ 0 \end{pmatrix}$

Example 2. $\begin{pmatrix} w & x \\ y & z \end{pmatrix} + \begin{pmatrix} w & -x \\ 2y & -2z \end{pmatrix} = \begin{pmatrix} 2w & 0 \\ 3y & -z \end{pmatrix}$

Exercise 3

Find the sums of the matrices in questions ***1*** to ***18***:

1 $\begin{pmatrix} 1 \\ 2 \end{pmatrix} + \begin{pmatrix} 3 \\ 4 \end{pmatrix}$ *2* $\begin{pmatrix} 0 \\ 1 \end{pmatrix} + \begin{pmatrix} -1 \\ 2 \end{pmatrix}$ *3* $\begin{pmatrix} 5 \\ -3 \end{pmatrix} + \begin{pmatrix} -2 \\ 1 \end{pmatrix}$

4 $\begin{pmatrix} a \\ b \end{pmatrix} + \begin{pmatrix} 5a \\ -3b \end{pmatrix}$ *5* $\begin{pmatrix} m \\ n \end{pmatrix} + \begin{pmatrix} 1 \\ 2 \end{pmatrix}$ *6* $\begin{pmatrix} p \\ q \end{pmatrix} + \begin{pmatrix} r \\ s \end{pmatrix}$

7 $(3 \quad 0) + (2 \quad 4)$ *8* $(2 \quad -3) + (-4 \quad 6)$

9 $(2u \quad 3v) + (u \quad -2v)$

10 $\begin{pmatrix} 2 \\ -1 \\ 3 \end{pmatrix} + \begin{pmatrix} 4 \\ 5 \\ 1 \end{pmatrix}$

11 $(2 \quad 3 \quad 1) + (4 \quad 1 \quad -3)$

12 $\begin{pmatrix} 1 & 0 \\ 0 & 1 \end{pmatrix} + \begin{pmatrix} 2 & 1 \\ 3 & 4 \end{pmatrix}$

13 $\begin{pmatrix} 3 & 4 \\ 1 & 2 \end{pmatrix} + \begin{pmatrix} 3 & 1 \\ 2 & 4 \end{pmatrix}$

14 $\begin{pmatrix} 6 & 5 \\ 3 & 2 \end{pmatrix} + \begin{pmatrix} 2 & -3 \\ 0 & 4 \end{pmatrix}$

15 $\begin{pmatrix} 3 & -2 \\ -1 & -4 \end{pmatrix} + \begin{pmatrix} -2 & 2 \\ 1 & 5 \end{pmatrix}$

16 $\begin{pmatrix} \frac{1}{2} & 1 \\ \frac{3}{4} & 0 \end{pmatrix} + \begin{pmatrix} \frac{1}{4} & -\frac{1}{2} \\ -\frac{1}{2} & 1 \end{pmatrix}$

17 $\begin{pmatrix} 2a & b \\ 3a & -b \end{pmatrix} + \begin{pmatrix} a & 2b \\ -4a & b \end{pmatrix}$

18 $\begin{pmatrix} x & -y \\ -x & 2y \end{pmatrix} + \begin{pmatrix} 2x & 3y \\ -x & 2y \end{pmatrix}$

19 $A = \begin{pmatrix} 2 & 1 \\ 3 & 4 \end{pmatrix}$ and $B = \begin{pmatrix} 3 & 2 \\ 4 & 5 \end{pmatrix}$. Find the matrices $A+B$ and $B+A$.

20a $P = \begin{pmatrix} a & b \\ c & d \end{pmatrix}$ and $Q = \begin{pmatrix} f & g \\ h & k \end{pmatrix}$. Find the matrices $P+Q$ and $Q+P$.

b Is it true that $P+Q = Q+P$? What law for matrix addition does this result suggest?

21a $A = \begin{pmatrix} 1 & 2 \\ 3 & 4 \end{pmatrix}$, $B = \begin{pmatrix} 5 & 6 \\ 7 & 8 \end{pmatrix}$ and $C = \begin{pmatrix} 3 & 4 \\ 9 & 10 \end{pmatrix}$. Find the following matrices:

(1) $A+B$ (2) $B+C$ (3) $(A+B)+C$ (4) $A+(B+C)$

b Is it true that $(A+B)+C = A+(B+C)$? What law for addition of matrices does this suggest?

A *zero matrix*, denoted by $\boldsymbol{O}$, is one whose elements are all zero. $\begin{pmatrix} 0 & 0 \\ 0 & 0 \end{pmatrix}$ is the 2×2 zero matrix.

22 Given that $\boldsymbol{O} = \begin{pmatrix} 0 & 0 \\ 0 & 0 \end{pmatrix}$ and $A = \begin{pmatrix} a & b \\ c & d \end{pmatrix}$, show that

$$\boldsymbol{O}+A = A+\boldsymbol{O} = A$$

23 Given that $A = \begin{pmatrix} 2 & 3 \\ 1 & 5 \end{pmatrix}$ and $B = \begin{pmatrix} -2 & -3 \\ -1 & -5 \end{pmatrix}$, find $A+B$ and $B+A$ and hence show that $A+B = B+A = \boldsymbol{O}$.

Note. Each entry in B is the negative of the corresponding entry in A. For this reason, B is called the *negative* of A and is written $-A$. We then have $A+(-A) = \boldsymbol{O}$.

24 Write down the negative of each of the following matrices:

a $\begin{pmatrix} 4 \\ 5 \end{pmatrix}$ *b* $\begin{pmatrix} 2 \\ 3 \\ -1 \end{pmatrix}$ *c* $\begin{pmatrix} 3 & -1 \\ 0 & -2 \end{pmatrix}$ *d* $\begin{pmatrix} -3 & 1 & 0 \\ 4 & -2 & -1 \end{pmatrix}$

e $(-m)$

4 Subtraction of matrices

We have seen that if a and b are two real numbers, then $a-b = a+(-b)$. In the same way, since each matrix has a negative, we can write $A+(-B)$ as $A-B$, and talk about *subtracting* one matrix from another.

$$A-B = A+(-B)$$

To subtract B from A, *add* the negative of B to A.

Example 1. If $P = \begin{pmatrix} 3 & 2 \\ -1 & 4 \end{pmatrix}$ and $Q = \begin{pmatrix} 2 & 0 \\ -1 & 5 \end{pmatrix}$, then

$$\begin{aligned} P-Q &= \begin{pmatrix} 3 & 2 \\ -1 & 4 \end{pmatrix} - \begin{pmatrix} 2 & 0 \\ -1 & 5 \end{pmatrix} \\ &= \begin{pmatrix} 3 & 2 \\ -1 & 4 \end{pmatrix} + \begin{pmatrix} -2 & 0 \\ 1 & -5 \end{pmatrix} \\ &= \begin{pmatrix} 1 & 2 \\ 0 & -1 \end{pmatrix} \end{aligned}$$

Example 2. Solve the equation $X + \begin{pmatrix} 2 \\ -1 \end{pmatrix} = \begin{pmatrix} -4 \\ 6 \end{pmatrix}$, given that X is a 2×1 matrix.

$$\begin{aligned} X + \begin{pmatrix} 2 \\ -1 \end{pmatrix} &= \begin{pmatrix} -4 \\ 6 \end{pmatrix} \\ \Leftrightarrow \quad X &= \begin{pmatrix} -4 \\ 6 \end{pmatrix} - \begin{pmatrix} 2 \\ -1 \end{pmatrix} \\ \Leftrightarrow \quad X &= \begin{pmatrix} -4 \\ 6 \end{pmatrix} + \begin{pmatrix} -2 \\ 1 \end{pmatrix} \\ \Leftrightarrow \quad X &= \begin{pmatrix} -6 \\ 7 \end{pmatrix} \end{aligned}$$

Exercise 4

1 Simplify each of the following:

a $\begin{pmatrix}4\\3\end{pmatrix}-\begin{pmatrix}1\\2\end{pmatrix}$ b $\begin{pmatrix}3\\4\end{pmatrix}-\begin{pmatrix}-2\\2\end{pmatrix}$

c $\begin{pmatrix}5\\-7\end{pmatrix}-\begin{pmatrix}4\\-3\end{pmatrix}$ d $\begin{pmatrix}-3\\0\end{pmatrix}-\begin{pmatrix}-2\\1\end{pmatrix}$

2 Find a 2×2 matrix equal to each of the following:

a $\begin{pmatrix}9&5\\4&2\end{pmatrix}-\begin{pmatrix}7&1\\3&0\end{pmatrix}$ b $\begin{pmatrix}2&-1\\3&7\end{pmatrix}-\begin{pmatrix}0&-1\\-2&0\end{pmatrix}$

c $\begin{pmatrix}1&2\\-3&1\end{pmatrix}-\begin{pmatrix}4&-1\\-2&3\end{pmatrix}$ d $\begin{pmatrix}3x&2\\4&5y\end{pmatrix}-\begin{pmatrix}x&-1\\2&-y\end{pmatrix}$

e $\begin{pmatrix}8&3\\6&4\end{pmatrix}+\begin{pmatrix}-1&2\\2&-6\end{pmatrix}-\begin{pmatrix}-3&2\\1&-4\end{pmatrix}$

3 Given $A=\begin{pmatrix}1&2\\3&4\end{pmatrix}$, $B=\begin{pmatrix}-2&3\\0&1\end{pmatrix}$ and $C=\begin{pmatrix}5&2\\-1&0\end{pmatrix}$, find in their simplest forms the following matrices:

a $A+B$ b $A+C$ c $(A+C)+(A+B)$ d $(A+C)-(A+B)$

e $A-B$ f $C-B$

Which of these are equal?

4 Solve each of the following equations for the 2×2 matrix X:

a $X+\begin{pmatrix}0&1\\1&0\end{pmatrix}=\begin{pmatrix}2&0\\0&2\end{pmatrix}$ b $\begin{pmatrix}2&3\\4&5\end{pmatrix}+X=\begin{pmatrix}4&-1\\3&2\end{pmatrix}$

c $X-\begin{pmatrix}3&5\\-2&1\end{pmatrix}=\begin{pmatrix}4&7\\5&0\end{pmatrix}$ d $\begin{pmatrix}3&-4\\2&7\end{pmatrix}-X=\begin{pmatrix}-1&-5\\3&-6\end{pmatrix}$

5 Find x, y and z in the following:

a $\begin{pmatrix}1\\2\\3\end{pmatrix}+\begin{pmatrix}x\\y\\z\end{pmatrix}=\begin{pmatrix}4\\1\\-2\end{pmatrix}$ b $(x\quad y\quad z)-(-4\quad 3\quad 1)=(-5\quad 1\quad 0)$

6 Find p, q, r, s such that:

$$\begin{pmatrix}p&q\\r&s\end{pmatrix}-\begin{pmatrix}2&3\\0&1\end{pmatrix}=\begin{pmatrix}3&5\\1&2\end{pmatrix}$$

5 *Multiplication of a matrix by a real number*

We know that $x+x = 2x$, and $x+x+x = 2x+x = 3x$. Now let $X = \begin{pmatrix} 3 & 4 \\ 5 & 6 \end{pmatrix}$. From the definition of addition of matrices,

$$X+X = \begin{pmatrix} 3 & 4 \\ 5 & 6 \end{pmatrix} + \begin{pmatrix} 3 & 4 \\ 5 & 6 \end{pmatrix} = \begin{pmatrix} 6 & 8 \\ 10 & 12 \end{pmatrix} = \begin{pmatrix} 2\times 3 & 2\times 4 \\ 2\times 5 & 2\times 6 \end{pmatrix},$$

and

$$X+X+X = \begin{pmatrix} 3 & 4 \\ 5 & 6 \end{pmatrix} + \begin{pmatrix} 3 & 4 \\ 5 & 6 \end{pmatrix} + \begin{pmatrix} 3 & 4 \\ 5 & 6 \end{pmatrix} = \begin{pmatrix} 9 & 12 \\ 15 & 18 \end{pmatrix}$$
$$= \begin{pmatrix} 3\times 3 & 3\times 4 \\ 3\times 5 & 3\times 6 \end{pmatrix}$$

If we now write

$$2\begin{pmatrix} 3 & 4 \\ 5 & 6 \end{pmatrix} = \begin{pmatrix} 2\times 3 & 2\times 4 \\ 2\times 5 & 2\times 6 \end{pmatrix}, \text{ and } 3\begin{pmatrix} 3 & 4 \\ 5 & 6 \end{pmatrix} = \begin{pmatrix} 3\times 3 & 3\times 4 \\ 3\times 5 & 3\times 6 \end{pmatrix},$$

it is reasonable to denote $X+X$ by $2X$ and $X+X+X$ by $3X$. Extending this idea, we make the following definition:

If k is a real number and A is a matrix, then kA is the matrix obtained by multiplying each entry of A by k.

Note. In matrix algebra, a real number is often called a *scalar*.

Example. Given that $A = \begin{pmatrix} 2 & 1 \\ 4 & 3 \end{pmatrix}$ and $B = \begin{pmatrix} 1 & 5 \\ 0 & 2 \end{pmatrix}$, find in their simplest forms the matrices ***a*** $2A$ and ***b*** $3A-2B$.

a $$2A = 2\begin{pmatrix} 2 & 1 \\ 4 & 3 \end{pmatrix} = \begin{pmatrix} 4 & 2 \\ 8 & 6 \end{pmatrix}$$

b $$3A-2B = 3\begin{pmatrix} 2 & 1 \\ 4 & 3 \end{pmatrix} - 2\begin{pmatrix} 1 & 5 \\ 0 & 2 \end{pmatrix} = \begin{pmatrix} 6 & 3 \\ 12 & 9 \end{pmatrix} - \begin{pmatrix} 2 & 10 \\ 0 & 4 \end{pmatrix} = \begin{pmatrix} 4 & -7 \\ 12 & 5 \end{pmatrix}$$

Exercise 5

1 Work out the following:

a $3\begin{pmatrix}1\\2\end{pmatrix}$ *b* $2\begin{pmatrix}5\\3\\2\end{pmatrix}$ *c* $5(3 \quad 1 \quad 2)$

d $\frac{1}{2}(6 \quad 8)$ *e* $2\begin{pmatrix}3 & 1\\4 & 2\end{pmatrix}$ *f* $-3\begin{pmatrix}1 & 0\\0 & -1\end{pmatrix}$

g $\frac{1}{2}\begin{pmatrix}4 & 2\\0 & 1\end{pmatrix}$ *h* $5\begin{pmatrix}a & 2b\\3a & 4b\end{pmatrix}$

2 $X = \begin{pmatrix}3 & 4\\1 & 2\end{pmatrix}$. Find the following 2×2 matrices:

a $2X$ *b* $3X$ *c* $5X$ *d* $-X$ *e* $(-1)X$

Which matrices are equal?

3 Using the results of your operations on the matrix X in question **2**, complete each of the following:

a $2X+3X$
$= \begin{pmatrix}6 & 8\\2 & 4\end{pmatrix} + \begin{pmatrix}9 & 12\\3 & 6\end{pmatrix}$
$= \begin{pmatrix}15 & .\\. & .\end{pmatrix}$
$= 5X$

b $5X-3X$
$= \begin{pmatrix}. & .\\. & .\end{pmatrix} - \begin{pmatrix}. & .\\. & .\end{pmatrix}$
$= \begin{pmatrix}. & .\\. & .\end{pmatrix}$
$= \ldots$

c $2X+X$
$= \begin{pmatrix}. & .\\. & .\end{pmatrix} + \begin{pmatrix}. & .\\. & .\end{pmatrix}$
$= \begin{pmatrix}. & .\\. & .\end{pmatrix}$
$= \ldots$

d $3X-X$
$= \begin{pmatrix}. & .\\. & .\end{pmatrix} - \begin{pmatrix}. & .\\. & .\end{pmatrix}$
$= \begin{pmatrix}. & .\\. & .\end{pmatrix}$
$= \ldots$

4 Fill in the entries in each of the following:

a $\begin{pmatrix}2 & 4\\6 & 8\end{pmatrix} = 2\begin{pmatrix}. & .\\. & .\end{pmatrix}$ *b* $\begin{pmatrix}6 & -3\\-9 & 0\end{pmatrix} = 3\begin{pmatrix}. & .\\. & .\end{pmatrix}$

c $\begin{pmatrix}-5 & 0\\15 & -10\end{pmatrix} = -5\begin{pmatrix}. & .\\. & .\end{pmatrix}$ *d* $\begin{pmatrix}2 & 3\\\frac{1}{2} & 1\end{pmatrix} = \frac{1}{2}\begin{pmatrix}. & .\\. & .\end{pmatrix}$

5 $A = \begin{pmatrix}2 & -3\\4 & 0\end{pmatrix}$ and $B = \begin{pmatrix}-3 & 4\\2 & -1\end{pmatrix}$. Simplify:

a $2A$ *b* $2B$ *c* $3A$

d $5A$ *e* $A+B$ *f* $2(A+B)$

g $A-B$ *h* $2(A-B)$ *i* $2A+2B$

j $2A-2B$ *k* $3A+2A$ *l* $3A+2B$

Which matrices are equal?

6 Simplify:

a $3\begin{pmatrix}2 & 1 & -3\\ 5 & 4 & 0\end{pmatrix}+2\begin{pmatrix}1 & -2 & 0\\ 3 & -5 & 2\end{pmatrix}$

b $3\begin{pmatrix}3 & -2 & 1\\ 1 & 3 & 4\end{pmatrix}-2\begin{pmatrix}1 & -3 & 2\\ -2 & 3 & 4\end{pmatrix}$

7 If $A = (3 \quad 1 \quad 2)$, $B = (4 \quad 3 \quad 1)$ and $C = (2 \quad 4 \quad 3)$, simplify:

a $3A+B$ *b* $2A+3B$ *c* $B-C$

d $2(B-C)$ *e* $A+B+C$ *f* $A+2B+3C$

8 Find the 3×1 matrix X in each of the following:

a $3X = \begin{pmatrix}6\\ -3\\ 9\end{pmatrix}$ *b* $2X+\begin{pmatrix}1\\ 5\\ 7\end{pmatrix}=\begin{pmatrix}5\\ 1\\ 9\end{pmatrix}$

c $4X-\begin{pmatrix}7\\ 3\\ 4\end{pmatrix}=\begin{pmatrix}1\\ 9\\ 0\end{pmatrix}$

9 Solve each of the following equations for the 2×2 matrix X:

a $3X=\begin{pmatrix}6 & -3\\ 9 & 12\end{pmatrix}$ *b* $2X+\begin{pmatrix}3 & 1\\ 4 & 2\end{pmatrix}=\begin{pmatrix}9 & 5\\ 2 & 8\end{pmatrix}$

c $4X-\begin{pmatrix}3 & 1\\ 4 & 7\end{pmatrix}=\begin{pmatrix}5 & 3\\ 0 & 13\end{pmatrix}$

6 Multiplication of two matrices

Can we multiply one matrix by another matrix? The following illustration will suggest an answer to this question.

Table 1 shows the purchases of fruit made by a housewife in two consecutive weeks, and *Table* 2 gives the cost of the fruit per kilogramme.

Table 1

Purchases (kg)	*Apples*	*Bananas*
First week	3	1
Second week	2	2

Table 2

Fruit	*Cost in p per kg*
Apples	25
Bananas	18

Multiplying the cost per kg by the number of kg bought, we obtain:
The total cost of the fruit for the first week
$= (3 \times 25)+(1 \times 18) = 75+18 = 93$p.

Setting out the information in *Tables* 1 and 2 in matrix form, the calculation may be shown as follows:

(i) $(3 \quad 1)\begin{pmatrix}25\\18\end{pmatrix} = (3\times 25 + 1\times 18) = (75+18) = (93)$,

giving the cost as 93p.

In effect, we have multiplied each entry in the 1×2 row matrix by the corresponding entry in the 2×1 column matrix, and found the sum of these products as a 1×1 cost matrix.

The total cost for the second week is given by:

(ii) $(2 \quad 2)\begin{pmatrix}25\\18\end{pmatrix} = (2\times 25 + 2\times 18) = (50+36) = (86)$,

giving the cost as 86p.

We can show the costs for both weeks as follows:

(iii) $\begin{pmatrix}3 & 1\\2 & 2\end{pmatrix}\begin{pmatrix}25\\18\end{pmatrix} = \begin{pmatrix}3\times 25 + 1\times 18\\2\times 25 + 2\times 18\end{pmatrix} = \begin{pmatrix}75+18\\50+36\end{pmatrix} = \begin{pmatrix}93\\86\end{pmatrix}$

This method of combining matrices is called *multiplication of matrices*. The rule is '*Multiply rows into columns, and add the resulting products*'.

If $A = \begin{pmatrix}a & b\\c & d\end{pmatrix}$ and $B = \begin{pmatrix}x\\y\end{pmatrix}$, then the product AB is defined by the equation

$$\begin{pmatrix}a & b\\c & d\end{pmatrix}\begin{pmatrix}x\\y\end{pmatrix} = \begin{pmatrix}ax+by\\cx+dy\end{pmatrix}.$$

The '*row into column*' process for matrix multiplication may be extended to an $m\times p$ matrix A and a $p\times 1$ matrix B, the product matrix AB being of order $m\times 1$. For example,

$$(a \quad b \quad c)\begin{pmatrix}x\\y\\z\end{pmatrix} = (ax+by+cz), \text{ and}$$

$$\begin{pmatrix} a & b & c \\ d & e & f \end{pmatrix}\begin{pmatrix} x \\ y \\ z \end{pmatrix} = \begin{pmatrix} ax+by+cz \\ dx+ey+fz \end{pmatrix}$$

It is clear that such products exist only if the number of *columns* in the left-hand matrix is the same as the number of *rows* in the right-hand matrix.

Note. In checking whether or not a product can be formed, and also in working out the order of the product matrix, a comparison with matching dominoes may be helpful, as shown in Figure 1.

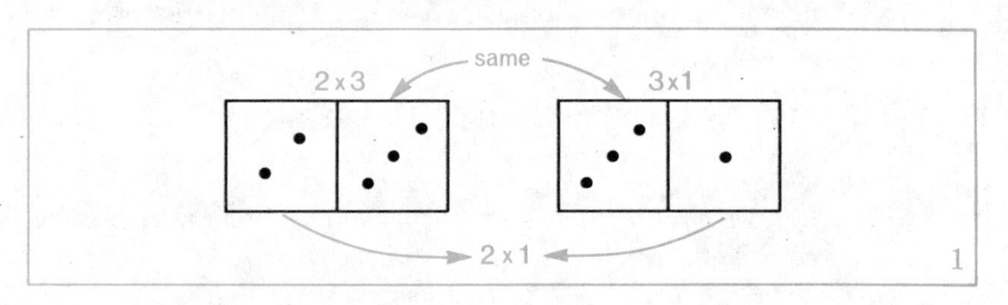

1

Example 1. Find the matrix product $(3 \quad 4)\begin{pmatrix} 5 \\ 6 \end{pmatrix}$.

Orders 1×2 and 2×1; order of product 1×1 (one element).

$$(3 \quad 4)\begin{pmatrix} 5 \\ 6 \end{pmatrix} = ((3\times5)+(4\times6)) = (15+24) = (39)$$

Example 2. Calculate, if possible, the following.

(i) $\begin{pmatrix} 1 & 2 & 3 \\ 4 & 5 & 6 \end{pmatrix}\begin{pmatrix} 7 \\ 8 \\ 9 \end{pmatrix}$ (ii) $\begin{pmatrix} 7 \\ 8 \\ 9 \end{pmatrix}\begin{pmatrix} 1 & 2 & 3 \\ 4 & 5 & 6 \end{pmatrix}$

(i) Orders 2×3 and 3×1; order of product 2×1.

$$\begin{pmatrix} 1 & 2 & 3 \\ 4 & 5 & 6 \end{pmatrix}\begin{pmatrix} 7 \\ 8 \\ 9 \end{pmatrix} = \begin{pmatrix} (1\times7)+(2\times8)+(3\times9) \\ (4\times7)+(5\times8)+(6\times9) \end{pmatrix} = \begin{pmatrix} 7+16+27 \\ 28+40+54 \end{pmatrix}$$

$$= \begin{pmatrix} 50 \\ 122 \end{pmatrix}$$

(ii) Orders 3×1 and 2×3. Since the number of columns in the left-hand matrix is not equal to the number of rows in the right-hand matrix, the product does not exist.

Exercise 6

1 Find the following matrix products:

a $(1 \quad 2)\begin{pmatrix}3\\4\end{pmatrix}$ b $(3 \quad 4)\begin{pmatrix}2\\5\end{pmatrix}$ c $(2 \quad 3)\begin{pmatrix}4\\1\end{pmatrix}$

d $(5 \quad 1)\begin{pmatrix}2\\3\end{pmatrix}$ e $(3 \quad 5)\begin{pmatrix}2\\4\end{pmatrix}$ f $(4 \quad 8)\begin{pmatrix}1\\0\end{pmatrix}$

g $(a \quad 2)\begin{pmatrix}1\\2\end{pmatrix}$ h $(4 \quad c)\begin{pmatrix}2\\5\end{pmatrix}$ i $(m \quad 1)\begin{pmatrix}2\\m\end{pmatrix}$

2 Simplify the following products:

a $(4 \quad 2)\begin{pmatrix}2\\-1\end{pmatrix}$ b $(5 \quad 3)\begin{pmatrix}-1\\6\end{pmatrix}$ c $(-2 \quad 3)\begin{pmatrix}4\\2\end{pmatrix}$

d $(2 \quad -1)\begin{pmatrix}5\\4\end{pmatrix}$ e $(3 \quad -2)\begin{pmatrix}3\\-4\end{pmatrix}$ f $(3 \quad -2)\begin{pmatrix}-3\\4\end{pmatrix}$

g $(-3 \quad -2)\begin{pmatrix}3\\4\end{pmatrix}$ h $(a \quad 5)\begin{pmatrix}a\\-2\end{pmatrix}$ i $(a \quad b)\begin{pmatrix}4\\-3\end{pmatrix}$

3 Find x in each of the following:

a $(x \quad 2)\begin{pmatrix}1\\3\end{pmatrix}=(10)$ b $(3 \quad x)\begin{pmatrix}5\\2\end{pmatrix}=(17)$

c $(x \quad 1)\begin{pmatrix}4\\x\end{pmatrix}=(15)$ d $(x \quad -3)\begin{pmatrix}2\\3\end{pmatrix}=(-1)$

4 Simplify the following products:

a $(1 \quad 2 \quad 3)\begin{pmatrix}1\\2\\3\end{pmatrix}$ b $(3 \quad 4 \quad 1)\begin{pmatrix}1\\-1\\2\end{pmatrix}$

c $(5 \quad 4 \quad 3)\begin{pmatrix}3\\-1\\-2\end{pmatrix}$ d $(3 \quad 1 \quad 5)\begin{pmatrix}x\\y\\z\end{pmatrix}$

5 Find each of the following products in its simplest form:

a $\begin{pmatrix}1 & 0\\0 & 1\end{pmatrix}\begin{pmatrix}3\\4\end{pmatrix}$ b $\begin{pmatrix}1 & 0\\0 & -1\end{pmatrix}\begin{pmatrix}3\\4\end{pmatrix}$ c $\begin{pmatrix}-1 & 0\\0 & 1\end{pmatrix}\begin{pmatrix}3\\4\end{pmatrix}$

d $\begin{pmatrix}-1 & 0\\0 & -1\end{pmatrix}\begin{pmatrix}3\\4\end{pmatrix}$ e $\begin{pmatrix}0 & -1\\1 & 0\end{pmatrix}\begin{pmatrix}3\\4\end{pmatrix}$ f $\begin{pmatrix}1 & -1\\1 & 1\end{pmatrix}\begin{pmatrix}3\\4\end{pmatrix}$

g $\begin{pmatrix}2 & -1\\1 & 2\end{pmatrix}\begin{pmatrix}2\\3\end{pmatrix}$ h $\begin{pmatrix}2 & 3\\4 & 5\end{pmatrix}\begin{pmatrix}1\\2\end{pmatrix}$ i $\begin{pmatrix}2 & 3\\1 & -2\end{pmatrix}\begin{pmatrix}1\\-3\end{pmatrix}$

6 By considering orders, determine which of the following products exist, and calculate those that do exist:

a $\begin{pmatrix}1 & 5\\2 & 4\end{pmatrix}\begin{pmatrix}3\\1\end{pmatrix}$ *b* $\begin{pmatrix}3\\2\end{pmatrix}\begin{pmatrix}7\\8\end{pmatrix}$ *c* $\begin{pmatrix}5\\4\end{pmatrix}\begin{pmatrix}1 & -1\\2 & 3\end{pmatrix}$

d $(3\quad 2)\begin{pmatrix}7\\8\end{pmatrix}$ *e* $\begin{pmatrix}1 & -3\\3 & 1\end{pmatrix}\begin{pmatrix}4\\0\end{pmatrix}$ *f* $\begin{pmatrix}\frac{1}{2} & -\frac{1}{2}\\\frac{1}{4} & \frac{2}{3}\end{pmatrix}\begin{pmatrix}12\\6\end{pmatrix}$

7 In each of the following, find by multiplication of matrices a system of equations in x and y. Hence find x and y.

a $\begin{pmatrix}3 & 0\\0 & 2\end{pmatrix}\begin{pmatrix}x\\y\end{pmatrix}=\begin{pmatrix}15\\6\end{pmatrix}$ *b* $\begin{pmatrix}x & 0\\1 & y\end{pmatrix}\begin{pmatrix}2\\3\end{pmatrix}=\begin{pmatrix}6\\-1\end{pmatrix}$

c $\begin{pmatrix}3 & 1\\2 & -1\end{pmatrix}\begin{pmatrix}x\\y\end{pmatrix}=\begin{pmatrix}9\\1\end{pmatrix}$ *d* $\begin{pmatrix}2 & 3\\2 & 1\end{pmatrix}\begin{pmatrix}x\\y\end{pmatrix}=\begin{pmatrix}6\\-2\end{pmatrix}$

e $\begin{pmatrix}2 & 1\\1 & 3\end{pmatrix}\begin{pmatrix}x\\y\end{pmatrix}=\begin{pmatrix}4\\-3\end{pmatrix}$ *f* $\begin{pmatrix}x & y\\y & x\end{pmatrix}\begin{pmatrix}2\\1\end{pmatrix}=\begin{pmatrix}7\\8\end{pmatrix}$

8 $A = (3\quad 4)$ and $B = \begin{pmatrix}2\\1\end{pmatrix}$.

a Find $5A$, and hence the product $(5A)B$.

b Find $5B$, and hence the product $A(5B)$.

c Find AB, and hence the product $5(AB)$.

Comment on your results.

7 *Using matrices for geometrical transformations*

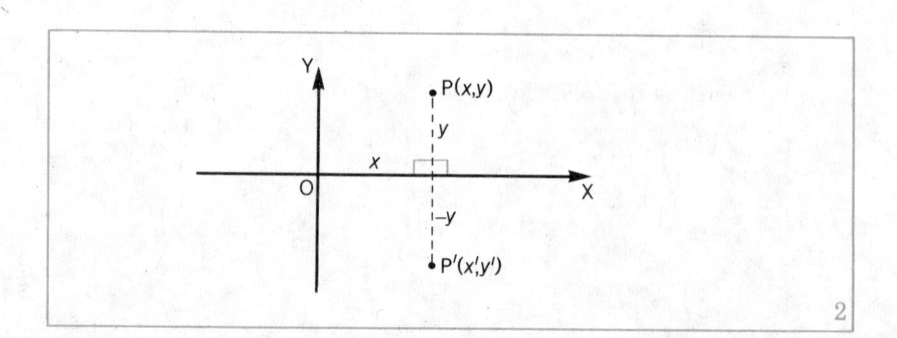

In Figure 2, the point $P'(x', y')$ is the image of the point $P(x, y)$ under reflection in the x-axis. So $x' = x$ and $y' = -y$. For example, the image of the point A(5, 3) is A′(5, −3).

Expressing x' and y' in terms of x and y, we get:

$$x' = \quad x = 1.x+0.y \quad \ldots\ldots \quad (1)$$

$$y' = -y = 0.x+(-1).y \quad \ldots\ldots \quad (2)$$

from which we can write

$$\begin{pmatrix} x' \\ y' \end{pmatrix} = \begin{pmatrix} 1.x+ & 0.y \\ 0.x+(-1).y \end{pmatrix} = \begin{pmatrix} 1 & 0 \\ 0 & -1 \end{pmatrix}\begin{pmatrix} x \\ y \end{pmatrix}$$

Hence the mapping $P(x, y) \to P'(x', y')$ is given by the *matrix equation*:

$$\begin{pmatrix} x' \\ y' \end{pmatrix} = \begin{pmatrix} 1 & 0 \\ 0 & -1 \end{pmatrix}\begin{pmatrix} x \\ y \end{pmatrix} \quad \ldots\ldots \quad (3)$$

The matrix $\begin{pmatrix} 1 & 0 \\ 0 & -1 \end{pmatrix}$ corresponds to the transformation of reflection in the x-axis.

Replacing x by 5 and y by 3 in equation (3),

$$\begin{pmatrix} x' \\ y' \end{pmatrix} = \begin{pmatrix} 1 & 0 \\ 0 & -1 \end{pmatrix}\begin{pmatrix} 5 \\ 3 \end{pmatrix} = \begin{pmatrix} 5 \\ -3 \end{pmatrix}$$

Therefore A′(5, −3) is the image of A(5, 3) as before.

Example 1. Under a mapping in which $P(x, y) \to P'(x', y')$, $x' = x+2y$ and $y' = 2x-3y$. Express the mapping in terms of matrices and use your result to find the image of P(7, −2).

We have $\begin{matrix} x' = \ x+2y \\ y' = 2x-3y \end{matrix}$ from which $\begin{pmatrix} x' \\ y' \end{pmatrix} = \begin{pmatrix} 1 & 2 \\ 2 & -3 \end{pmatrix}\begin{pmatrix} x \\ y \end{pmatrix}$.

Hence $\begin{pmatrix} x' \\ y' \end{pmatrix} = \begin{pmatrix} 1 & 2 \\ 2 & -3 \end{pmatrix}\begin{pmatrix} 7 \\ -2 \end{pmatrix} = \begin{pmatrix} 7-4 \\ 14+6 \end{pmatrix} = \begin{pmatrix} 3 \\ 20 \end{pmatrix}$, and so the image of P(7, −2) is P′(3, 20).

Example 2. With what geometrical transformation is the matrix $\begin{pmatrix} 0 & -1 \\ -1 & 0 \end{pmatrix}$ associated?

Suppose that under the transformation, $P(x, y) \to P'(x', y')$.

Then $\begin{pmatrix} x' \\ y' \end{pmatrix} = \begin{pmatrix} 0 & -1 \\ -1 & 0 \end{pmatrix}\begin{pmatrix} x \\ y \end{pmatrix} = \begin{pmatrix} -y \\ -x \end{pmatrix}$

So $x' = -y$, $y' = -x$,
and $P(x, y) \to P'(-y, -x)$.
Three points whose images are easy to find are A(1, 0), B(0, 1) and O(0, 0). These images A′(0, −1), B′(−1, 0) and O(0, 0) are shown in Figure 3 from which we can see that the transformation is the reflection of all points of the plane in the line with equation $y = -x$.

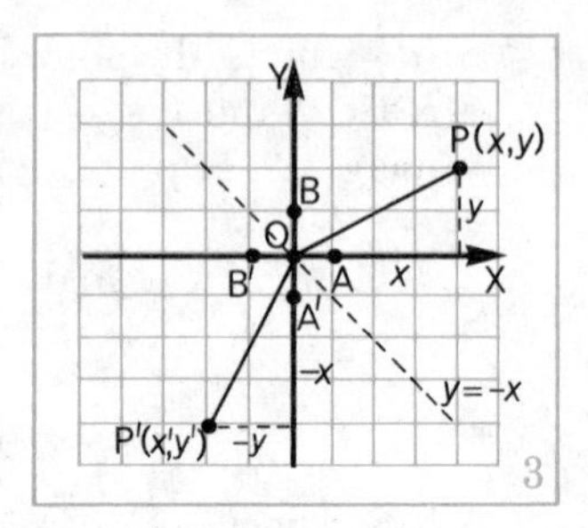

3

Exercise 7

In questions *1–4* answer the following questions for each of the given transformations.

1 Reflection in the y-axis, so that $x' = -x$ and $y' = y$ (see Figure 4).

a Express x' and y' in terms of x and y as in equations (1) and (2) on page 33.

b Hence write down the matrix equation for this transformation, corresponding to (3) on page 33.

c Use the matrix equation to find the image of P(6, 2) under the transformation. Illustrate in a sketch.

d Write down the matrix that corresponds to the transformation.

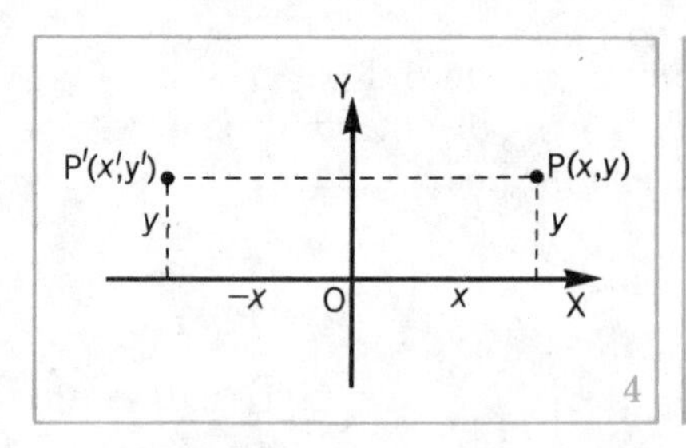

4

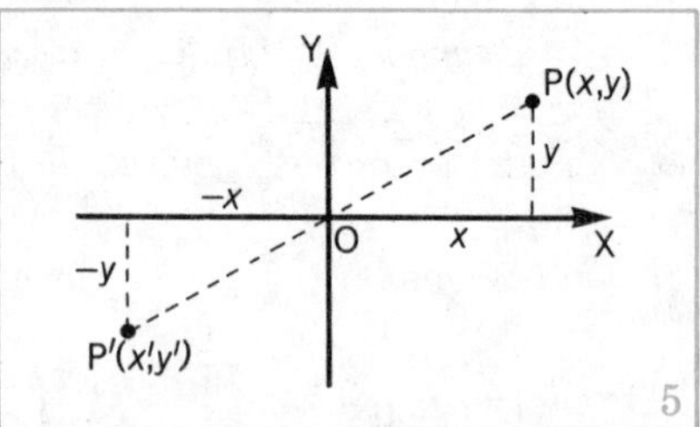

5

2 Half turn about the origin, so that $x' = -x$ and $y' = -y$ (see Figure 5).

3 Rotation of $+90°$ about the origin, so that $x' = -y$ and $y' = x$ (see Figure 6).

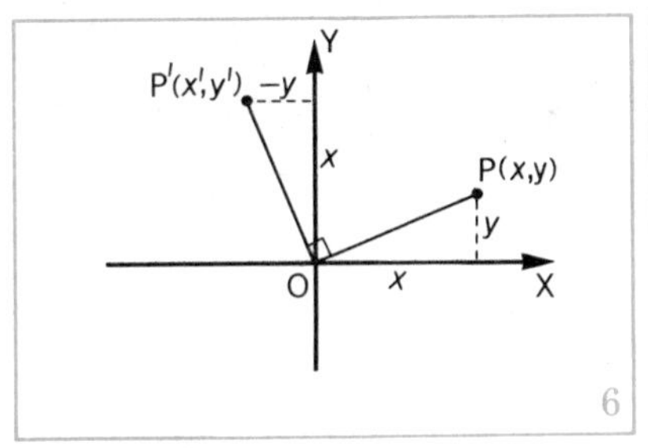

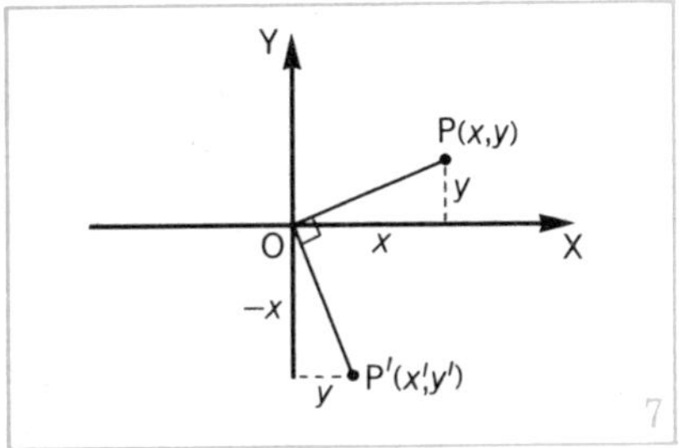

4 Rotation of $-90°$ about the origin, so that $x' = y$ and $y' = -x$ (see Figure 7).

5 Using the method of the Worked Example on page 33, find the geometrical transformation of all points of the plane associated with each of the following matrices:

a $\begin{pmatrix} 1 & 0 \\ 0 & 1 \end{pmatrix}$ *b* $\begin{pmatrix} 1 & 0 \\ 0 & -1 \end{pmatrix}$ *c* $\begin{pmatrix} 2 & 0 \\ 0 & 2 \end{pmatrix}$ *d* $\begin{pmatrix} 0 & 1 \\ 1 & 0 \end{pmatrix}$

6 Under a mapping in which $P(x, y) \rightarrow P'(x', y')$, $x' = 2x - y$ and $y' = x + 2y$.

a Write down the matrix equation for this mapping.

b Use this equation to find the image of the point A(1, 2).

c In the same way, find the image of each of the points B(−2, 1), C(−1, −2) and D(2, −1).

7 A mapping of $P(x, y) \rightarrow P'(x', y')$ is such that

$$\begin{pmatrix} x' \\ y' \end{pmatrix} = \begin{pmatrix} a & b \\ c & d \end{pmatrix} \begin{pmatrix} x \\ y \end{pmatrix}.$$

Given that (4, 2), (6, 3) are the respective images of (1, 0), (0, 3), find a, b, c and d.

8 A transformation of the plane maps $P(x, y) \rightarrow P'(x', y')$ such that

$$\begin{pmatrix} x' \\ y' \end{pmatrix} = \begin{pmatrix} a & b \\ c & d \end{pmatrix} \begin{pmatrix} x \\ y \end{pmatrix}.$$

Under this mapping, the images of the points (1, 0) and (3, 2) are (1, −2) and (5, −4) respectively. Find the matrix $\begin{pmatrix} a & b \\ c & d \end{pmatrix}$.

8 More about multiplication of matrices

Consider the system of equations:

$$x' = 2x + y$$
$$y' = 3x + 2y$$

Since $\begin{pmatrix} x' \\ y' \end{pmatrix} = \begin{pmatrix} 2x+y \\ 3x+2y \end{pmatrix} = \begin{pmatrix} 2 & 1 \\ 3 & 2 \end{pmatrix}\begin{pmatrix} x \\ y \end{pmatrix}$, we can write

$$\begin{pmatrix} x' \\ y' \end{pmatrix} = \begin{pmatrix} 2 & 1 \\ 3 & 2 \end{pmatrix}\begin{pmatrix} x \\ y \end{pmatrix},$$

which defines a mapping $(x, y) \to (x', y')$.

For example, the image A′ of A(−3, 5) is found by calculating $\begin{pmatrix} 2 & 1 \\ 3 & 2 \end{pmatrix}\begin{pmatrix} -3 \\ 5 \end{pmatrix} = \begin{pmatrix} -6+5 \\ -9+10 \end{pmatrix} = \begin{pmatrix} -1 \\ 1 \end{pmatrix}$, which gives A′(−1, 1).

In the same way, the image B′ of B(3, −2) is found by calculating $\begin{pmatrix} 2 & 1 \\ 3 & 2 \end{pmatrix}\begin{pmatrix} 3 \\ -2 \end{pmatrix} = \begin{pmatrix} 6-2 \\ 9-4 \end{pmatrix} = \begin{pmatrix} 4 \\ 5 \end{pmatrix}$, which gives B′(4, 5).

The *row into column* process of matrix multiplication suggests that the coordinates of A′ and B′ can be found by *combining* the two calculations as follows:

$$\begin{pmatrix} 2 & 1 \\ 3 & 2 \end{pmatrix}\begin{pmatrix} -3 & 3 \\ 5 & -2 \end{pmatrix} = \begin{pmatrix} 2\times(-3)+1\times5 & 2\times3+1\times(-2) \\ 3\times(-3)+2\times5 & 3\times3+2\times(-2) \end{pmatrix}$$
$$= \begin{pmatrix} -6+5 & 6-2 \\ -9+10 & 9-4 \end{pmatrix} = \begin{pmatrix} -1 & 4 \\ 1 & 5 \end{pmatrix}$$

This leads to the definition of the product of two 2×2 matrices:

$$\begin{pmatrix} a & b \\ c & d \end{pmatrix}\begin{pmatrix} p & q \\ r & s \end{pmatrix} = \begin{pmatrix} ap+br & aq+bs \\ cp+dr & cq+ds \end{pmatrix}$$

The 2×2 matrix on the right-hand side can be obtained by 'diving' each row of the first matrix down each column of the second matrix and adding products as indicated below:

First column ↓   First entry

$$\text{First row} \to \begin{pmatrix} a & b \\ \cdot & \cdot \end{pmatrix}\begin{pmatrix} p & \cdot \\ r & \cdot \end{pmatrix} = \begin{pmatrix} ap+br & \cdot \\ \cdot & \cdot \end{pmatrix}$$

Similarly,

$$\begin{pmatrix} a & b \\ \cdot & \cdot \end{pmatrix}\begin{pmatrix} \cdot & q \\ \cdot & s \end{pmatrix} = \begin{pmatrix} \cdot & aq+bs \\ \cdot & \cdot \end{pmatrix}$$

$$\begin{pmatrix} \cdot & \cdot \\ c & d \end{pmatrix}\begin{pmatrix} p & \cdot \\ r & \cdot \end{pmatrix} = \begin{pmatrix} \cdot & \cdot \\ cp+dr & \cdot \end{pmatrix}$$

$$\begin{pmatrix} \cdot & \cdot \\ c & d \end{pmatrix}\begin{pmatrix} \cdot & q \\ \cdot & s \end{pmatrix} = \begin{pmatrix} \cdot & \cdot \\ \cdot & cq+ds \end{pmatrix}$$

Example. Given $P = \begin{pmatrix} 1 & 2 \\ 3 & 1 \end{pmatrix}$ and $Q = \begin{pmatrix} 4 & 5 \\ 2 & 0 \end{pmatrix}$, find PQ and QP.

$$PQ = \begin{pmatrix} 1 & 2 \\ 3 & 1 \end{pmatrix}\begin{pmatrix} 4 & 5 \\ 2 & 0 \end{pmatrix} = \begin{pmatrix} 4+4 & 5+0 \\ 12+2 & 15+0 \end{pmatrix} = \begin{pmatrix} 8 & 5 \\ 14 & 15 \end{pmatrix}$$

$$QP = \begin{pmatrix} 4 & 5 \\ 2 & 0 \end{pmatrix}\begin{pmatrix} 1 & 2 \\ 3 & 1 \end{pmatrix} = \begin{pmatrix} 4+15 & 8+5 \\ 2+0 & 4+0 \end{pmatrix} = \begin{pmatrix} 19 & 13 \\ 2 & 4 \end{pmatrix}$$

Notice that $PQ \neq QP$, so multiplication of matrices in general is not commutative.

To avoid ambiguity in the multiplication of matrices, PQ may be described as P post-multiplied by Q, or Q pre-multiplied by P.

Exercise 8

1 Calculate:

a $\begin{pmatrix} 1 & 2 \\ 3 & 1 \end{pmatrix}\begin{pmatrix} 2 & 1 \\ 1 & 5 \end{pmatrix}$ b $\begin{pmatrix} 2 & 1 \\ 1 & 5 \end{pmatrix}\begin{pmatrix} 1 & 2 \\ 3 & 1 \end{pmatrix}$ c $\begin{pmatrix} 1 & 2 \\ 2 & 4 \end{pmatrix}\begin{pmatrix} 3 & 2 \\ 1 & 4 \end{pmatrix}$

d $\begin{pmatrix} 3 & 2 \\ 1 & 4 \end{pmatrix}\begin{pmatrix} 1 & 2 \\ 2 & 4 \end{pmatrix}$ e $\begin{pmatrix} 1 & 0 \\ 2 & 3 \end{pmatrix}\begin{pmatrix} 2 & 3 \\ 3 & 4 \end{pmatrix}$ f $\begin{pmatrix} 2 & 3 \\ 3 & 4 \end{pmatrix}\begin{pmatrix} 1 & 0 \\ 2 & 3 \end{pmatrix}$

2 Find the following products in their simplest form:

a $\begin{pmatrix} 1 & 0 \\ 0 & -1 \end{pmatrix}\begin{pmatrix} 2 & 3 \\ 4 & 5 \end{pmatrix}$ b $\begin{pmatrix} 3 & 1 \\ 5 & 2 \end{pmatrix}\begin{pmatrix} 2 & -1 \\ 1 & 2 \end{pmatrix}$

c $\begin{pmatrix} 0 & 2 \\ 3 & 1 \end{pmatrix}\begin{pmatrix} 4 & -1 \\ -2 & 3 \end{pmatrix}$ d $\begin{pmatrix} -1 & 2 \\ 0 & 3 \end{pmatrix}\begin{pmatrix} -1 & 2 \\ 0 & 3 \end{pmatrix}$

e $\begin{pmatrix} 3 & 1 \\ 5 & 2 \end{pmatrix}\begin{pmatrix} 2 & -1 \\ -5 & 3 \end{pmatrix}$ f $\begin{pmatrix} -2 & 3 \\ -1 & 5 \end{pmatrix}\begin{pmatrix} 3 & 1 \\ 2 & 4 \end{pmatrix}$

3 Given that $P = \begin{pmatrix} 1 & 2 \\ 3 & 4 \end{pmatrix}$, $Q = \begin{pmatrix} 3 & 2 \\ 1 & 4 \end{pmatrix}$, $R = \begin{pmatrix} 2 & 5 \\ 3 & 1 \end{pmatrix}$, find:

a PQ b QP c QR d RQ e PR f RP

4 Work out the following products:

a $\begin{pmatrix}1 & 2\\3 & 4\end{pmatrix}\begin{pmatrix}1 & 0\\0 & 1\end{pmatrix}$ b $\begin{pmatrix}3 & -2\\-4 & 5\end{pmatrix}\begin{pmatrix}1 & 0\\0 & 1\end{pmatrix}$ c $\begin{pmatrix}a & b\\c & d\end{pmatrix}\begin{pmatrix}1 & 0\\0 & 1\end{pmatrix}$

d $\begin{pmatrix}1 & 0\\0 & 1\end{pmatrix}\begin{pmatrix}1 & 2\\3 & 4\end{pmatrix}$ e $\begin{pmatrix}1 & 0\\0 & 1\end{pmatrix}\begin{pmatrix}3 & -2\\-4 & 5\end{pmatrix}$ f $\begin{pmatrix}1 & 0\\0 & 1\end{pmatrix}\begin{pmatrix}a & b\\c & d\end{pmatrix}$

$\begin{pmatrix}1 & 0\\0 & 1\end{pmatrix}$ is called the *unit matrix* of order 2, and is denoted by I. It behaves like unity in the real number system.

If A is a 2×2 matrix, then $AI = IA = A$.

5 Work out the following products:

a $\begin{pmatrix}1 & 2\\3 & 4\end{pmatrix}\begin{pmatrix}0 & 0\\0 & 0\end{pmatrix}$ b $\begin{pmatrix}a & b\\c & d\end{pmatrix}\begin{pmatrix}0 & 0\\0 & 0\end{pmatrix}$ c $\begin{pmatrix}0 & 0\\0 & 0\end{pmatrix}\begin{pmatrix}p & q\\r & s\end{pmatrix}$

6 Find a single matrix equal to each of the following:

a $\begin{pmatrix}1 & 1\\1 & -1\end{pmatrix}\begin{pmatrix}a & b\\c & d\end{pmatrix}$ b $\begin{pmatrix}1 & k\\0 & 1\end{pmatrix}\begin{pmatrix}a & b\\c & d\end{pmatrix}$ c $\begin{pmatrix}1 & 0\\0 & k\end{pmatrix}\begin{pmatrix}a & b\\c & d\end{pmatrix}$

d $\begin{pmatrix}a & b\\c & d\end{pmatrix}\begin{pmatrix}1 & 1\\1 & -1\end{pmatrix}$ e $\begin{pmatrix}a & b\\c & d\end{pmatrix}\begin{pmatrix}1 & k\\0 & 1\end{pmatrix}$ f $\begin{pmatrix}a & b\\c & d\end{pmatrix}\begin{pmatrix}1 & 0\\0 & k\end{pmatrix}$

7 If $A = \begin{pmatrix}0 & 0\\1 & 0\end{pmatrix}$ and $B = \begin{pmatrix}1 & 0\\0 & 0\end{pmatrix}$, show that $AB \neq \boldsymbol{O}$ but $BA = \boldsymbol{O}$.

8 Show that if $X = \begin{pmatrix}3 & -4\\1 & -1\end{pmatrix}$, then $X.X - 2X + I = \boldsymbol{O}$.

Exercise 8B

1 Given $A = \begin{pmatrix}0 & 2\\4 & 3\end{pmatrix}$, $B = \begin{pmatrix}1 & -1\\3 & 2\end{pmatrix}$ and $C = \begin{pmatrix}5 & 0\\1 & -2\end{pmatrix}$, find:

a $B+C$ b $A(B+C)$ c AB d AC e $AB+AC$

What law connecting matrix multiplication and matrix addition is suggested by your answers to ***b*** and ***e***?

2 $A = \begin{pmatrix}1 & 2\\3 & 4\end{pmatrix}$, $B = \begin{pmatrix}2 & 5\\3 & 4\end{pmatrix}$, $C = \begin{pmatrix}1 & -1\\-1 & 1\end{pmatrix}$. Simplify:

a AB b BC c $(AB)C$ d $A(BC)$

What law is suggested by your answers to ***c*** and ***d***?

3 a Given $P = \begin{pmatrix}1 & 3\\4 & 2\end{pmatrix}$ and $Q = \begin{pmatrix}4 & -2\\0 & 3\end{pmatrix}$, find PQ and QP.

Is multiplication of matrices a commutative operation?

b If k is a non-zero real number, verify that:

(1) $(kP)Q = k(PQ)$ (2) $k(PQ) = P(kQ)$

4 Powers of a square matrix A are defined as follows:

$A^2 = A.A$, $A^3 = A.A^2$, $A^4 = A.A^3$, and so on.

Given that $A = \begin{pmatrix} 1 & 2 \\ 3 & 4 \end{pmatrix}$ and $B = \begin{pmatrix} 1 & -1 \\ 2 & 3 \end{pmatrix}$, calculate:

a A^2 *b* B^2 *c* A^3 *d* B^3

5 Show that if $X = \begin{pmatrix} 1 & 3 \\ 0 & -2 \end{pmatrix}$, $X^2 + 3X + 4I = 2\begin{pmatrix} 4 & 3 \\ 0 & 1 \end{pmatrix}$

6 *a* If $\left.\begin{aligned} x &= 3a - 2b \\ y &= a + b \end{aligned}\right\}$ and $\left.\begin{aligned} a &= 2p + 3q \\ b &= 5p - 2q \end{aligned}\right\}$ express x and y in terms of p and q by direct substitution.

b By writing these equations in matrix form, $\begin{pmatrix} x \\ y \end{pmatrix} = \begin{pmatrix} 3 & -2 \\ 1 & 1 \end{pmatrix}\begin{pmatrix} a \\ b \end{pmatrix}$ and $\begin{pmatrix} a \\ b \end{pmatrix} = \begin{pmatrix} 2 & 3 \\ 5 & -2 \end{pmatrix}\begin{pmatrix} p \\ q \end{pmatrix}$, show how to derive result ***a*** by matrix multiplication.

Application to geometrical transformations

In Section 7 we found that the following matrices represent the geometrical transformations of all points of the plane as shown:

$\begin{pmatrix} 1 & 0 \\ 0 & 1 \end{pmatrix}$ ... no change (identity transformation)

$\begin{pmatrix} 1 & 0 \\ 0 & -1 \end{pmatrix}$ ... reflection in x-axis

$\begin{pmatrix} -1 & 0 \\ 0 & 1 \end{pmatrix}$ ... reflection in y-axis

$\begin{pmatrix} -1 & 0 \\ 0 & -1 \end{pmatrix}$ ... reflection in origin (or half turn about origin)

$\begin{pmatrix} 0 & 1 \\ 1 & 0 \end{pmatrix}$ ... reflection in $y = x$

$\begin{pmatrix} 0 & 1 \\ -1 & 0 \end{pmatrix}$ ... rotation $-90°$ about O

$\begin{pmatrix} 0 & -1 \\ 1 & 0 \end{pmatrix}$ ... rotation $+90°$ about O

$\begin{pmatrix} 0 & -1 \\ -1 & 0 \end{pmatrix}$ ... reflection in $y = -x$

In using geometrical transformations we find it useful to extend the multiplication of matrices to obtain the product of a 2×2 matrix and a $2 \times n$ matrix (where n is an integer greater than 2), in the form of a $2 \times n$ product matrix.

Example. $\triangle$ABC has vertices A$(-2, 1)$, B$(4, 2)$, C$(-1, 5)$. Find the coordinates of the vertices of the image of $\triangle$ABC under reflection in the x-axis.

Using $\begin{pmatrix} x' \\ y' \end{pmatrix} = \begin{pmatrix} 1 & 0 \\ 0 & -1 \end{pmatrix}\begin{pmatrix} x \\ y \end{pmatrix}$, we have

$$\begin{pmatrix} 1 & 0 \\ 0 & -1 \end{pmatrix}\overset{\begin{matrix}\text{A} & \text{B} & \text{C}\end{matrix}}{\begin{pmatrix} -2 & 4 & -1 \\ 1 & 2 & 5 \end{pmatrix}} = \overset{\begin{matrix}\text{A}' & \text{B}' & \text{C}'\end{matrix}}{\begin{pmatrix} -2 & 4 & -1 \\ -1 & -2 & -5 \end{pmatrix}}$$

So A$'$ is $(-2, -1)$, B$'$ is $(4, -2)$ and C$'$ is $(-1, -5)$

Exercise 9

Use matrix products to find the images of the triangles formed by joining the points in questions ***1–6*** under the given transformations. In each case finish by sketching the triangle and its image.

1 A(2, 2), B(4, 0), C(6, 5) under reflection in the x-axis.

2 D(0, 1), E(1, 3), F(0, 5) under reflection in the y-axis.

3 G(4, 4), H(2, 0), K(1, 3) under reflection in the line $y = x$.

4 L$(-3, 5)$, M(0, 5), N$(-1, -1)$ under reflection in the origin.

5 O(0, 0), P(6, 1), Q$(6, -1)$ under a rotation of $-90°$ about O.

6 R(2, 2), S$(-3, 1)$, T$(-1, -4)$ under reflection in the line $y = -x$.

7 Find the image of the unit square O(0, 0), A(1, 0), B(1, 1), C(0, 1) under the transformation given by each of the following matrices; illustrate each in a sketch.

a $\begin{pmatrix} 3 & 0 \\ 0 & 3 \end{pmatrix}$ *b* $\begin{pmatrix} 2 & 1 \\ 1 & 2 \end{pmatrix}$ *c* $\begin{pmatrix} 2 & 2 \\ 1 & 3 \end{pmatrix}$ *d* $\begin{pmatrix} -2 & 0 \\ 0 & -2 \end{pmatrix}$

8 Quadrilateral OPQR has vertices O(0, 0), P(2, 2), Q$(4, -1)$, R$(0, -5)$. Find the coordinates of the vertices of its image under the transformation given by the matrix $\begin{pmatrix} 2 & 1 \\ 3 & -2 \end{pmatrix}$.

9 Under a certain mapping P$(x, y) \rightarrow$ P$'(x', y')$ such that $x' = 3x-4y$ and $y' = 4x+3y$. Form a matrix equation, and hence find the image of the set $\{(4, 2), (-1, 3), (1, 6)\}$.

10 Repeat question ***9*** for the mapping given by $x' = x+y$, $y' = 2x+2y$, and the set of points $\{(0, 0), (3, -1), (-2, 2)\}$.

11*a* Find the images of the points (3, 4), (2, −1), (−3, 5), (−7, −3) under the mapping with matrix $\begin{pmatrix} 4 & 2 \\ 2 & 1 \end{pmatrix}$.

b Show that the images lie on a straight line, and find its equation.

9 The inverse of a square matrix of order 2

Let $A = \begin{pmatrix} a & b \\ c & d \end{pmatrix}$ and $I = \begin{pmatrix} 1 & 0 \\ 0 & 1 \end{pmatrix}$.

Pre-multiplying A by I, $IA = \begin{pmatrix} 1 & 0 \\ 0 & 1 \end{pmatrix}\begin{pmatrix} a & b \\ c & d \end{pmatrix} = \begin{pmatrix} a & b \\ c & d \end{pmatrix} = A$.

Post-multiplying A by I, $AI = \begin{pmatrix} a & b \\ c & d \end{pmatrix}\begin{pmatrix} 1 & 0 \\ 0 & 1 \end{pmatrix} = \begin{pmatrix} a & b \\ c & d \end{pmatrix} = A$.

Therefore $IA = AI = A$.

For this reason, the unit 2×2 matrix I is called the *identity matrix* for multiplication of 2×2 matrices. Note that A commutes with I, i.e. $IA = AI$.

Now let $P = \begin{pmatrix} 3 & 2 \\ 7 & 5 \end{pmatrix}$ and $Q = \begin{pmatrix} 5 & -2 \\ -7 & 3 \end{pmatrix}$.

Pre-multiplying Q by P, $PQ = \begin{pmatrix} 3 & 2 \\ 7 & 5 \end{pmatrix}\begin{pmatrix} 5 & -2 \\ -7 & 3 \end{pmatrix} = \begin{pmatrix} 1 & 0 \\ 0 & 1 \end{pmatrix} = I$

Post-multiplying Q by P, $QP = \begin{pmatrix} 5 & -2 \\ -7 & 3 \end{pmatrix}\begin{pmatrix} 3 & 2 \\ 7 & 5 \end{pmatrix} = \begin{pmatrix} 1 & 0 \\ 0 & 1 \end{pmatrix} = I$

Therefore $PQ = QP = I$.

For this reason Q is called the *multiplicative inverse* of P and is denoted by P^{-1}. Also, P is the multiplicative inverse of Q and is therefore denoted by Q^{-1}.

It is customary to use the phrase *inverse of a matrix* to refer to its multiplicative inverse, since its additive inverse is usually called its negative. We now make the following definition.

If A and B are square matrices of the same order such that $AB = BA = I$, then B is an inverse of A and A is an inverse of B.

Example 1. If $A = \begin{pmatrix} 5 & -2 \\ 3 & -1 \end{pmatrix}$ and $B = \begin{pmatrix} -1 & 2 \\ -3 & 5 \end{pmatrix}$, show that A and B are inverses of each other.

We have to show that $AB = I = BA$.

$$AB = \begin{pmatrix} 5 & -2 \\ 3 & -1 \end{pmatrix}\begin{pmatrix} -1 & 2 \\ -3 & 5 \end{pmatrix} = \begin{pmatrix} 1 & 0 \\ 0 & 1 \end{pmatrix} = I$$

$$BA = \begin{pmatrix} -1 & 2 \\ -3 & 5 \end{pmatrix}\begin{pmatrix} 5 & -2 \\ 3 & -1 \end{pmatrix} = \begin{pmatrix} 1 & 0 \\ 0 & 1 \end{pmatrix} = I$$

Since $AB = I = BA$, A and B are inverses of each other.

Example 2. Show that $\begin{pmatrix} -1 & 0 \\ 0 & 1 \end{pmatrix}$ is its own inverse. Explain the result geometrically.

$\begin{pmatrix} -1 & 0 \\ 0 & 1 \end{pmatrix}\begin{pmatrix} -1 & 0 \\ 0 & 1 \end{pmatrix} = \begin{pmatrix} 1 & 0 \\ 0 & 1 \end{pmatrix}$, so $\begin{pmatrix} -1 & 0 \\ 0 & 1 \end{pmatrix}$ is its own inverse.

The matrix represents reflection in the y-axis. The product represents reflection in the y-axis, followed by reflection in the y-axis, which gives the identity transformation. We can see this as follows:

$\begin{pmatrix} -1 & 0 \\ 0 & 1 \end{pmatrix}\begin{pmatrix} x \\ y \end{pmatrix} = \begin{pmatrix} -x \\ y \end{pmatrix}$, and $\begin{pmatrix} -1 & 0 \\ 0 & 1 \end{pmatrix}\begin{pmatrix} -x \\ y \end{pmatrix} = \begin{pmatrix} x \\ y \end{pmatrix}$.

So $(x, y) \rightarrow (-x, y) \rightarrow (x, y)$.

Exercise 10

In questions ***1*** to ***6***, show that each matrix is the inverse of the other.

1 $\begin{pmatrix} 3 & 2 \\ 1 & 1 \end{pmatrix}$ and $\begin{pmatrix} 1 & -2 \\ -1 & 3 \end{pmatrix}$

2 $\begin{pmatrix} 1 & 1 \\ 1 & 2 \end{pmatrix}$ and $\begin{pmatrix} 2 & -1 \\ -1 & 1 \end{pmatrix}$

3 $\begin{pmatrix} 3 & 4 \\ 5 & 7 \end{pmatrix}$ and $\begin{pmatrix} 7 & -4 \\ -5 & 3 \end{pmatrix}$

4 $\begin{pmatrix} 5 & -7 \\ -2 & 3 \end{pmatrix}$ and $\begin{pmatrix} 3 & 7 \\ 2 & 5 \end{pmatrix}$

5 $\begin{pmatrix} 3 & -2 \\ -4 & 3 \end{pmatrix}$ and $\begin{pmatrix} 3 & 2 \\ 4 & 3 \end{pmatrix}$

6 $\begin{pmatrix} 2 & 5 \\ 3 & 8 \end{pmatrix}$ and $\begin{pmatrix} 8 & -5 \\ -3 & 2 \end{pmatrix}$

7 Study the pattern in the entries of the pairs of matrices in questions ***1*** to ***6***. Use this pattern to write down a possible inverse of each of the following matrices, and check by multiplication:

a $\begin{pmatrix} 2 & 1 \\ 1 & 1 \end{pmatrix}$ *b* $\begin{pmatrix} 2 & 3 \\ 3 & 5 \end{pmatrix}$ *c* $\begin{pmatrix} 4 & 3 \\ 9 & 7 \end{pmatrix}$

d $\begin{pmatrix} 2 & -3 \\ -1 & 2 \end{pmatrix}$ *e* $\begin{pmatrix} 9 & -5 \\ -7 & 4 \end{pmatrix}$ *f* $\begin{pmatrix} 5 & -7 \\ 3 & -4 \end{pmatrix}$

8 *a* $I = \begin{pmatrix} 1 & 0 \\ 0 & 1 \end{pmatrix}$. Show that I is its own inverse, i.e. $I.I = I$.

b $A = \begin{pmatrix} 2 & 0 \\ 0 & 2 \end{pmatrix}$. Write down A^{-1}, the inverse of A.

[*Hint.* $\begin{pmatrix} 2 & 0 \\ 0 & 2 \end{pmatrix} = 2\begin{pmatrix} 1 & 0 \\ 0 & 1 \end{pmatrix}$]

9 $M = \begin{pmatrix} 3 & 5 \\ 1 & 2 \end{pmatrix}$. Find, as in question 7, M^{-1}.
Investigate whether or not the *squares* of M and M^{-1} are also inverses of one another ($M^2 = M.M$, etc.).

10 Show that each of the following matrices is its own inverse, and explain the results geometrically.

a $\begin{pmatrix} 1 & 0 \\ 0 & -1 \end{pmatrix}$ *b* $\begin{pmatrix} -1 & 0 \\ 0 & -1 \end{pmatrix}$ *c* $\begin{pmatrix} 0 & 1 \\ 1 & 0 \end{pmatrix}$ *d* $\begin{pmatrix} 0 & -1 \\ -1 & 0 \end{pmatrix}$

11 Show that each of the matrices $\begin{pmatrix} 0 & -1 \\ 1 & 0 \end{pmatrix}$ and $\begin{pmatrix} 0 & 1 \\ -1 & 0 \end{pmatrix}$ is the inverse of the other, and explain this geometrically.

12*a* Find the coordinates of the images A′, B′, C′ of the points A(1, 4), B(3, −2), C(−5, 7) respectively under the mapping whose matrix is $X = \begin{pmatrix} 2 & 1 \\ 1 & 1 \end{pmatrix}$.

b Write down the matrix X^{-1}.

c Find the images of the points A′, B′, C′ under the mapping whose matrix is X^{-1}.

d What is the effect of the mapping whose matrix is X^{-1}?

10 *More about inverses of square matrices of order 2*

Given a square matrix A of order 2, can we always find another matrix B such that $AB = BA = I$?

From your answers to question 7 of Exercise 10, did you notice that

(i) *the difference of the 'cross-products' of the entries was always 1?*

For example, from $\begin{pmatrix} 2 & 3 \\ 3 & 5 \end{pmatrix}$, $(2\times 5)-(3\times 3) = 10-9 = 1$

Note the order—the main diagonal product first.

(ii) *the inverse matrix could be found by interchanging the entries in the main diagonal, and changing the signs of the entries in the other diagonal?*

For example, $\begin{pmatrix} 2 & 1 \\ 1 & 1 \end{pmatrix}$ has inverse $\begin{pmatrix} 1 & -1 \\ -1 & 2 \end{pmatrix}$,

and $\begin{pmatrix} 5 & -7 \\ 3 & -4 \end{pmatrix}$ has inverse $\begin{pmatrix} -4 & 7 \\ -3 & 5 \end{pmatrix}$.

If we do this for the matrix $\begin{pmatrix} 5 & 3 \\ 6 & 4 \end{pmatrix}$ we obtain $\begin{pmatrix} 4 & -3 \\ -6 & 5 \end{pmatrix}$.

Pre-multiplying, $\begin{pmatrix} 4 & -3 \\ -6 & 5 \end{pmatrix}\begin{pmatrix} 5 & 3 \\ 6 & 4 \end{pmatrix} = \begin{pmatrix} 2 & 0 \\ 0 & 2 \end{pmatrix} = 2\begin{pmatrix} 1 & 0 \\ 0 & 1 \end{pmatrix}$

Post-multiplying, $\begin{pmatrix} 5 & 3 \\ 6 & 4 \end{pmatrix}\begin{pmatrix} 4 & -3 \\ -6 & 5 \end{pmatrix} = \begin{pmatrix} 2 & 0 \\ 0 & 2 \end{pmatrix} = 2\begin{pmatrix} 1 & 0 \\ 0 & 1 \end{pmatrix}$

Hence $\frac{1}{2}\begin{pmatrix} 4 & -3 \\ -6 & 5 \end{pmatrix}\begin{pmatrix} 5 & 3 \\ 6 & 4 \end{pmatrix} = \begin{pmatrix} 1 & 0 \\ 0 & 1 \end{pmatrix}$, so that $\frac{1}{2}\begin{pmatrix} 4 & -3 \\ -6 & 5 \end{pmatrix}$, or

$\begin{pmatrix} 2 & -\frac{3}{2} \\ -3 & \frac{5}{2} \end{pmatrix}$, is the inverse of $\begin{pmatrix} 5 & 3 \\ 6 & 4 \end{pmatrix}$.

Does every 2 × 2 matrix have an inverse?

The general 2×2 matrix is $A = \begin{pmatrix} a & b \\ c & d \end{pmatrix}$.

Pre-multiplying A by $\begin{pmatrix} d & -b \\ -c & a \end{pmatrix}$, we have

$$\begin{pmatrix} d & -b \\ -c & a \end{pmatrix}\begin{pmatrix} a & b \\ c & d \end{pmatrix} = \begin{pmatrix} ad-bc & 0 \\ 0 & ad-bc \end{pmatrix} = (ad-bc)\begin{pmatrix} 1 & 0 \\ 0 & 1 \end{pmatrix}$$

Hence $\left[\dfrac{1}{ad-bc}\begin{pmatrix} d & -b \\ -c & a \end{pmatrix}\right]\begin{pmatrix} a & b \\ c & d \end{pmatrix} = \begin{pmatrix} 1 & 0 \\ 0 & 1 \end{pmatrix}$

Similarly, post-multiplying A by $\dfrac{1}{ad-bc}\begin{pmatrix} d & -b \\ -c & a \end{pmatrix}$ we obtain

$$\begin{pmatrix} a & b \\ c & d \end{pmatrix}\left[\frac{1}{ad-bc}\begin{pmatrix} d & -b \\ -c & a \end{pmatrix}\right] = \begin{pmatrix} 1 & 0 \\ 0 & 1 \end{pmatrix}$$

It follows that if $ad-bc \neq 0$, the matrix $A = \begin{pmatrix} a & b \\ c & d \end{pmatrix}$ has inverse

$A^{-1} = \dfrac{1}{ad-bc}\begin{pmatrix} d & -b \\ -c & a \end{pmatrix}$.

$ad-bc$ is called the *determinant* of the matrix A, and is written det A. If det $A = 0$, A does not have an inverse, and is called a *singular* matrix.

Example. Given $P = \begin{pmatrix} 4 & 3 \\ -1 & -2 \end{pmatrix}$, find, if possible, P^{-1}.

det $P = 4.(-2)-(-1).3 = -8+3 = -5 \neq 0$.
Hence P^{-1} exists.

$$P^{-1} = \frac{1}{\det P}\begin{pmatrix} -2 & -3 \\ 1 & 4 \end{pmatrix} = \frac{1}{-5}\begin{pmatrix} -2 & -3 \\ 1 & 4 \end{pmatrix} = \begin{pmatrix} \frac{2}{5} & \frac{3}{5} \\ -\frac{1}{5} & -\frac{4}{5} \end{pmatrix}$$

Exercise 11

State whether each of the matrices in questions *1* to *15* has an inverse. If the inverse exists, find it.

1 $A = \begin{pmatrix} 4 & 3 \\ 2 & 2 \end{pmatrix}$ *2* $A = \begin{pmatrix} 2 & 1 \\ 4 & 3 \end{pmatrix}$ *3* $A = \begin{pmatrix} 3 & 2 \\ 1 & 2 \end{pmatrix}$

4 $A = \begin{pmatrix} 4 & 2 \\ 10 & 5 \end{pmatrix}$ *5* $A = \begin{pmatrix} 7 & 4 \\ 16 & 9 \end{pmatrix}$ *6* $A = \begin{pmatrix} 2 & 3 \\ 1 & 4 \end{pmatrix}$

7 $A = \begin{pmatrix} 2 & 1 \\ 4 & 2 \end{pmatrix}$ *8* $A = \begin{pmatrix} 5 & 7 \\ 6 & 9 \end{pmatrix}$ *9* $A = \begin{pmatrix} 1 & 1 \\ 1 & 0 \end{pmatrix}$

10 $A = \begin{pmatrix} 2 & -3 \\ 1 & 5 \end{pmatrix}$ *11* $A = \begin{pmatrix} 1 & 6 \\ 3 & 4 \end{pmatrix}$ *12* $A = \begin{pmatrix} 2 & 3 \\ -1 & 2 \end{pmatrix}$

13 $A = \begin{pmatrix} 3 & -1 \\ -1 & 2 \end{pmatrix}$ *14* $A = \begin{pmatrix} -2 & 4 \\ 1 & -1 \end{pmatrix}$ *15* $A = \begin{pmatrix} -1 & 2 \\ 0 & 3 \end{pmatrix}$

16 Given the matrices $A = \begin{pmatrix} 2 & 3 \\ 0 & 1 \end{pmatrix}$ and $B = \begin{pmatrix} 2 & 5 \\ 1 & 3 \end{pmatrix}$, calculate:

a AB *b* BA *c* A^{-1} *d* B^{-1}

e $(AB)^{-1}$ *f* $A^{-1}B^{-1}$ *g* $B^{-1}A^{-1}$ *h* $(BA)^{-1}$

State which products are equal.

11 Using matrices to solve systems of linear equations

Consider the following system of equations in which x and y are variables on the set of real numbers.

$$\left.\begin{aligned} 3x + y &= 9 \\ 3x + 2y &= 12 \end{aligned}\right\} \quad \ldots \ldots \quad (1)$$

Since $\begin{pmatrix} 3x + y \\ 3x + 2y \end{pmatrix} = \begin{pmatrix} 3 & 1 \\ 3 & 2 \end{pmatrix}\begin{pmatrix} x \\ y \end{pmatrix}$, system (1) may be written as a single-matrix equation:

$$\begin{pmatrix} 3 & 1 \\ 3 & 2 \end{pmatrix}\begin{pmatrix} x \\ y \end{pmatrix} = \begin{pmatrix} 9 \\ 12 \end{pmatrix} \quad \ldots \ldots \quad (2)$$

If we can find an equation equivalent to (2) of the form $\begin{pmatrix} x \\ y \end{pmatrix} = \begin{pmatrix} a \\ b \end{pmatrix}$, the solution of the system can be written down at once. To do this, we make use of the fact that the product of a matrix and its inverse is I and proceed as follows.

For $\begin{pmatrix} 3 & 1 \\ 3 & 2 \end{pmatrix}$, det $= (3 \times 2) - (1 \times 3) = 3$

and so $\begin{pmatrix} 3 & 1 \\ 3 & 2 \end{pmatrix}^{-1} = \frac{1}{3}\begin{pmatrix} 2 & -1 \\ -3 & 3 \end{pmatrix}$.

Pre-multiplying both sides of (2) by the inverse $\frac{1}{3}\begin{pmatrix} 2 & -1 \\ -3 & 3 \end{pmatrix}$,

$$\frac{1}{3}\begin{pmatrix} 2 & -1 \\ -3 & 3 \end{pmatrix}\begin{pmatrix} 3 & 1 \\ 3 & 2 \end{pmatrix}\begin{pmatrix} x \\ y \end{pmatrix} = \frac{1}{3}\begin{pmatrix} 2 & -1 \\ -3 & 3 \end{pmatrix}\begin{pmatrix} 9 \\ 12 \end{pmatrix}$$

$$\Leftrightarrow \quad \begin{pmatrix}1 & 0\\0 & 1\end{pmatrix}\begin{pmatrix}x\\y\end{pmatrix} = \tfrac{1}{3}\begin{pmatrix}6\\9\end{pmatrix}$$

$$\Leftrightarrow \quad \begin{pmatrix}x\\y\end{pmatrix} = \begin{pmatrix}2\\3\end{pmatrix}$$

Hence $x = 2$ and $y = 3$, which gives $\{(2, 3)\}$ as the solution set of the system.

Replacing x by 2 and y by 3 in (1) readily verifies that $\{(2, 3)\}$ is the solution set of the system. This check is always worth making.

Exercise 12

Find the solution sets of the systems of equations in questions ***1–15*** by matrix methods; the variables are on the set of real numbers.

1 $x-y = 5$, $x+y = 11$

2 $x-y = 0$, $x+y = 8$

3 $x+y = -1$, $x-y = 3$

4 $3x+y = 7$, $3x+2y = 5$

5 $3x-y = 5$, $2x+y = 15$

6 $2x+y = 5$, $2x+3y = -1$

7 $3x-4y = 18$, $5x+y = 7$

8 $2x+y = 12$, $3x-2y = 25$

9 $x+2y = 11$, $2x-y = 2$

10 $2x+3y = 9$, $3x+2y = 16$

11 $3x+y = 4$, $7x+2y = 9$

12 $2x+3y = 5$, $4x-5y = 21$

13 $5x+y = 13$, $x+5y = 17$

14 $5x-3y = 9$, $7x-6y = 9$

15 $10x+5y+3 = 0$, $5x+10y+9 = 0$

16 Try to solve the following systems of equations by matrices. Explain, with the aid of a Cartesian diagram, why you failed.

a $x+y = 4$, $3x+3y = 12$

b $2x-y = 3$, $6x-3y = 15$

17 Find the inverse of the matrix $\begin{pmatrix}2 & -1\\1 & 3\end{pmatrix}$, and use it to solve the following systems:

a $2x-y = 0$, $x+3y = 7$

b $2x-y = 9$, $x+3y = 1$

c $4x-2y-5 = 0$, $2x+6y+1 = 0$

Summary

1 *A matrix is a rectangular array of numbers arranged in rows and columns.*

The numbers or letters are called *entries* or *elements.*

2 The *order of a matrix* is given by the number of rows followed by the number of columns,

e.g. $\begin{pmatrix} 3 & 1 & 7 \\ 4 & 2 & 5 \end{pmatrix}$ *Order*: 2×3

$\begin{pmatrix} 3 & 7 \\ 9 & 4 \end{pmatrix}$ *Order*: 2×2, *or* a square matrix of order 2.

3 *Two matrices are equal* only if they are of the same order and if their corresponding entries are equal.

4 *A zero matrix*, $\boldsymbol{O}$, is a matrix whose elements are all zero.

5 *A unit matrix*, I, is a square matrix whose elements in the main diagonal are unity and whose other elements are all zero,

e.g. $I = \begin{pmatrix} 1 & 0 \\ 0 & 1 \end{pmatrix}$ and $I = \begin{pmatrix} 1 & 0 & 0 \\ 0 & 1 & 0 \\ 0 & 0 & 1 \end{pmatrix}$

6 *Addition of matrices*

If A and B are two matrices of the same order, the sum of A and B, denoted by $A+B$, is the matrix obtained by adding the entries of A to the corresponding entries of B,

$$\text{e.g.} \quad \underset{A}{\begin{pmatrix} a & b \\ c & d \end{pmatrix}} + \underset{B}{\begin{pmatrix} p & q \\ r & s \end{pmatrix}} = \underset{A+B}{\begin{pmatrix} a+p & b+q \\ c+r & d+s \end{pmatrix}}$$

7 *Subtraction of matrices*

If A and B are two matrices of the same order, to subtract B from A, the negative of B is added to A, i.e.

$$A - B = A + (-B).$$

The negative of B is the matrix whose entries are the negatives of the entries in B, so that $B + (-B) = \boldsymbol{O}$.

$$\text{e.g.}\ \overset{A}{\begin{pmatrix} a & b \\ c & d \end{pmatrix}} - \overset{B}{\begin{pmatrix} p & q \\ r & s \end{pmatrix}} = \overset{A}{\begin{pmatrix} a & b \\ c & d \end{pmatrix}} + \overset{(-B)}{\begin{pmatrix} -p & -q \\ -r & -s \end{pmatrix}}$$

$$\overset{A-B}{= \begin{pmatrix} a-p & b-q \\ c-r & d-s \end{pmatrix}}$$

8 *Multiplication of matrices by real numbers*

To multiply a matrix by a number k, we multiply each entry by that number,

$$\text{e.g.}\ k\begin{pmatrix} a & b \\ c & d \end{pmatrix} = \begin{pmatrix} ka & kb \\ kc & kd \end{pmatrix}$$

9 *Multiplication of two matrices*

a $\begin{pmatrix} a & b \\ c & d \end{pmatrix}\begin{pmatrix} x \\ y \end{pmatrix} = \begin{pmatrix} ax+by \\ cx+dy \end{pmatrix}$

b $\begin{pmatrix} a & b \\ c & d \end{pmatrix}\begin{pmatrix} p & q \\ r & s \end{pmatrix} = \begin{pmatrix} ap+br & aq+bs \\ cp+dr & cq+ds \end{pmatrix}$

Rule. Multiply 'row into column and add the products'. Multiplication of two matrices is not in general commutative.

10 *Inverse of a* 2×2 *matrix*

The *inverse* of the matrix $A = \begin{pmatrix} a & b \\ c & d \end{pmatrix}$ is

$A^{-1} = \dfrac{1}{ad-bc}\begin{pmatrix} d & -b \\ -c & a \end{pmatrix}$, provided that $ad-bc \neq 0$.

$ad-bc$ is the *determinant* of matrix A. If $\det A = 0$, A has no inverse, and is a *singular* matrix.

Property. $A^{-1}A = AA^{-1} = I$

11 2×2 *matrices correspond to geometrical transformations.*

For example:

$\begin{pmatrix} 1 & 0 \\ 0 & -1 \end{pmatrix}$ ... reflection in x-axis

$\begin{pmatrix} -1 & 0 \\ 0 & 1 \end{pmatrix}$ ... reflection in y-axis

$\begin{pmatrix} -1 & 0 \\ 0 & -1 \end{pmatrix}$ ... reflection in origin

$\begin{pmatrix} 0 & -1 \\ 1 & 0 \end{pmatrix}$ ... rotation $+90°$ about O

Functions; the Quadratic Function and its Graph

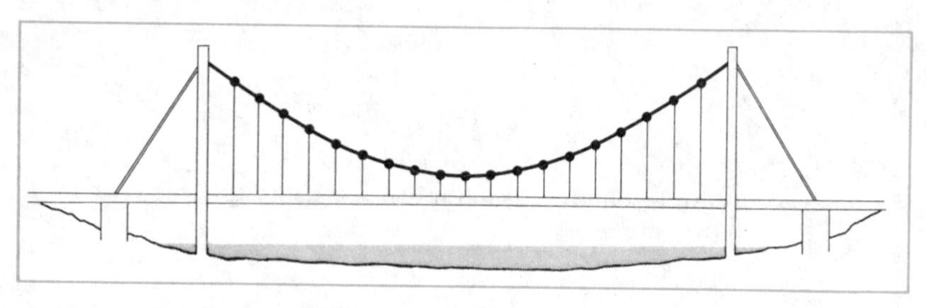

1 Mappings, or functions

In Book 3, we looked at some relations and mappings from one set to another set. In the present chapter, we develop these ideas further and study the important concept of *function*.

Example 1. Let $A = \{1, 4, 9\}$ and $B = \{1, 2, 3, 4\}$. Show in an arrow diagram the following relations from A to B:

(i) *is greater than* (ii) *is the square of*

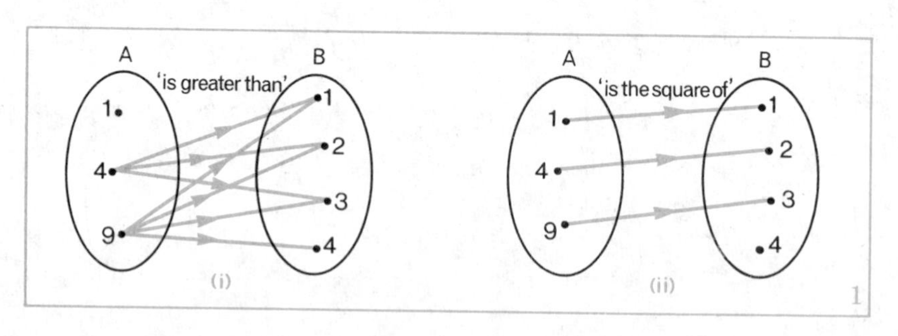

1

In (ii), the relation is a mapping since each element of A is related to exactly one element of B as is shown in Figure 1 (ii) where one arrow leaves each element of A.

Relations which are mappings are of prime importance in mathematics and are often referred to as *functional relations*, or simply *functions*. A function is therefore another name for a mapping. Both terms are useful; the idea of a mapping as a kind of operation helps to give a picture of a function.

A *function*, or *mapping*, from a set A to a set B is a relation in which each element of A is related to exactly one element of B. We write $A \rightarrow B$ (A maps to B).

Notation: If a function f maps an element x of set A to an element y of set B, we write $f: x \rightarrow y$, which may be read 'f maps x to y'.

y is called the *image* of x under f, and the set of images form the *range* of the function.

Thus for a function we require:
(i) a set A, called the *domain* of the function
(ii) a relation which assigns each element of A to exactly one element of B. The set of images in B is called the *range* of the function.

In Worked Example 1 (ii), the *domain* is $A = \{1, 4, 9\}$, the *function* is *is the square of*, and the *range* is $\{1, 2, 3\}$.

Example 2. Show the function $f: x \rightarrow 2x, x \in N$, as:
(i) an arrow diagram (ii) a Cartesian graph.
The domain of f is $N = \{1, 2, 3, 4, \ldots\}$
Replacing x by 1, 2, 3, 4, . . . ,
$f: 1 \rightarrow 2, f: 2 \rightarrow 4, f: 3 \rightarrow 6, f: 4 \rightarrow 8, \ldots$
The set of images, i.e. the range, is the set $\{2, 4, 6, 8, \ldots\}$. The function maps the set of natural numbers to the set of even numbers as shown in Figure 2.

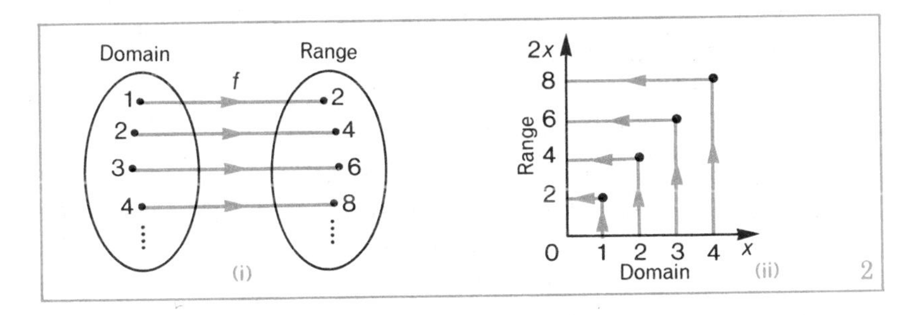

2

Exercise 1

1 Relations from the set $A = \{a, b, c\}$ to the set $B = \{p, q, r\}$ are shown in Figure 3. Which of these relations are mappings?

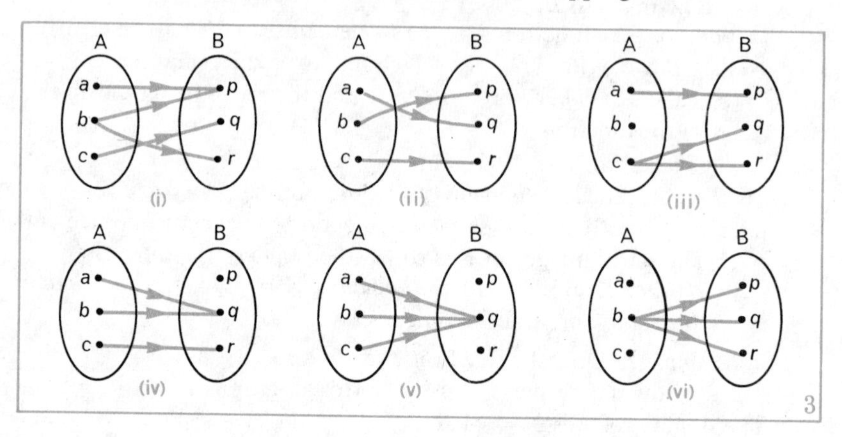

2 A function f is defined by $f: 1 \to 2, 2 \to 4, 3 \to 4, 4 \to 2, 5 \to 8$.

a List the elements of the domain and also of the range (the image set).

b Illustrate by means of (*1*) an arrow diagram (*2*) a Cartesian graph.

3 Mr X has a son Jim and a daughter Mary, and Mr Y has three sons, Bob, Ian, and Peter. Let $A = \{\text{Jim, Mary, Bob, Ian, Peter}\}$, $B = \{\text{Mr } X, \text{Mr } Y\}$, and let f denote the relation '*has as father*' from set A to set B.

a Show the relation f in an arrow diagram.

b Is f a function? Give a reason for your answer.

4 George, Tom and Bill live in North Street, Ian and David in South Street, and Alan lives in Market Street.

a List the set B of boys and the set S of streets.

b Show the mapping '*lives in*' from B to S in an arrow diagram.

5 Let $X = \{\text{Paris, London, Oslo, Copenhagen}\}$ and let $Y = \{\text{Norway, Denmark, England, France}\}$.

a State a relation which gives a mapping from the set X of capitals to the set Y of countries.

b Show your relation in an arrow diagram.

6 The function $f:x \to x+1$ has domain $\{0, 1, 2, 3, 4\}$.

a Find the range of f, i.e. the image set.

b Illustrate f by means of a Cartesian graph.

7 The domain of the function $x \to 3x$, $x \in N$, is $\{x:x < 5\}$.

a List the elements of the domain.

b List the corresponding elements of the range.

c Show the function in a Cartesian graph.

8 Find the range of the function $g:x \to x^2$ with the domain $\{3, 2, 1, 0, -1, -2, -3\}$. Illustrate g by means of a Cartesian graph.

9 $S = \{p, q, r\}$ and $T = \{1, 2\}$. Figure 4 shows three different functions from S to T. Use arrow diagrams to show five more functions from S to T.

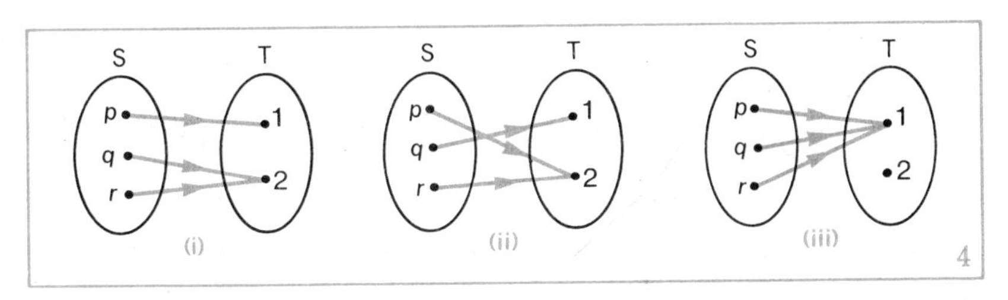

4

10 The domain of the function $h:x \to 4$ is $\{5, 10, 15, 20\}$. Show h in an arrow diagram and state the range of h.

11 For the function $f:x \to x^3$, the domain is $\{3, 2, 1, 0, -1, -2, -3\}$.

a List the elements of the range.

b Illustrate f in a Cartesian diagram.

12 For the function $g:x \to x^2-1$, copy and complete the following table:

Element of domain	3	2	1	0	−1	−2	−3
Image	8						

a Show g in a Cartesian graph.

b By drawing a smooth curve through the set of points, illustrate the mapping $g : x \to x^2-1$, $x \in R$.

2 Functional notation

Let f be a function from set A to set B. Under f, the image of an element x of set A is denoted by $f(x)$, which is read 'f of x'.

For example, from $f: x \to x^2+x+1$, $f(x) = x^2+x+1$.
$f(x) = x^2+x+1$ is called a *formula for the function* f.

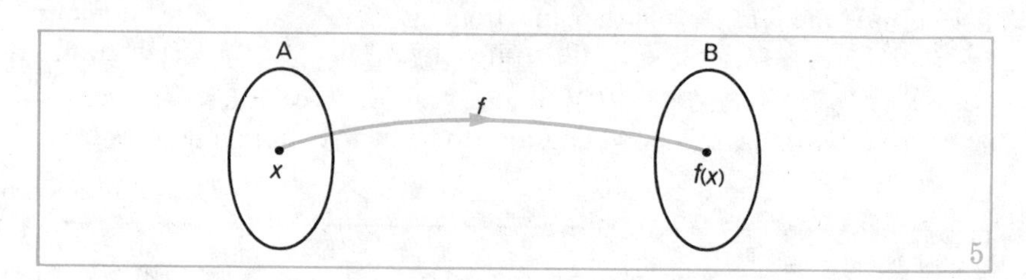

Example 1. Find the images of 3 and -1 under the function $f: x \to x^2+x+1$, $x \in R$.

Since for each $x \in R$, the image of x is $f(x) = x^2+x+1$,
the image of 3 is $f(3) = 3^2+3+1 = 13$, and
the image of -1 is $f(-1) = (-1)^2+(-1)+1 = 1$.

Note. Since the *value* of x^2+x+1 is 13 when x is replaced by 3, we say that $f(3)$ is the *value* of f at 3.

In general, $f(a) = a^2+a+1$ is the *value* of f at a.

Example 2. A function g is defined by the formula $g(x) = x^3$, $x \in R$. Find:

a the image of 2 under g

b the value of g at -3

c the number a, given that $g(a) = -1$

Since for each $x \in R$, $g(x) = x^3$,

a $g(2) = 2^3 = 8$

b $g(-3) = (-3)^3 = -27$

c $g(a) = a^3$, so $a^3 = -1$, and $a = -1$

Example 3. A function f on R is defined by the formula $f(x) = px+q$, where p and q are integers. Given that $f(1) = 4$ and $f(-2) = 1$, find p and q.

Since $f(x) = px+q$,

$$f(1) = p+q = 4$$
$$f(-2) = -2p+q = 1$$

we have a system of equations in p and q.

$$\left.\begin{array}{r} p+q = 4 \\ -2p+q = 1 \end{array}\right\}$$

Subtract $3p = 3$

$\Leftrightarrow$ $p = 1$

and so $q = 3$

Exercise 2

The functions in this Exercise are on the set of real numbers.

1 For $f: x \rightarrow x+3$, write down $f(x)$ and find:

a $f(1)$ *b* $f(2)$ *c* $f(-3)$ *d* $f(0)$ *e* $f(\frac{1}{2})$

2 For $f: x \rightarrow 3-4x$, write down $f(x)$ and find:

a $f(1)$ *b* $f(3)$ *c* $f(-2)$ *d* $f(\frac{1}{2})$

3 The formula $g(x) = x^2+1$ defines a function g.

a Calculate $g(2)$, $g(-1)$, $g(4)$, $g(-3)$.

b Given $g(a) = 50$, form an equation in a, and solve it.

4 $f: x \rightarrow 2x-5$.

a Find the images of $3, \frac{1}{2}, 0, -4, 4$.

b If $f(a) = 99$, find a.

5 For $g: x \rightarrow \frac{1}{2}(x+5)$, find:

a the values of g at $3, 0, -3$, i.e. find $g(3)$, $g(0)$, $g(-3)$.

b x if $g(x) = 0$.

6 For the function $h: x \rightarrow 3x-1$, find x such that

a $h(x) = 20$ *b* $h(x) = -16$

7 Function f is described by the formula $f(x) = ax+b$, where a and b are integers. Given $f(3) = 11$ and $f(1) = 7$, find a and b.

8 For $f: x \rightarrow px+q$, where p and q are integers, $f(1) = 3$ and $f(5) = 7$. Find p and q.

9 For $g: x \rightarrow ax+b$, where a and b are integers, $g(0) = -3$ and $g(2) = 1$. Find a and b, and hence obtain $g(4)$.

10 For $h:x \to 2^x$,

a What are the images of 3, 2, 1, 0, -1, -2?

b What element of the domain has 64 as its image?

Exercise 2B

Unless otherwise stated, the functions in this Exercise are on the set of real numbers.

1 Which of the following statements are true?

a The image of -3 under $f:x \to 5-x$ is 8.

b The image of -3 under $f:x \to x^2+9$ is zero.

c The functions $f:x \to 2x-1$ and $h:x \to x^2-1$ have the same value when $x = 2$.

d $f:x \to 3x+4$; $f(a) = a$ implies that $a = -2$.

e $f(2) = 5$ and $g(2) = 5$ mean that f and g are the same function.

f $f:x \to x+3$ and $g:t \to t+3$ define the same function ($x, t \in R$).

2 For the function f defined by the arrow diagram in Figure 6, state the images $f(a)$, $f(b)$ and $f(c)$. State the range of f.

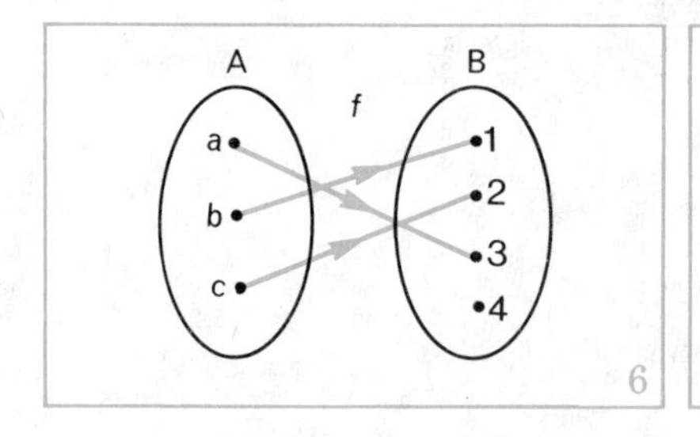

6

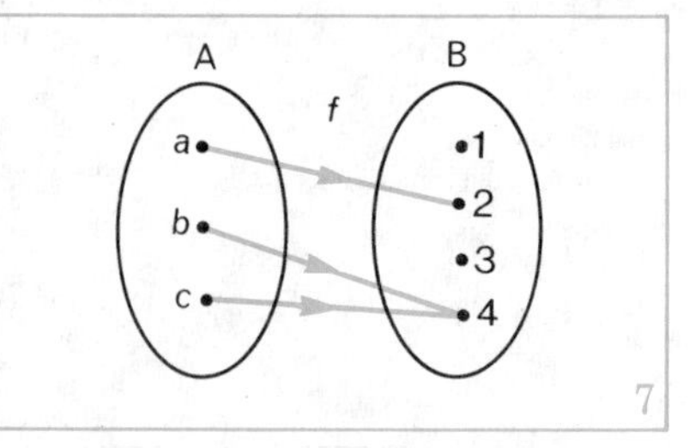

7

3 Repeat question **2** for the function f defined in Figure 7.

4 For the function $f:x \to x^2+x$, find the image set of $A = \{-2, -1, 0, 1, 2\}$, by finding the image of each element of A.

5 The domain of a function L is the set of positive integers, and L is defined by $L:x \to \log_{10}x$. Use your tables to find, to three significant figures:

a the images of 5, 25 and 50

b the element of the domain which has 2·5 as its image.

6 The sine function s is defined by $s:x \to \sin x°$ and so $s(x) = \sin x°$.

a Use your tables to find $s(30)$, $s(60)$, $s(90)$ to three significant figures.

b Given that $s(x) = 0{\cdot}750$, find x to the nearest $0{\cdot}1°$ $(0 < x < 90)$.

c For the domain $\{x : 0 \leqslant x \leqslant 360, x \in R\}$, what is the range of s?

7 A function S is defined on the set of positive integers as follows: '$S(n)$ is the sum of all possible divisors of n'. For example, $S(6) = 1+2+3+6 = 12$.

a Find the values $S(2)$, $S(4)$, and $S(18)$.

b Show that (*1*) $S(14) = S(15)$ (2) $S(3).S(5) = S(15)$.

8 The number of diagonals of a polygon of n sides is $\frac{1}{2}n(n-3)$. Define a function d by the formula $d(n) = \frac{1}{2}n(n-3)$.

a Find $d(6)$, $d(8)$, $d(10)$, $d(12)$.

b How many diagonals will a polygon of 20 sides have?

c Show, by listing elements, a suitable domain for d and a corresponding range.

9 The height in metres of a projectile after t seconds is $40t-5t^2$. Define a height function h on the set of non-negative real numbers by formula $h(t) = 40t-5t^2$. Calculate the values $h(0)$, $h(1)$, $h(4)$, $h(1\frac{1}{2})$, $h(8)$. Interpret the meaning of the last result.

10 For the function $f:x \to \dfrac{1}{x^2-4}$, $x \neq 2$ or -2, find:

a the values $f(1)$, $f(-3)$, $f(0)$, $f(\frac{1}{2})$

b the number a such that $f(a) = \dfrac{1}{12}$.

Note that this function has no value at $x = 2$ or $x = -2$. Why is this?

3 *The quadratic function and its graph*

A function f on R, defined by $f(x) = ax^2+bx+c$, where $a, b, c \in R$ and $a \neq 0$, is called a *quadratic function*.

The simplest quadratic function is given by $f(x) = x^2$. Some elements of f are $(0, 0)$, $(1, 1)$, $(2, 4)$, $(3, 9)$, $(-1, 1)$, $(-2, 4)$, $(-3, 9)$.

Notice that if $(h, k) \in f$, then $(-h, k) \in f$. The graph of f is therefore symmetrical about the $f(x)$-axis, and is shown in Figure 8. The *domain* of f is R, and the *range* of f is the set of non-negative real numbers. The graph is called a *parabola*.

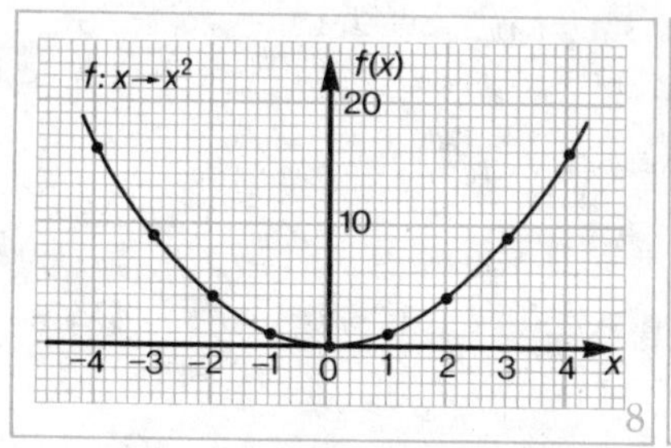

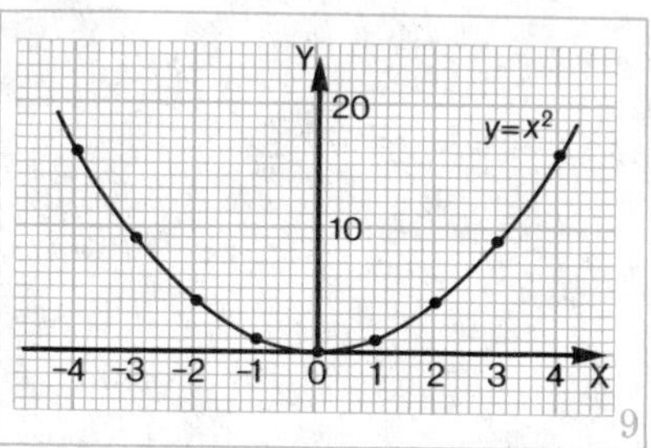

Note on the equation of a graph
The coordinate plane is the set of points defined by $\{(x, y): x \in R, y \in R\}$. Since a function is defined by the set of ordered pairs $\{(x, f(x)): x \in D\}$, where D is a subset of R, we consider the set of points given by $\{(x, y): y = f(x), x \in D\}$ to be the graph of the function f. $y = f(x)$ is called the *equation of the graph* of f.

In the above example, $y = x^2$ is the equation of the graph of the function f for which $f(x) = x^2$, as shown in Figure 9.

Example 1. Figure 10 shows the graph of the quadratic function f given by $f(x) = x^2 - 4x$, with domain $\{x: -2 \leqslant x \leqslant 6, x \in R\}$, i.e. the set of real numbers from -2 to 6 inclusive.

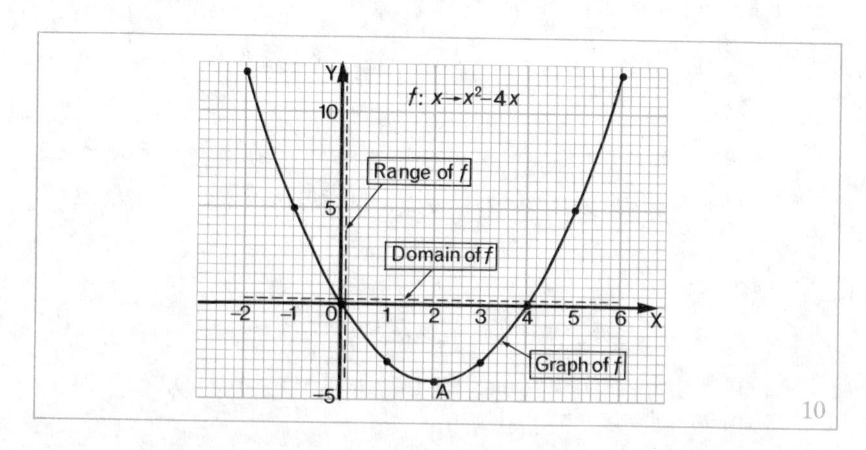

To draw the graph, we first select several suitable replacements for x and calculate the corresponding values of f. The following table shows a convenient way of setting out the calculation of values of f:

x	-2	-1	0	1	2	3	4	5	6
x^2	4	1	0	1	4	9	16	25	36
$-4x$	8	4	0	-4	-8	-12	-16	-20	-24
$f(x)$	12	5	0	-3	-4	-3	0	5	12

We now plot the points $(-2, 12)$, $(-1, 5)$, $(0, 0)$, $(1, -3)$, $(2, -4)$, $(3, -3)$, $(4, 0)$, $(5, 5)$, $(6, 12)$ and draw a smooth curve through these points as shown. The resulting curve is a parabola whose axis of symmetry has equation $x = 2$. The range of f is $\{y: -4 \leqslant y \leqslant 12, y \in R\}$.

The graph shows that as x increases from -2 to 6, *the value of the function decreases* from 12 to -4, and then *increases* from -4 to 12. The point A$(2, -4)$ at which this change from decreasing to increasing values takes place is called the *turning point*, or *vertex*, of the parabola. Since no other point on the curve has its y-coordinate less than -4, A is said to be a *minimum turning point*. The corresponding value -4 of f at $x = 2$ is called the *minimum value* of the function, since it is the least value taken by f.

Example 2. Draw the graph of the quadratic function f defined by $f(x) = 6+4x-2x^2$ with domain $\{x: -3 \leqslant x \leqslant 4, x \in R\}$.

Table of values of f

x	-3	-2	-1	0	1	2	3	4
6	6	6	6	6	6	6	6	6
$4x$	-12	-8	-4	0	4	8	12	16
$-2x^2$	-18	-8	-2	0	-2	-8	-18	-32
$f(x)$	-24	-10	0	6	8	6	0	-10

We now plot the points $(-3, -24)$, $(-2, -10)$, $(-1, 0)$, $(0, 6)$, $(1, 8)$, $(2, 6)$, $(3, 0)$, $(4, -10)$ on squared paper and draw a smooth curve through these points as shown in Figure 11. The curve is again a parabola whose axis of symmetry has equation $x = 1$.

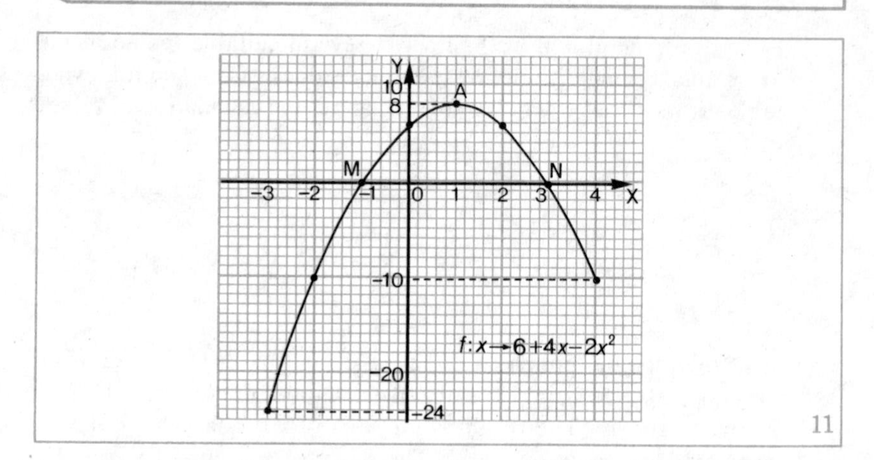

11

The range of f is $\{y: -24 \leqslant y \leqslant 8, y \in R\}$.
The graph of f shows that as x increases from -3 to 4, the value of the function *increases* from -24 to 8, and then *decreases* from 8 to -10. In this case, the point A(1, 8) is said to be a *maximum turning point* since no other point on the curve has its y-coordinate greater than 8. The value 8 of f at $x = 1$ is called the *maximum value* of the function, since it is the greatest value taken by f.

Notice that $f(-1) = 0$ and $f(3) = 0$. For this reason, -1 and 3 are called *zeros* of the function. The graph of f crosses the x-axis at the corresponding points M(-1, 0) and N(3, 0). The zeros of a function are very important elements of the domain.

Exercise 3

(In this Exercise, it will be assumed that x is a variable on R.)

1 Figure 12 shows the graph of the function f, defined by $f(x) = 4 - x^2$, for domain $\{x: -3 \leqslant x \leqslant 3\}$. Write down:

a the coordinates of the maximum turning point

b the maximum value of f

c the zeros of the function, i.e. the elements x of the domain such that $f(x) = 0$

d the range of f

e the equation of the axis of symmetry of the parabola.

2 Figure 13 shows the graph of the function f such that $f(x) = x^2+2x-3$ for domain $\{x: -5 \leqslant x \leqslant 3\}$. Write down:

a the coordinates of the minimum turning point

b the minimum value of f

c the zeros of f

d the range of f

e the equation of the axis of symmetry of the parabola.

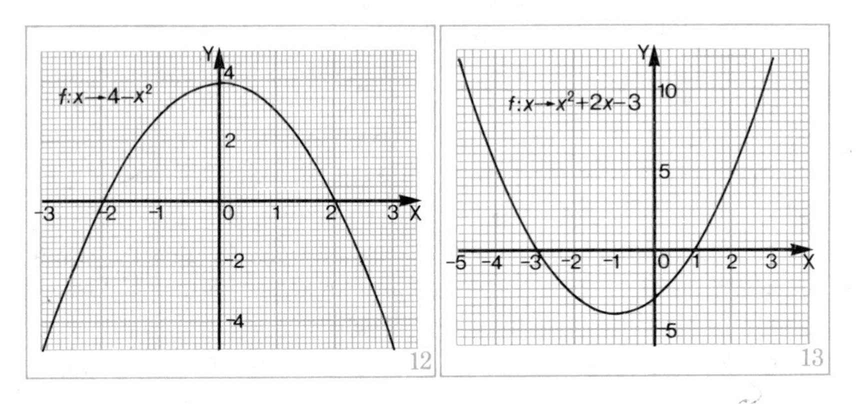

12

13

3 *a* Complete the table for each of the functions whose formula is given and whose domain is $\{x: -4 \leqslant x \leqslant 4\}$.

(*1*) $f(x) = x^2+4$

x	-4	-3	-2	-1	0	1	2	3	4
$f(x)$	20								

(*2*) $f(x) = x^2-4$

x	-4	-3	-2	-1	0	1	2	3	4
$f(x)$	12								

b Taking a scale of 1 cm to 1 unit for x and 1 cm to 5 units for $f(x)$, draw the graph of each function on 2-mm squared paper. Compare with the graph in question ***1***.

c Which function has no zeros?

In questions ***4–9***, draw the graph of each function on 2-mm squared paper. Answer the five parts of questions ***1*** or ***2*** above for each graph.

	Formula	Scale for x	Scale for $f(x)$	Domain
4	$f(x) = 2x^2$	1 cm per unit	2 cm per 5 units	$\{x: -3 \leqslant x \leqslant 3\}$
5	$f(x) = x^2 - 6x$	1 cm per unit	2 cm per 5 units	$\{x: -1 \leqslant x \leqslant 7\}$
6	$f(x) = x^2 - 4x + 4$	1 cm per unit	2 cm per 5 units	$\{x: -2 \leqslant x \leqslant 6\}$
7	$f(x) = 3 + 2x - x^2$	2 cm per unit	1 cm per unit	$\{x: -2 \leqslant x \leqslant 4\}$
8	$f(x) = x^2 + 2x + 5$	2 cm per unit	2 cm per 5 units	$\{x: -4 \leqslant x \leqslant 2\}$
9	$f(x) = 8 - 2x - x^2$	1 cm per unit	2 cm per 5 units	$\{x: -6 \leqslant x \leqslant 4\}$

Sometimes difficulty may be experienced in completing the graph near a turning point. Calculation of extra values for the neighbourhood of the turning point may be helpful.

10*a* Copy and complete the table of values of the function f for which $f(x) = x^2 - 5x + 4$, the domain being $\{x: 0 \leqslant x \leqslant 5\}$. Hence draw the graph of f, using a scale of 2 cm per unit for both x and $f(x)$. The extra values are shown in colour.

x	0	1	2	3	4	5	1·5	2·5	3·5
x^2				9			2·25	6·25	12·25
$-5x$				−15			−7·50	−12·50	−17·50
4				4			4	4	4
$f(x)$				−2			−1·25	−2·25	−1·25

b On what subset of the domain is $f(x)$ positive?

c What is the minimum value of f?

11*a* Copy and complete the table of values of the function f whose formula is $f(x) = 6 - x - 2x^2$ and whose domain is $\{x: -3 \leqslant x \leqslant 3\}$. Hence draw the graph of f, using a scale of 2 cm per unit for x and 2 cm per 5 units for $f(x)$.

x	-3	-2	-1	0	1	2	3	$\frac{1}{2}$	$-\frac{1}{2}$
6 $-x$ $-2x^2$									
$f(x)$									

b Use your graph to find:

(*1*) $f(2{\cdot}5)$

(*2*) the subset of the domain on which $f(x) > 0$

(*3*) $\{x : f(x) = 3\}$

12 Figure 14 shows the graph of function f on R such that $f(x) = 25 - x^2$.

a State the maximum value of f.

b Given that P is the point $(2, k)$, find k.

c Write down the zeros of f.

d If Q is the point $(h, -11)$, where $h < 0$, find h.

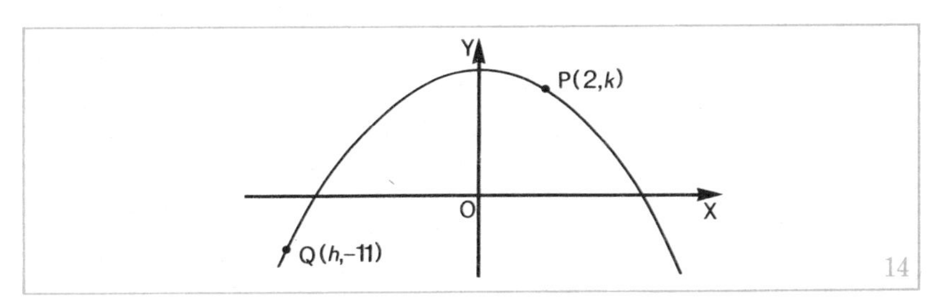

14

Each of the graphs in Exercise 3 is *similar* to all the others. Every graph of the quadratic function f defined by $f(x) = ax^2 + bx + c$ is a *parabola* with a maximum or minimum *turning point* which indicates the maximum or minimum value of the function.

Notice that the parabola has an axis of bilateral symmetry which passes through the turning point. This is called the *axis* of the parabola.

4 *Problems solved by the quadratic function and its graph*

The quadratic function and its graph have many practical applications, a few of which are shown in Figure 15, and in the diagram of the suspension bridge at the beginning of the chapter.

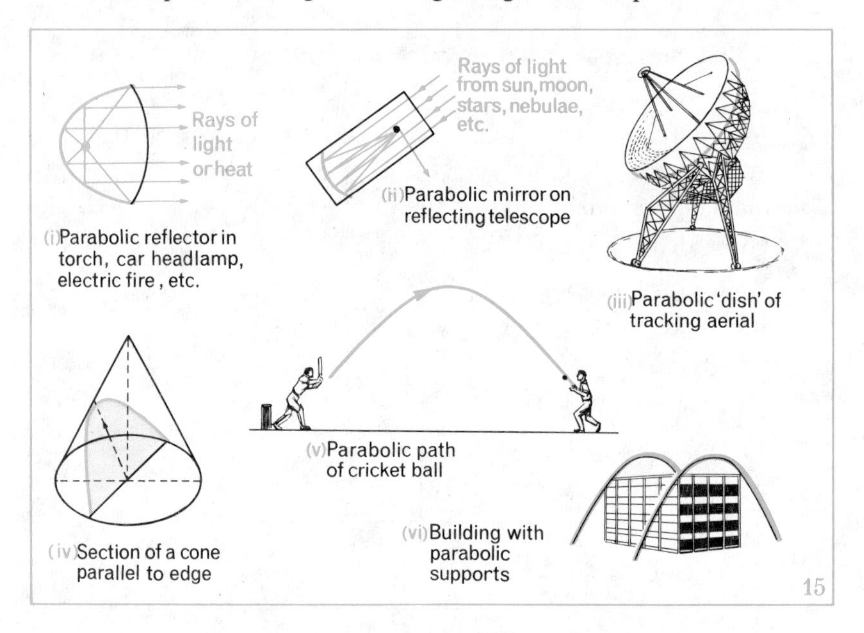

The formula $A = x^2$ for the area of a square associates a unique positive real number A with each positive real number x, and so defines a function on the positive real numbers. If this function is denoted by f then $f(x) = x^2$. To avoid introducing a new letter we often denote the function by the same letter A as in the formula, and so obtain an 'area function' A defined on the positive real numbers by the equation $A(x) = x^2$; the range of A is also the positive real numbers.

Exercise 4

1 A rectangle is x metres long and $(5-x)$ metres broad. The formula $A(x) = x(5-x)$ defines an area function A whose domain is $\{x:0 \leqslant x \leqslant 5\}$.

a Draw a graph of the function A by first calculating values of A at 0, 1, 2, 3, 4, 5, $\frac{1}{2}$, $2\frac{1}{2}$, $4\frac{1}{2}$. Use a scale of 2 cm per unit for both axes.

b From the graph, deduce:

(*1*) the area of the rectangle when $x = 2{\cdot}4$

(*2*) possible dimensions for the rectangle when its area is $5{\cdot}3 \text{ m}^2$

(*3*) the maximum area of the rectangle and the corresponding length and breadth. Comment on the result.

2 The height of a projectile above its point of projection after t seconds is $30t-5t^2$ metres. The formula for the height function h is $h(t) = 30t-5t^2$, the domain of h being $\{t:0 \leqslant t \leqslant 6\}$.

Draw the graph of the function h, using a scale of 2 cm per unit for t and 2 cm per 10 units for $h(t)$.

From your graph find:

(*1*) the time taken by the projectile to reach its maximum height and what this height was,

(*2*) the interval of time during which the projectile was over 20 metres above its point of projection.

3 In Figure 16, PQRS is a rectangle 10 cm long and 6 cm wide. W, X, Y, Z are points on PQ, QR, RS, SP such that PW = QX = RY = SZ = x cm.

a Write down the area of rectangle PQRS.

b Write down the area of each of the four triangles at the corners of the rectangle.

c Show that the area of quadrilateral WXYZ is $60-16x+2x^2 \text{ cm}^2$.

d Draw the graph of the area function A defined by formula $A(x) = 60-16x+2x^2$ by first calculating values of A at 0, 1, 2, 3, 4, 5, 6. Use a scale of 2 cm to 1 unit for x and 2 cm to 10 units for A.

e Find the minimum area of quadrilateral WXYZ.

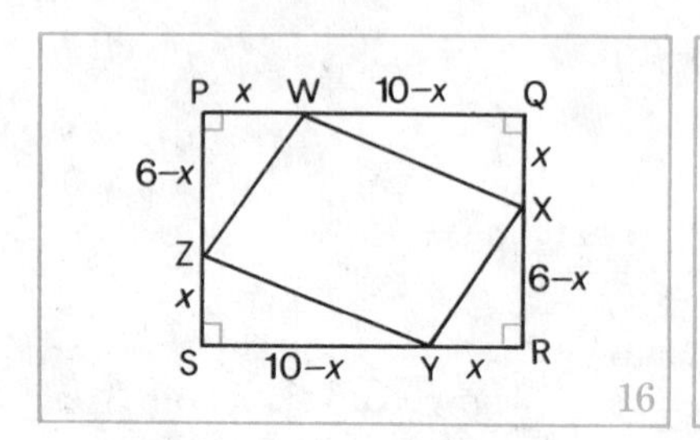

16

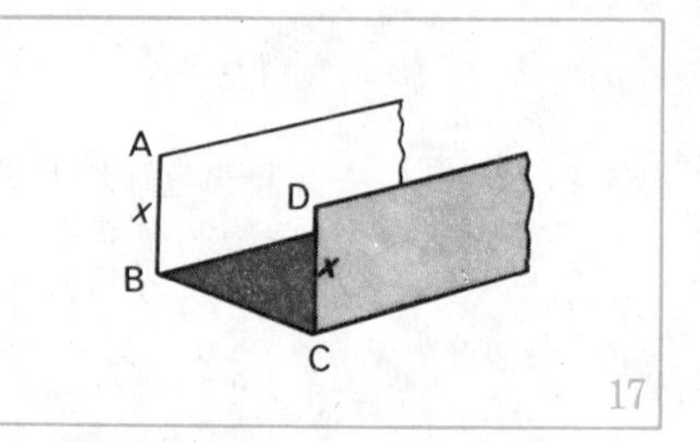

17

4 Figure 17 shows part of a long rectangular sheet of roofing metal, 40 cm wide, which is bent to form an open rectangular rain gutter.

a If AB = CD = x cm, show that the formula for the cross-sectional area function A is $A(x) = 40x - 2x^2$.

b Draw a graph of the function A by first calculating values of A at 0, 2, 4, 6, 8, . . ., 20. Use a scale of 1 cm per 2 units for x and 1 cm per 20 units for $A(x)$.

c What should be the height of the cross-section and the width of the base so as to let as much water as possible flow through?

5 Two boys played the following game. The first said 'Think of a number, double it, subtract this from 8, and then multiply by the original number. The one who gets the largest answer is the winner.'

The second boy found the following formula for the product $P(n)$, where n represents the first number: $P(n) = (4-n)2n$. Can you obtain this formula?

Then by calculating P at -1, 0, 1, 2, 3, 4, 5 he drew the graph of the function P, and found which number gave the greatest product. Can you?

Summary

1. A *relation* from a set A to a set B is a pairing of elements of A with elements of B.

2. A *function*, or *mapping*, f from a set A to a set B is a special kind of relation in which every element of A is paired with exactly one element of B.

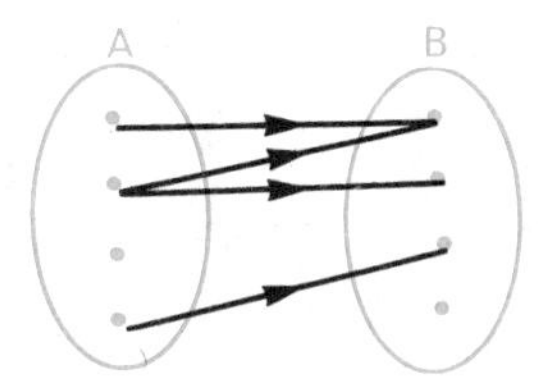

A relation which is not a function

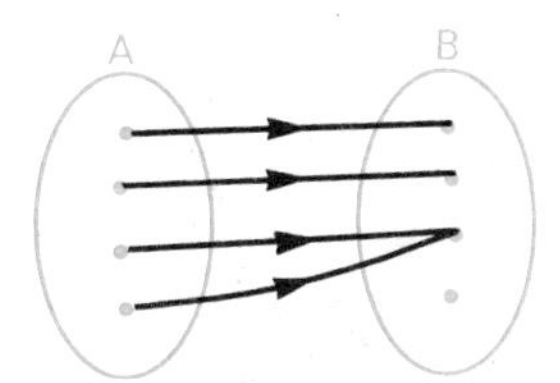

A relation which is a function

3. The set A is called the *domain* of the function, and the set of elements in B that are paired with the elements of A is called the *range* of the function.

4. If a is an element of the domain, the corresponding element of the range is called the *image of a* under the function f and is denoted by $f(a)$; the set of images form the range of the function.
$f(a)$ is also called the *value of the function* at a.

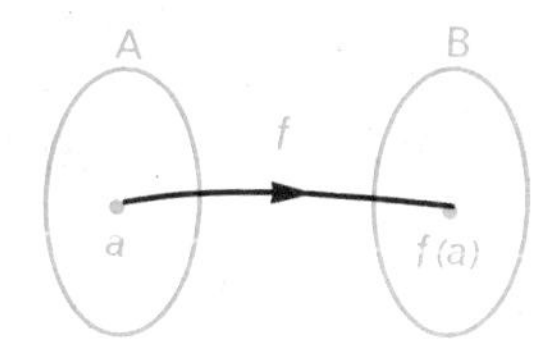

$f: a \rightarrow f(a)$.

5. A function f for which $f(x) = ax^2+bx+c$, where a, b, $c \in R$ and $a \neq 0$ is called a *quadratic function.*
$f(x) = ax^2+bx+c$ is a *formula* for the function.
$y = ax^2+bx+c$ is the *equation of the graph* of the function.

6 Every Cartesian graph of a quadratic function is a *parabola* with a *maximum*, or *minimum*, *turning point* through which the *axis of bilateral symmetry* passes.
If $(a, f(a))$ are the coordinates of a turning point, $f(a)$ is called a *turning value* of the function.

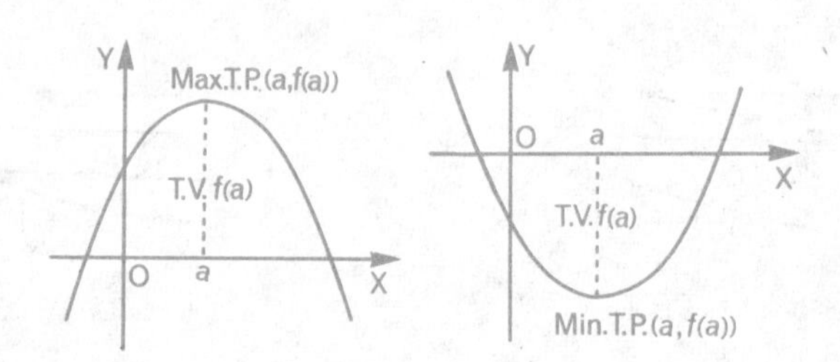

7 If when $x = a$, $f(a) = 0$, then a is called a *zero of the function.*
The graph of f cuts the x-axis at the point $(a, 0)$.

8 In this chapter a function has been defined:

(i) in words, e.g. *is the square of* from set A to set B

(ii) as a mapping, e.g. $f: x \rightarrow x^2 + 3x - 5$

(iii) by a formula, e.g. $f(x) = x^2 + 3x - 5$

(iv) by a set of ordered pairs, e.g. $\{(x, f(x)): f(x) = x^2\}$

(v) by a graph

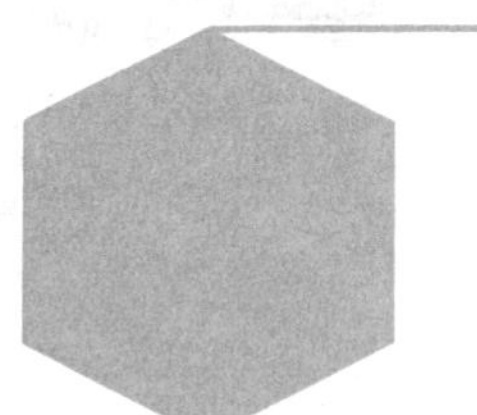

Revision Exercises

Revision Exercises on Chapter 1
Sums, Products and Squares

Revision Exercise 1A

1 Simplify the following:

a $5x^2-3x^2$ b $2y^2-6y^2$ c $-z^2-z^2$ d k^2-k^2

e $2x+3y-x-3y$ f $p^2-3q^2-3p^2+6q^2$

g $3(x+y)+2(x-y)$ h $5(m-n)-3(m+n)$

2 Find the sum of:

a $2x+3y$ and $-6x+12y$ b $p-3q-r$ and $5p+3q+r$

c $4c-6d+5e$ and $-2c-3d-7e$ d x^2-x-4 and $3x^2-x+5$

3 Subtract:

a $2x+3y$ from $-6x+12y$ b $-7a-8b$ from $3a-9b$

c $p-3q-r$ from $5p+3q+r$ d $4c-6d+5e$ from $-2c-3d-2e$

4 Simplify:

a $(3a+4b)+2(a-5b)$ b $3(x+y)-(x-y)$

5 Find the solution sets of the following, where $x \in R$:

a $5x-2(x-2)=1$ b $6-(3-2x)>7$

c $3(x-5)+2(x-1)=3$ d $4(1+2x)-3(3x+1) \leqslant 0$

6 Express the following products as sums:

a $(x+1)(x+2)$ b $(x+3)(2x+1)$ c $(2x+3)(3x+2)$

d $(y-2)(y-4)$ e $(y-1)(2y-1)$ f $(2y-3)(4y-1)$

g $(a+3)(a-2)$ h $(b-5)(b+1)$ i $(c+6)(c-6)$

j $(2x+1)(3x-1)$ k $(5y-2)(2y-5)$ l $(4z-3)(2z+7)$

7 Expand the following squares:

a $(c+d)^2$ b $(m+n)^2$ c $(u-v)^2$ d $(x-y)^2$

e $(x+6)^2$ f $(y-3)^2$ g $(2x+5)^2$ h $(3y-2)^2$

8 Solve the following equations, where $x \in R$:

a $(x+5)^2 = x^2+30$ b $(x+4)(x-2) = x^2$

c $(x+4)(x+7) = (x+10)(x+3)$ d $(x-3)(x+3) = (x-5)^2$

e $x(2x+6) = 2(x^2-5)$ f $(2x-3)(4x+1) = 2(2x-1)^2$

9 Prove that each of the following equations is an identity:

a $(x+1)^2-1 = x(x+2)$ b $(a+b)^2+(a-b)^2 = 2(a^2+b^2)$

c $(3k+6)(3k-6) = 9(k^2-4)$ d $a(b-c)+b(c-a)+c(a-b) = 0$

10 A rectangular lawn is a metres long and b metres wide. A concrete path 1 metre wide is laid round the outside of the lawn. Show this in a diagram. Write down the length and breadth of the path. By subtracting the areas of two rectangles find an expression for the area of the path.

Revision Exercise 1B

1 Simplify the following:

a $4x^2-4x-2x^2-2x$ b $a^3-b^3-a^3-b^3$

c $(x^2-2x-3)+(2x^2+2x-3)$ d $(2y^2+3y-4)-(y^2-3y-4)$

e $i+j-k-(2i-3j+5k)$ f $k(2k-1)-2k(k+2)$

g $2(x^2-x)+3(x^2+x)-(x^2+2x)$ h $x(1-2x)-2x(x+2)+(x^2+x)$

2 Find the solution sets of the following, where $x \in R$:

a $3(1-x)-2(x-1) = 0$ b $3x-2(3x-1) > 4$

c $2x(x-1)-x(4x+1) = 2(1-x^2)$ d $x(x^2-3)-(x^3+2x) \leqslant 1$

3 Express the following products as sums:

a $(a+b)(a-b)$ b $(2a+b)(a-3b)$

c $(5a-2b)(3a-2b)$ d $(2+x)(5+x)$

e $(1-x)(4-x)$ f $(2+3x)(2-3x)$

g $(y+2)(3-y)$ h $(x+4)(1-x)$

i $(z-3)(2-z)$

4 Expand the following, and simplify where possible:

a $(a+2)(b+5)$ b $(x-1)(y-4)$

c $(x^2+y^2)(x^2-y^2)$ d $(x+3)(2x^2-x-3)$

e $(5y-2)(3y^2-4y+5)$

5 Expand and simplify:

a $(a+b)^2-(a-b)^2$

b $(p-q)^2-(p+q)^2$

c $(2x+3)^2-(3x+2)^2$

d $\left(x+\frac{1}{x}\right)^2+\left(x-\frac{1}{x}\right)^2$

6 Solve the following equations, where $x \in R$:

a $(2x+3)(x-2) = (2x-1)(x-1)$

b $x(x-9)+(x-2)^2 = 2x(x-6)$

c $4x^2-(2x-7)^2 = 7$

d $2(x+3)^2-(x-1)^2 = x(x-3)$

7 Prove that:

a $(a+b)(a^2-ab+b^2) = a^3+b^3$

b $(a^2+b^2)^2-(a^2-b^2)^2 = 4a^2b^2$

c $(x+y)(x+z)-x^2 = (y+z)(y+x)-y^2$

d $(mq-sp)^2-m(p-sq)^2 = (s^2-m)(p^2-mq^2)$

8 Writing 2·01 as $2+0{\cdot}01$, show that $2{\cdot}01^2 = 4{\cdot}04$ to two decimal places.

9 With the aid of a diagram explain why $(1+x)^2 \doteqdot 1+2x$ when x is small. Hence calculate $1{\cdot}006^2$ to three decimal places.

10 In △ABC, the lengths of AB, BC and CA are 13 cm, 14 cm and 15 cm respectively. AD is the altitude from A to BC, and is h cm long. CD is x cm long. Use Pythagoras' theorem in triangles ADC and ADB to express h^2 in two ways; hence obtain an equation in x, and solve it. Then calculate h, and the area of △ABC.

Revision Exercises on Chapter 2
Matrices

Revision Exercise 2A

1 Write down the number of rows, the number of columns, the order, and the element in the second row and first column of each of the following matrices:

a $\begin{pmatrix} 5 & 6 \\ 7 & 8 \end{pmatrix}$

b $\begin{pmatrix} -1 & 0 & 1 & 2 \\ -4 & 0 & 4 & 8 \end{pmatrix}$

c $\begin{pmatrix} 1 \\ 3 \\ 5 \\ 7 \end{pmatrix}$

d $\begin{pmatrix} -1 & -1 & -1 \\ 2 & 0 & -2 \\ 3 & 6 & 9 \end{pmatrix}$

2 a In question *1*, which are square matrices, and which is a column matrix?

b How many entries are there in (*1*) a 5×3 matrix (*2*) a $p \times q$ matrix?

3 Write down the matrix defined as follows. It is a 3×2 matrix; the first row contains the first two odd numbers, in order; the second row contains the first two whole numbers, in order; the third row consists of elements which are the sum of the corresponding elements in the first two rows.

4 $A = \begin{pmatrix} 3 & 1 \\ 0 & -2 \end{pmatrix}$ and $B = \begin{pmatrix} -3 & 2 \\ 2 & 0 \end{pmatrix}$. Verify that $A+B = B+A$.

5 $P = \begin{pmatrix} 1 & 0 \\ 0 & -1 \end{pmatrix}$, $Q = \begin{pmatrix} -1 & 0 \\ 0 & 1 \end{pmatrix}$, $R = \begin{pmatrix} -1 & 0 \\ 0 & -1 \end{pmatrix}$. Verify that $(P+Q)+R = P+(Q+R)$.

6 Find x and y given by each of the following matrix equations:

a $(2x \quad 3y) = (12 \quad -6)$ *b* $\begin{pmatrix} 5x \\ 4y \end{pmatrix} = \begin{pmatrix} 0 \\ 2 \end{pmatrix}$ *c* $\begin{pmatrix} x-3 \\ y+7 \end{pmatrix} = \begin{pmatrix} 1 \\ 2 \end{pmatrix}$

7 $X = (3 \quad -1 \quad 4)$, $Y = (-3 \quad 0 \quad -2)$. Find $X+Y$, $X-Y$, $Y+X$, $Y-X$, $2X+3Y$ and $6X-2Y$.

8 Solve for x and y:

a $\begin{pmatrix} x \\ y \end{pmatrix} + \begin{pmatrix} 3 \\ -4 \end{pmatrix} = \begin{pmatrix} 5 \\ -7 \end{pmatrix}$ *b* $\begin{pmatrix} x \\ y \end{pmatrix} - 3\begin{pmatrix} -2 \\ 1 \end{pmatrix} = 2\begin{pmatrix} 1 \\ 0 \end{pmatrix}$

9 Find a, b, c, d given that:

$$\begin{pmatrix} a & b \\ c & d \end{pmatrix} + \begin{pmatrix} 5 & -3 \\ 3 & 2 \end{pmatrix} = \begin{pmatrix} 2 & -1 \\ 3 & 1 \end{pmatrix}.$$

10 Given that $A = \begin{pmatrix} 1 & -1 \\ -1 & 1 \end{pmatrix}$ and $B = \begin{pmatrix} 1 & 1 \\ 2 & 2 \end{pmatrix}$, find AB and BA.

11 Find the coordinates of the images P′, Q′ of the points P(1, −1), Q(−3, 4) under the mapping whose matrix is $\begin{pmatrix} 3 & -2 \\ 1 & 4 \end{pmatrix}$.

12 Find x in each of the following matrix equations:

a $(x \quad 2)\begin{pmatrix} 3 \\ 4 \end{pmatrix} = (20)$ *b* $(3 \quad x)\begin{pmatrix} -2 \\ 5 \end{pmatrix} = (24)$

c $(x \quad 3 \quad 1)\begin{pmatrix} 2 \\ x \\ 4 \end{pmatrix} = (39)$

13 $A = \begin{pmatrix} 5 & 2 \\ 7 & 3 \end{pmatrix}$, $B = \begin{pmatrix} 2 & -1 \\ 5 & -3 \end{pmatrix}$ and $C = \begin{pmatrix} 4 & 2 \\ 1 & 1 \end{pmatrix}$. Find:

a AB, BC and CA *b* A^2, B^2 and C^2

c A^{-1}, B^{-1} and C^{-1} *d* $A(BC)$, $(AB)C$ and $A(B+C)$

14 Using squared paper, draw the triangle with vertices (2, 1), (6, 1) and (6, 3).
By first calculating the coordinates of the image of each vertex, construct the image of the given triangle under each of the transformations whose matrices are:

a $\begin{pmatrix} 1 & 0 \\ 0 & -1 \end{pmatrix}$ *b* $\begin{pmatrix} -1 & 0 \\ 0 & 1 \end{pmatrix}$ *c* $\begin{pmatrix} 0 & -1 \\ -1 & 0 \end{pmatrix}$

15 Using a matrix method, solve the following systems of equations:

a $3x+4y = 6$, $x+2y = 4$ *b* $5x-2y = 12$, $2x+y = 3$ *c* $2x+y = 7$, $5x-3y = 12$

Revision Exercise 2B

1 $A = \begin{pmatrix} 2 & 4 \\ 6 & 8 \end{pmatrix}$, $B = \begin{pmatrix} -2 & -4 \\ -6 & -8 \end{pmatrix}$, $C = \begin{pmatrix} 2 & 6 \\ 4 & 8 \end{pmatrix}$, $D = \begin{pmatrix} 0 & -4 \\ -6 & 0 \end{pmatrix}$.

a Which matrices are the transposes of each other?

b Which matrices are the negatives of each other?

c Find in simplest form:

(1) $A+B$ (2) $C+D$ (3) $A-C$ (4) $B-D$

(5) $2A$ (6) $2D$ (7) $2A+2D$ (8) $2A-2D$

2 $P = \begin{pmatrix} 1 \\ 0 \end{pmatrix}$, $Q = \begin{pmatrix} 0 \\ 1 \end{pmatrix}$. Find $3P+4Q$ and $2P-Q$.

3 $X = (a \quad -b \quad c)$, $Y = (-a \quad b \quad -c)$. Find $X+Y$, $X-Y$, $Y+X$, $Y-X$, $2X+3Y$ and $6X-2Y$.

4 $\begin{pmatrix} p & q \\ r & s \end{pmatrix} + \begin{pmatrix} 1 & -1 \\ 0 & 2 \end{pmatrix} = \begin{pmatrix} 3 & 4 \\ 5 & 6 \end{pmatrix}$. Find p, q, r, s.

5 $\begin{pmatrix} 3 & 7 & -1 \\ -2 & 0 & 4 \end{pmatrix} + \begin{pmatrix} 2 & 1 & 1 \\ 2 & -1 & -5 \end{pmatrix} = \begin{pmatrix} a & b & c \\ d & e & f \end{pmatrix}$. Find a, b, c, d, e, f.

6 In each of the following, form a system of two equations in x and y and solve them:

a $x\begin{pmatrix}5\\2\end{pmatrix}+y\begin{pmatrix}3\\3\end{pmatrix}=\begin{pmatrix}7\\1\end{pmatrix}$ *b* $x\begin{pmatrix}3\\2\end{pmatrix}+y\begin{pmatrix}1\\-1\end{pmatrix}=\begin{pmatrix}4\\6\end{pmatrix}$

7 Simplify the following products:

a $\begin{pmatrix}2&1\\-1&1\end{pmatrix}\begin{pmatrix}3&4\\-3&1\end{pmatrix}$ *b* $\begin{pmatrix}4&1\\-2&3\end{pmatrix}\begin{pmatrix}2&1&-4\\1&-3&2\end{pmatrix}$

c $\begin{pmatrix}a&b\\-b&a\end{pmatrix}\begin{pmatrix}a&-b\\-b&a\end{pmatrix}$ *d* $\begin{pmatrix}a&b\\-b&a\end{pmatrix}^2$

8 Under a mapping $(x, y) \to (x', y')$ defined by
$\begin{pmatrix}x'\\y'\end{pmatrix}=\begin{pmatrix}3&-2\\-2&3\end{pmatrix}\begin{pmatrix}x\\y\end{pmatrix}$.

Find the coordinates of the images of the points $(4, 1)$, $(2, -3)$, $(-3, 4)$, $(-1, -5)$.

9 $A=\begin{pmatrix}1&2\\2&1\end{pmatrix}$ and $I=\begin{pmatrix}1&0\\0&1\end{pmatrix}$. Simplify $A-kI$, where k is a real number.
Find k for which $\det(A-kI)=0$.

10 Under a mapping $(x, y) \to (x', y')$ defined by $\begin{pmatrix}x'\\y'\end{pmatrix}=\begin{pmatrix}a&b\\c&d\end{pmatrix}\begin{pmatrix}x\\y\end{pmatrix}$, the images of the points $(1, 0)$ and $(1, -1)$ are $(1, 4)$ and $(-1, 7)$. Find a, b, c and d.

11 Write down the inverse of the matrix $\begin{pmatrix}3&2\\1&1\end{pmatrix}$.
Use your result to find the 2×2 matrix X, given that

$$\begin{pmatrix}3&2\\1&1\end{pmatrix}X=\begin{pmatrix}4&5\\-2&3\end{pmatrix}.$$

12 Find, if possible, the inverse of each of the following:

a $\begin{pmatrix}1&1\\-1&2\end{pmatrix}$ *b* $\begin{pmatrix}2&4\\-1&2\end{pmatrix}$ *c* $\begin{pmatrix}2&4\\1&2\end{pmatrix}$ *d* $\begin{pmatrix}1&-3\\-2&6\end{pmatrix}$

13 X is a 2×2 matrix, and $\begin{pmatrix}3&-\frac{1}{2}\\1&\frac{1}{2}\end{pmatrix}X=\begin{pmatrix}4&2\\0&6\end{pmatrix}$. Find X.

14 Use a matrix method to solve the following systems of equations:

a $3x+4y=4$
$2x+3y=2$

b $x+4y=5$
$3y=x+9$

c $4x+6y=9$
$x+2y=2$

15 Figure 1 shows a network of roads connecting four towns *A*, *B*, *C*, *D*.

a Work out the total number of routes from *A* to *C* via *B* and via *D*.

b The tables summarize the number of routes from *A* to *B* and to *D*, and from *B* and *D* to *C*. Extract matrices from these, and interpret their product in terms of your answer to ***a***.

	B	*D*
A	2	3

	C
B	3
D	1

c If a new road is made from *D* to *C*, calculate by means of a matrix product the total number of routes from *A* to *C* now.

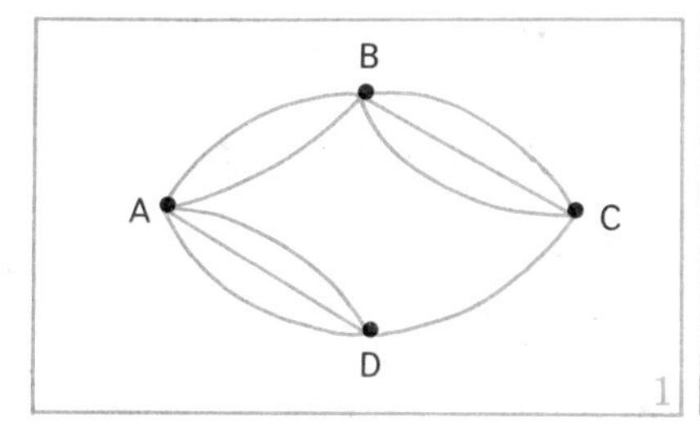

1

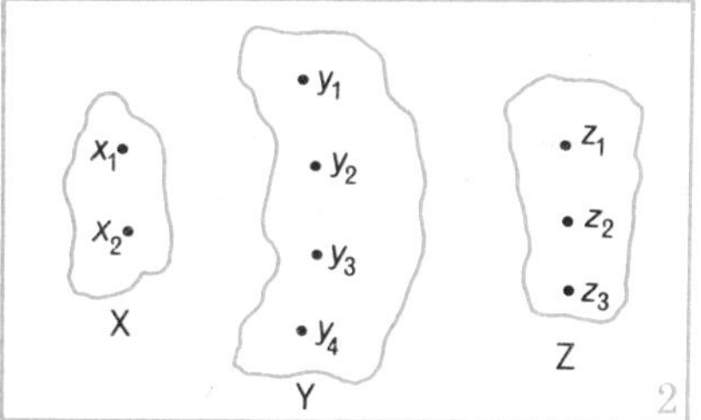

2

16 Figure 2 shows three countries *X*, *Y* and *Z* in which there are two, four and three airports. Travel agents supply the following tables of information about the number of routes connecting the airports.

	y_1	y_2	y_3	y_4
x_1	3	1	0	0
x_2	0	0	2	1

	z_1	z_2	z_3
y_1	1	1	0
y_2	1	1	0
y_3	0	1	2
y_4	0	0	1

a Copy Figure 2, and complete the network of routes between the airports.

b Calculate the number of routes from airports x_1 and x_2 in *X* to airports in *Z*:

(*1*) from your diagram (*2*) using matrices.

Revision Exercises on Chapter 3 Functions; the Quadratic Function and its Graph

Revision Exercise 3A

1 $A = \{\text{Mr Smith, Mr Jones}\}$, $B = \{\text{Alan Smith, Jean Smith, Tom Jones, Mary Jones}\}$ define some members of two families. Show in arrow diagrams the relations:

a *is the father of* from set A to set B

b *is the child of* from set B to set A.

Which relation is a function? Why?

2 The arrow diagram in Figure 3 defines a function g from A to A.

a Write down the image of 2 under g.

b What elements of the domain have 2 as their image?

c Write down the range of g.

d List the ordered pairs of g, and illustrate g in a Cartesian graph.

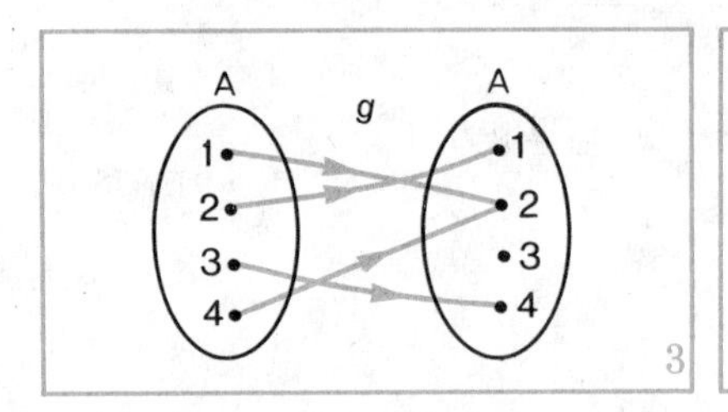

3

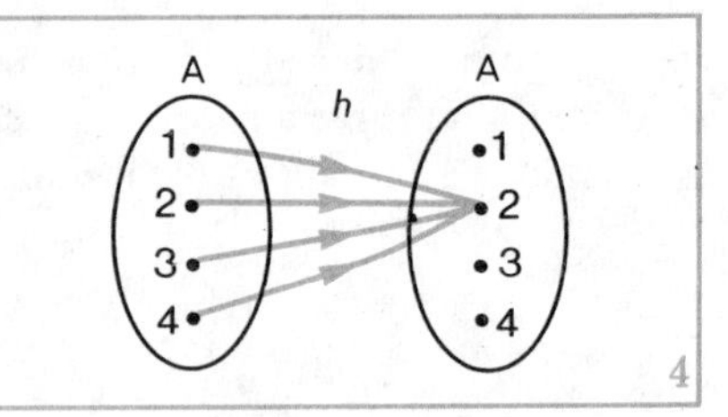

4

3 Repeat question **2** for the function h defined in Figure 4.

4 A function f is defined by $f(x) = x^4$ and the domain of f is the set $A = \{-2, -1, 0, 1, 2\}$. Find the range of the function and illustrate f in an arrow diagram.

5 Redefine the following functions from R to R by means of a formula:

a f maps each real number x to its square plus 1.

b g maps each real number x to its double less 1.

c h maps each real number x to half its square.

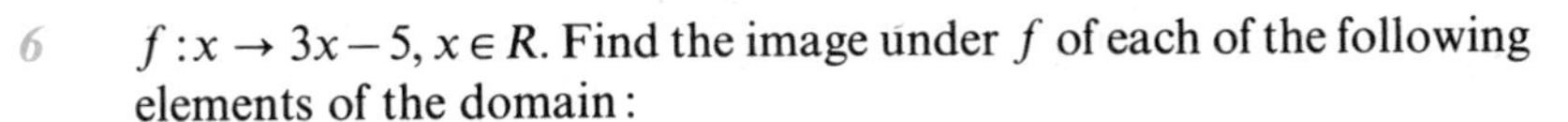

6 $f: x \to 3x-5, x \in R$. Find the image under f of each of the following elements of the domain:

a 0 b 4 c -2 d 0·5

Which element of the domain has 40 as its image?

7 A function h is defined by $h(x) = 2x^2, x \in Z$.

a Find the image of each of the following elements:

(1) 1 (2) 5 (3) -3 (4) $2t, \quad t \in Z$

b Given that $h(a) = 8$, find a.

8 A function f on R is defined by $f(x) = ax+b$, where a and b are integers. Given that $f(3) = 7$ and $f(1) = 1$, find a and b.

9 A function g on R is such that $g(x) = px+q$, where p and q are rational numbers. Given that $g(2) = 6$ and $g(-2) = 4$, find p and q. Hence find the image of 14 under g.

10 A quadratic function f is defined by $f(x) = x^2+3x-4$, the domain of f being the set $\{x: -6 \leqslant x \leqslant 3, x \in R\}$.

a Copy and complete the following table of values of f:

x	-6	-5	-4	-3	-2	-1	0	1	2	3
x^2	36	25								9
$3x$	-18	-15								9
-4	-4	-4								-4
$f(x)$	14	6								14

b Using these values of f, draw the graph of f. Scales: 1 cm per unit for x and 2 cm per 5 units for $f(x)$.

c Estimate the coordinates of the turning point, and hence state the turning value of f. Is it a maximum or a minimum?

d Use your graph to obtain as accurately as you can the values of:

(1) $f(1{\cdot}8)$ (2) $f(-2{\cdot}4)$

11 Draw the graph of the quadratic function f, for which $f(x) = x^2+2$, on 2-mm squared paper. Use the following information:
Scales: 2 cm per unit for the x-axis and 1 cm per unit for the $f(x)$-axis.
Domain: $\{x: -3 \leqslant x \leqslant 3, x \in R\}$. *Range*: $\{y: 2 \leqslant y \leqslant 11\}$. Write down the coordinates of the turning point of the graph of f and state its nature. Is there an x such that $f(x) = 0$?

12 The domain of function f is $\{x: -4 \leqslant x \leqslant 6\}$ and $f(x) = 8+2x-x^2$. Draw the graph of f using a scale of 1 cm to 1 unit for x and 2 cm to 5 units for $f(x)$.

a Write down the zeros of f.

b Write down the coordinates of the turning point, and state its nature.

c What is the equation of the axis of symmetry of the curve?

d State the restrictions on x such that $f(x) < 0$.

13 A farmer has 40 metres of wire-netting to make a sheep pen in the form of a rectangle. Show that he could enclose an area $A(x) = 20x-x^2$, where x metres is the length of the pen. (A sketch will help.)

Draw a graph of the area function A by calculating values at 7, 8, 9, 10, 11, 12, 13. Use a scale of 2 cm to 1 unit for x and 1 cm to 1 unit for $A(x)$, taking (7, 91) as origin.

Estimate from your graph the greatest area he could enclose, and the corresponding dimensions of the sheep pen.

14 On the same diagram draw the graphs of the functions f and g for which $f(x) = x^2$ and $g(x) = 2x+3$, the domain of each being $\{x: -4 \leqslant x \leqslant 4, x \in R\}$.

Write down *two* elements of the domain for which $f(x) = g(x)$.

15 The range of the function f for which $f(x) = 5-2x$ is the set $\{-1, -5, 3, 7\}$. Find the domain of f.

Revision Exercise 3B

1 $P = \{-1, 0, 1, 8, 27\}$ and $Q = \{-1, 0, 1, 2, 3\}$. Find a suitable relation from P to Q and illustrate in an arrow diagram. Explain why the relation is a function.

2 $f(x) = 2x^2+3x-2, x \in R$. Find:

a $f(2)$ *b* $f(-1)$ *c* $f(0)$ *d* $f(\frac{1}{2})$ *e* $f(-\frac{1}{2})$

3 A function $g = \{(-3, -3), (-2, -1), (-1, 1), (0, 3), (1, 5), (2, 7), (3, 9)\}$

a List the elements of the domain and range of g.

b Draw a Cartesian graph of g using a scale of 1 cm to 1 unit on both axes.

c Given $g(x) = ax+b$, where $a, b \in Z$, use the fact that $g(1) = 5$ and $g(2) = 7$ to find a and b.

4 A function m, with domain $\{x: -3 \leqslant x \leqslant 3, x \in R\}$, is defined as follows:

$$m(x) = x, \text{ for } x \geqslant 0$$
$$m(x) = -x, \text{ for } x < 0$$

a Write down the values of m at $-3, -2, -1, 0, 1, 2, 3$.

b List the ordered pairs of m given by these replacements for x.

c Draw a Cartesian graph of m using a scale of 2 cm to 1 unit on each axis.

Note. Function m is called a modulus function.

5 $f(x) = \dfrac{2x}{1-x}$, $x \in R$ but $x \neq 1$.

a Find $f(3)$, $f(0)$, $f(-1)$ and $f(5)$.

b Find a, given that the image of a is 1.

c Why is $x \neq 1$?

6 For a function $f: x \to ax^2+bx$, where a and b are integers, $f(2) = -2$ and $f(-2) = 10$. Find a and b. Draw a graph of the function, the domain being $\{x: -2 \leqslant x \leqslant 5\}$.

7 The distance in metres travelled by a car after t seconds, when accelerating, is $16t+\frac{1}{2}t^2$.

a Write down a formula for the distance function s in terms of t, the domain being the set of non-negative real numbers.

b Calculate the distance travelled after (*1*) 10 seconds (*2*) 12 seconds.

c Hence estimate the average speed of the car over the 10 to 12 second interval.

8 Draw the graphs of the functions f and g defined by $f(x) = x^2$ and $g(x) = 6-x$ with domain $\{x: -4 \leqslant x \leqslant 3, x \in R\}$. Write down *two* elements of the domain for which the functions have the same value.

9 For the quadratic function f defined by $f(x) = ax^2+bx+c$, with domain R, $a, b, c \in R$ and $a \neq 0$, which of the following are true and which are false?

a The graph of f is a parabola.

b The range of the function is R.

c The axis of symmetry of the graph always passes through the turning point.

d The graph of f always has an axis of symmetry perpendicular to the x-axis.

e $f(0)$ is a zero of f.

10 The function f is such that $f(x) = x - \frac{1}{2}x^2$, and its domain is $\{x: -2 \leqslant x \leqslant 4, x \in R\}$.

a Find the values of f at -2, -1, 0, 1, 2, 3, 4, and hence draw the graph of f, using a scale of 2 cm to 1 unit on both axes.

b State the maximum value of f.

c Use the graph to find $f(3{\cdot}5)$.

d Write down the condition that $f(x) > 0$.

11 Figure 5 shows the graph of the quadratic function f for which $f(x) = x^2 + px + q$, and the domain of f is $\{x: -3 \leqslant x \leqslant 6, x \in R\}$. Use the graph to answer the following:

a What are the zeros of f?

b Find $f(0)$ and hence obtain q.

c Find the image of 4 and hence obtain p.

d Write down the set of all x for which the value of $f(x)$ is increasing.

e Estimate the minimum value of f.

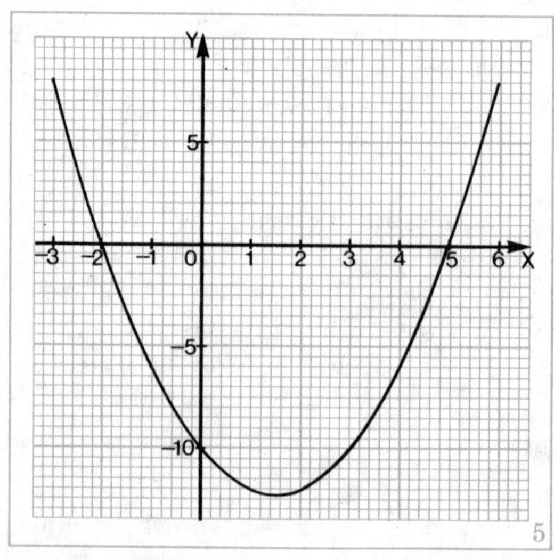

5

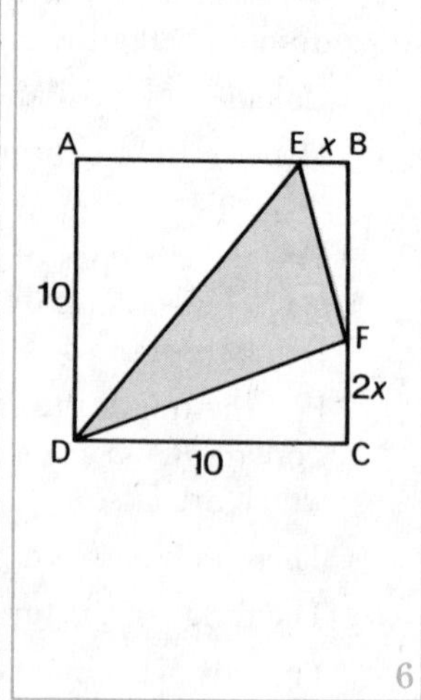

6

12 The quadratic function f is defined by $f(x) = 4x^2 - 4x - 15$, and the domain is $\{x: -2 \leqslant x \leqslant 3, x \in R\}$.

a Copy and complete the following table of values of f:

x	-2	-1	0	1	2	3	$\frac{1}{2}$	$1\frac{1}{2}$	$-\frac{1}{2}$	$-1\frac{1}{2}$
$4x^2$										
$-4x$										
-15										
$f(x)$										

b Draw the graph of f using a scale of 2 cm to 1 unit for x and 2 cm to 5 units for $f(x)$.

c Use the graph of f to answer the following:

(1) State the coordinates of the turning point and say whether it is a maximum or minimum.

(2) Find x such that $f(x) = 0$.

(3) Find $f(1{\cdot}2)$.

13 Repeat question ***12*** for function f defined by $f(x) = 15 + 4x - 4x^2$.

14 In Figure 6, ABCD is a square of side 10 cm. BE is x cm long and CF is $2x$ cm long. Write down the lengths of AE and BF in terms of x.
Show that the area A cm^2 of triangle DEF is given by formula $A(x) = 50 - 10x + x^2$.
Using a graphical method, find x such that A is a minimum.

Geometry

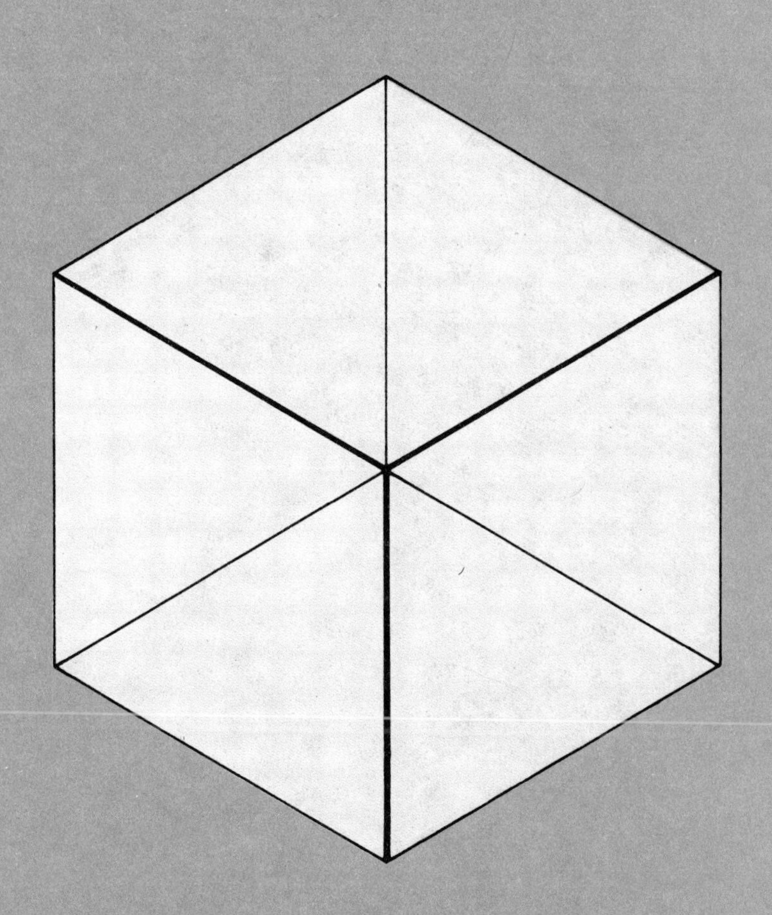

Geometry

Similar Shapes

1 Scale drawings

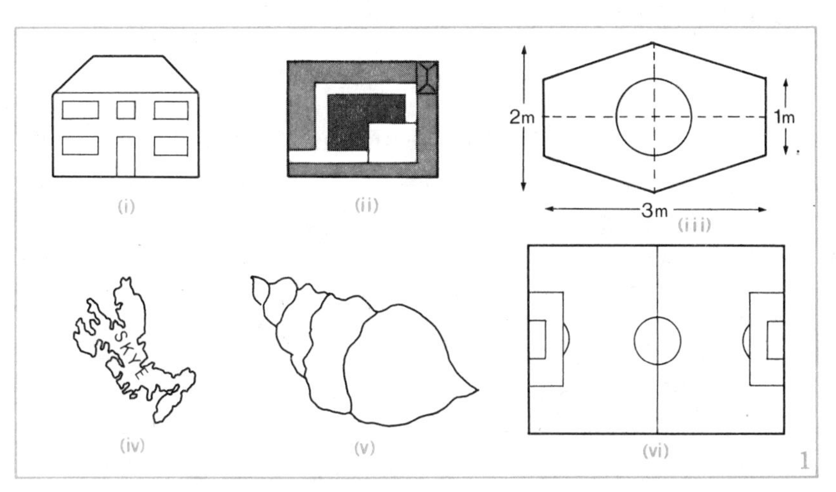

Figure 1 illustrates some of the ways in which people use scale drawings to give information about the shapes and sizes of objects. If exact information is wanted, the diagrams are not merely pictures, but are two-dimensional drawings made accurately to scale.

Exercise 1

1 Figure 1 (i) shows the *front elevation* of a house to a scale of 1 cm representing 5 m.

Use your ruler to measure the length of the front, and the height of the house in the drawing. Hence calculate the actual length of the front, and the height of the house.

2 Figure 1 (ii) shows the plan of a garden, drawn to a scale of 1 cm representing 8 m. Measure the length and breadth of the plan, and calculate the actual dimensions of the garden.

3 Figure 1 (iii) shows a scale drawing of a metal plate. To what scale has it been drawn?

4 Figure 1 (iv) shows a map of the Isle of Skye, which lies off the west coast of Scotland. The island is actually 66 km long from the furthest north cape to the furthest south. What scale has been used?

5 Figure 1 (v) is a drawing of a whelk with its dimensions reduced to $\frac{3}{4}$ life size. What is the true length of the whelk from end to end?

6 Figure 1 (vi) shows the plan of a football pitch. If the scale used is 1 cm representing 30 m, calculate the length and breadth of the pitch, and the diameter of the centre circle.

2 *Photographs and scale models*

Figure 2 (i) shows a photograph of the front of a house. In the photograph the height of the door is 0·5 cm and the height of the house is 1·8 cm. Figure 2 (ii) shows the actual dimensions of the door.

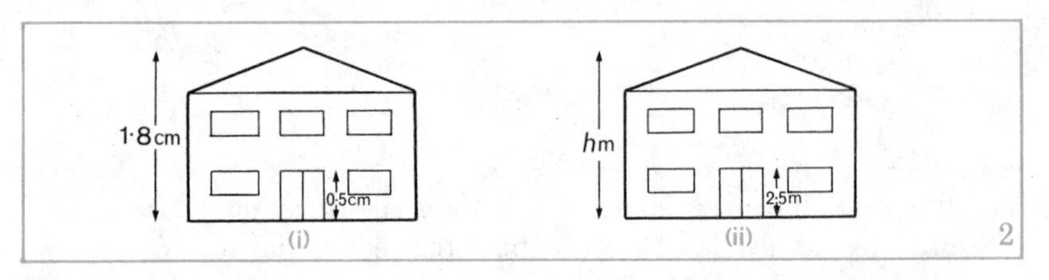

(i) (ii) 2

If the actual door is 2·5 m high, we can calculate the height of the house, h metres, as follows. We expect that in the photograph *all the dimensions of the house will be reduced in the same ratio.*

Compare the actual height of the house with the height in the photograph, and do the same for the door.

$$\frac{h}{1{\cdot}8} = \frac{2{\cdot}5}{0{\cdot}5}$$

$$\Leftrightarrow \quad 0{\cdot}5h = 1{\cdot}8 \times 2{\cdot}5 \text{ (by cross-multiplication)}$$

$$\Leftrightarrow \quad h = \frac{1{\cdot}8 \times 2{\cdot}5}{0{\cdot}5}$$

$$= 9$$

The height of the house is 9 metres.

Note. Slide rules are useful in calculations of this kind.

Exercise 2

1. Suppose that in Figure 2 the actual door is 3 m high. What is the actual height of the house?

2. Suppose that in Figure 2 the height of the door and the house in the photograph were 0·6 cm and 2·4 cm respectively, and that the actual door was 2·2 m high. Calculate the height of the house.

3. An aircraft has a length of 24 m and a wing span of 32 m. A scale model is to be made with a wing span of 12 cm. Calculate the length of the model.

4. Figure 3 is a sketch of a motor-car with its actual dimensions shown.
   a If a scale model of the car is 30 cm long, calculate its height.
   b If the distance between the front and back axles of the car is 3 m, find the corresponding distance in the model.

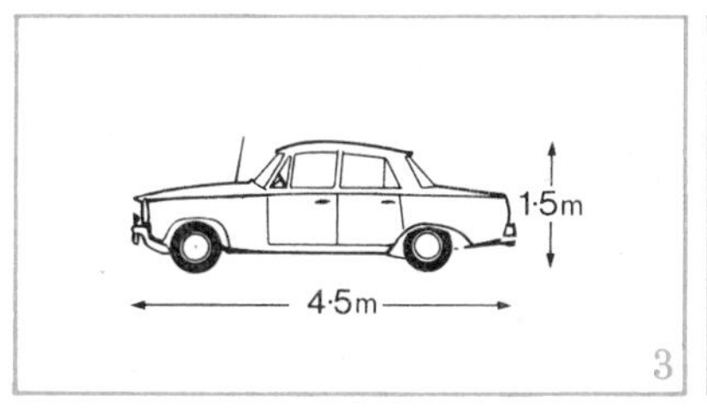

3

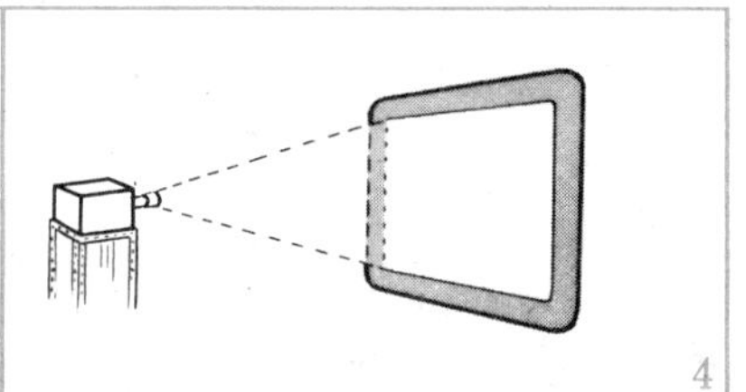
4

5. Figure 4 shows the image of a slide measuring 34 mm wide and 22 mm high. If the picture on the screen is 85 cm wide, what is the height of the picture?

6. The same slide as in question ***5*** gives a picture 44 cm high when the screen is moved nearer. Find the width of the picture.

7. A warship is 250 m long and 40 m broad at its widest point. A model is to be made 50 cm long. Find its breadth.
   If the mast is 20 m high, what height would it be on the model?

8. A church 25 m high and 20 m wide appears on a television screen. If the screen image is 12 cm wide, what is its height?

9. A shop window 3 m high has an image on a television screen 18 cm high and 25 cm wide. Calculate the actual width of the window, to the nearest tenth of a metre.

10. A photograph 8 cm high and 6 cm wide is enlarged to give a print 12 cm wide. What is the height of the print?
    Calculate the ratio of the area of the photograph to the area of the print.

3 Similar figures

In Sections 1 and 2 we took it for granted that a scale drawing (or model) had the same *shape* as the original, but was of a different size; each length measured in the drawing (or model) was a certain ratio of the original length.

We now look more closely at the idea of 'same shape'.

Exercise 3

1 Which, if any, of the following could be regarded as being of the same shape as a football pitch measuring 100 metres by 60 metres?

a A square 8 cm by 8 cm.

b A rectangle 5 mm by 3 mm.

c A parallelogram with sides of 10 cm and 6 cm and one angle 85°.

2 Which, if any, of the following could be regarded as being of the same shape as a boxing ring 5 metres square?

a A square carpet measuring 6 metres by 6 metres.

b A page of your textbook.

c A square field of side 280 metres.

d One of the smallest squares on a sheet of graph paper.

Note. In geometry we use the word 'similar' with a special meaning. We want to restrict its use to the kind of resemblance between figures which is illustrated by a photograph of an object and an enlargement of the same photograph.

As far as *figures with straight sides* are concerned we shall say they are *similar* if

(i) they are equiangular, i.e. pairs of angles taken in order are equal
(ii) pairs of corresponding sides are in the same ratio.

You will have to reach a clear idea about the meaning of 'equiangular' and 'corresponding sides'.

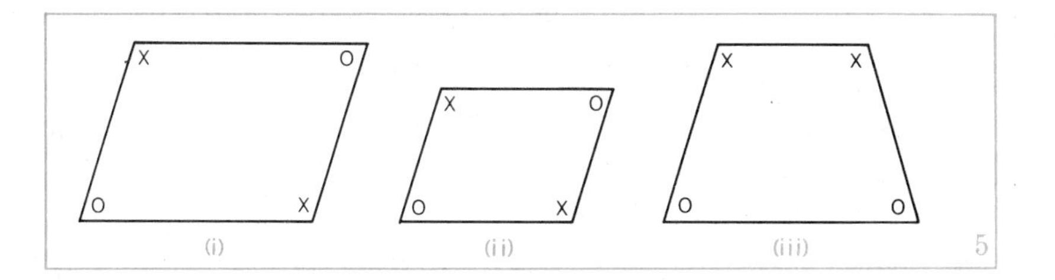

(i) (ii) (iii) 5

In Figure 5, (i) and (ii) are equiangular, but (iii) is not equiangular to (i) and (ii). Even though the angles are the same sizes, they are not in the right order. Once the relation between the angles is settled, corresponding sides are then easily picked out.

3 In Figure 6, lengths of sides are given in centimetres. Which of the quadrilaterals (ii), (iii), (iv) and (v) might be, and which cannot be, similar to (i)?

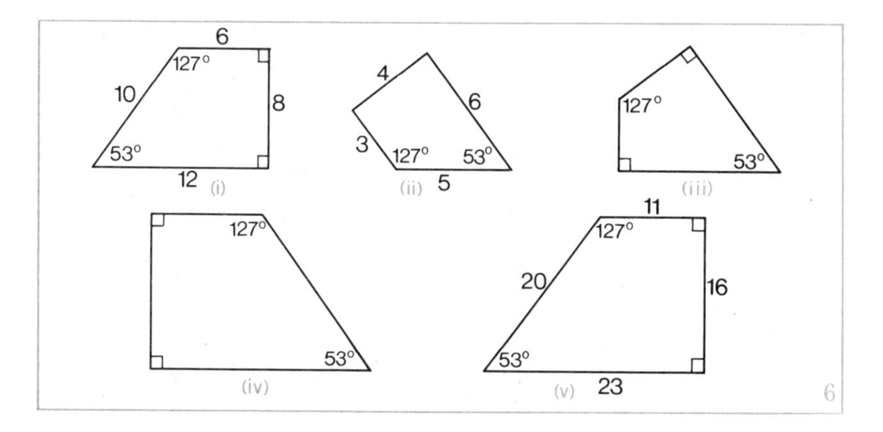

(i) (ii) (iii) (iv) (v) 6

4 A rectangular photographic mount measures 20 cm by 30 cm. A photograph placed centrally on the mount leaves a uniform margin $2\frac{1}{2}$ cm all round. Is the rectangle of the photograph *similar* to the rectangle of the mount?

5 A photograph is mounted on a card which is 20 cm wide and 30 cm tall, so as to leave a margin 3 cm wide at the top and at each side. If the photograph is *similar* to the mount, what width of margin will be left at the foot of the mount?

6 In a football pitch, certain dimensions are fixed by the laws of the game, e.g. the centre circle is 9 m in diameter, the penalty area extends to 16 m from the goal line and the goals are 7 m wide.

Two pitches are fully marked out. One measures 96 m by 64 m and the other 105 m by 70 m.

a Are the rectangular playing areas similar?

b Are the figures consisting of all the marking lines similar?

7 Which of the following *must* be similar to each other?

a Two equilateral triangles

b Two isosceles triangles

c Two squares

d Two regular hexagons

e Two kites

f Two circles

4 Enlarging and reducing

Exercise 4B

1 Figure 7 illustrates an embroidery stitch known as 'wave stitch filling'. Copy it on squared paper, some members of the class using 5-mm squares and others 2-mm squares. There will then be in the room diagrams of the pattern in three different sizes. Will they all be similar?

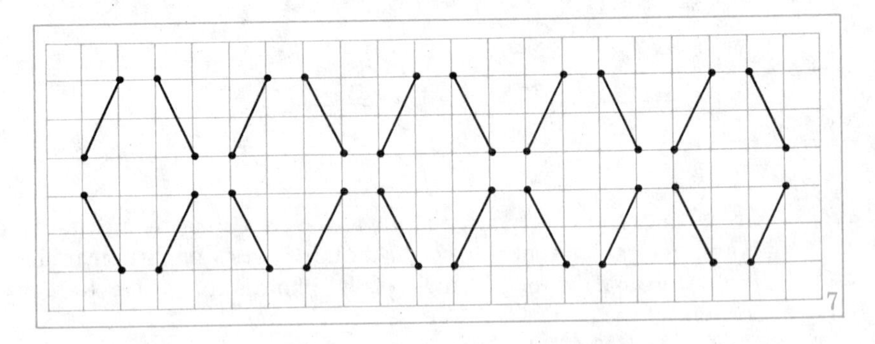

2 Figure 8 bears some resemblance to a face in profile. Copy it on 5-mm squared paper, inserting an ear and an eye in reasonable positions.

On the same sheet of paper make a drawing which is similar to your first one, magnified either two-fold or three-fold.

Draw it again half-size.

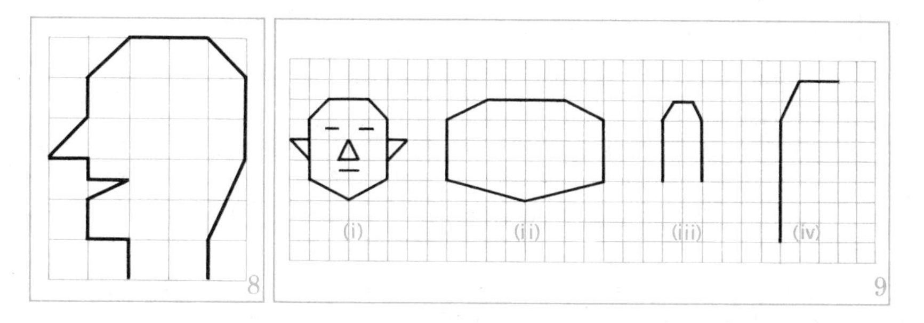

3 Figure 9 (i) is a crude full-face drawing. In Figure 9 (ii) all widths have been doubled, heights remaining unchanged. Copy and complete 9 (ii).

Copy and complete 9 (iii) and 9 (iv) which are variations on the same drawing. Are any of the drawings similar?

4 Draw a shape on squared paper using straight or curved lines or both. Make a drawing which is similar to the original, enlarged or reduced in some ratio you choose.

5 The square grid drawn in Figure 10 (i) may be regarded as a net like a fishing net. Such a net can be pulled out of shape as in Figure 10 (ii) to give a grid of rhombuses. The sides of the rhombuses are equal in length to the sides of the square.

The two **L**-shapes shown in (i) are intended to be similar. The second one has been drawn incorrectly. How would you correct it?

They appear distorted in (ii).

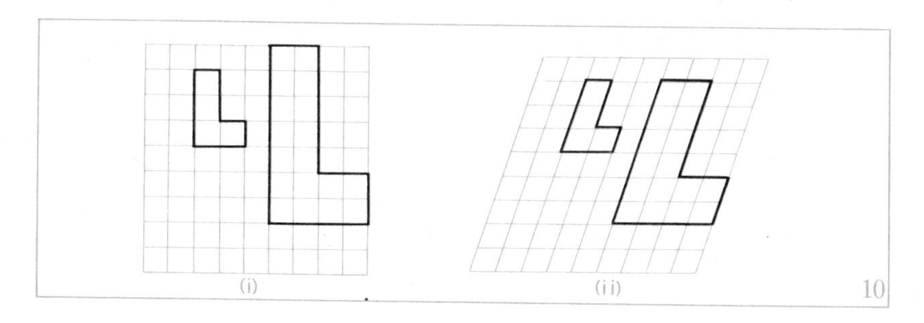

Of the four **L**-shapes, including the corrected one,

a which are equiangular?

b which have corresponding sides in proportion (or equal)?

c which are similar?

Find the ratios of corresponding sides, perimeters and areas of the two **L**s in (ii).

Note. In both (i) and (ii) the two **L**s can be regarded as having been built up, each in the same way, from four square tiles of two different sizes in (i) and four rhombus tiles of two different sizes in (ii).

6 A child's building set contains cubes in two sizes. One has an edge 3 cm long, the other 6 cm long. Calculate the ratios of:

a the edges of the two kinds of brick *b* the areas of their faces

c their volumes.

7 A standard building brick measures 24 cm by 12 cm by 8 cm. A child's building set contains model bricks, similar to the real one, and 6 cm long. What are the dimensions of a model brick?

Calculate the ratios of:

a the total lengths of the edges of the two bricks

b the total surface area of the two bricks

c the volumes of the two bricks.

Repeat for bricks measuring x cm by y cm by z cm, and mx cm by my cm by mz cm.

5 *Similar triangles*

Figure 11 shows a tiling of triangles formed by three sets of parallel lines.

Exercise 5

1 a Why are the two triangles in Figure 11 (i) equiangular to each other (i.e. having the angles of one respectively equal to the angles of the other)?

b What is the ratio of the lengths of corresponding sides in these triangles?

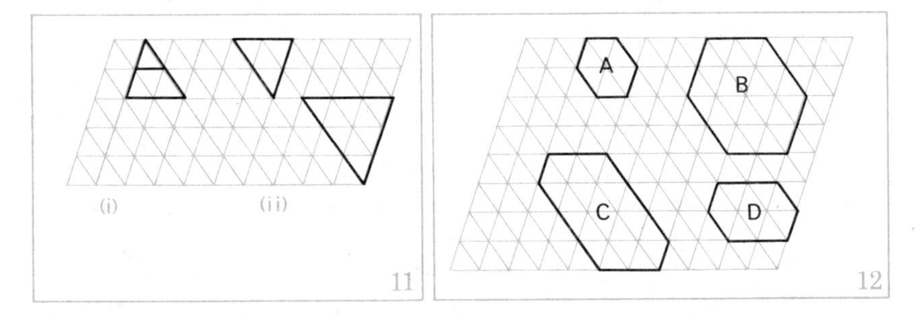

2 Repeat question ***1*** for the pair of triangles in Figure 11 (ii).

3 Can you find two triangles in the tiling of Figure 11 which are equiangular to each other and whose side lengths are in the ratio:

a 4:1 *b* 5:2?

Make a tracing of the triangular tiling and draw the triangles on it.

Note. The triangular tiling contains numerous triangles that are similar to each other.

4 Can you find in the tiling two triangles which are equiangular to each other and whose sides are *not* in proportion?

Note. If two triangles are equiangular to each other, *then* their corresponding sides are in proportion.

5 Figure 12 shows a number of hexagons which are equiangular to one another. Are any of these hexagons similar, i.e. are their corresponding sides in proportion as well?

6 Draw a tiling of congruent triangles (or make a tracing of the tiling in Figure 11). Mark two triangles PQR and XYZ such that the sides of XYZ are each $\frac{5}{3}$ of the sides of PQR.

Repeat for other fractions of your own choice.

In the tiling of congruent triangles in Figure 13, triangles ABC and DEF are equiangular. Each side of $\triangle$DEF is double the length of the corresponding side of $\triangle$ABC.

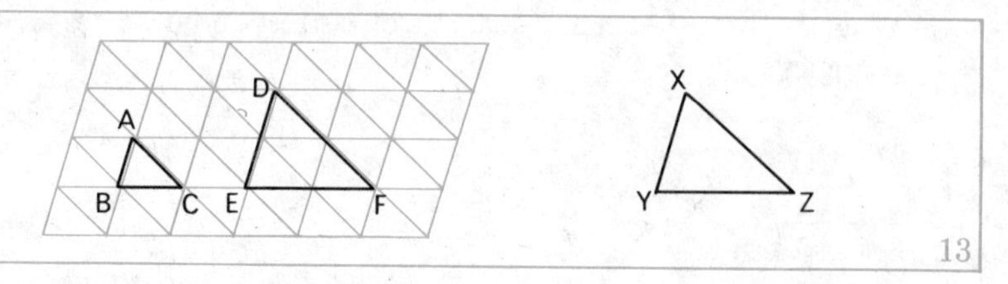

13

Suppose we draw $\triangle XYZ$ with its sides respectively double the sides of $\triangle ABC$. Can we show that $\triangle XYZ$ must be equiangular to $\triangle ABC$?

The sides of $\triangle XYZ$ are equal to the corresponding sides of $\triangle DEF$.

It follows that $\triangle XYZ$ is congruent to $\triangle DEF$, and therefore that $\triangle XYZ$ is equiangular to $\triangle DEF$.

So $\triangle XYZ$ is equiangular to $\triangle ABC$.

This argument could be repeated for every possible given ratio of sides and, with a different tiling, for every size and shape of triangle.

If two triangles have their sides in proportion, the triangles are equiangular.

Note. If two figures are to be similar, we must be able to say that:

(i) they are equiangular to each other,
(ii) their corresponding sides are in proportion.

But triangles are very special figures. We have seen in Exercise 5 that:

if two triangles are equiangular to each other,
then their corresponding sides are in proportion; and
if two triangles have their sides in proportion,
then they are equiangular (corresponding angles being equal).

In the case of equiangular triangles we no longer need to check the order of the angles.

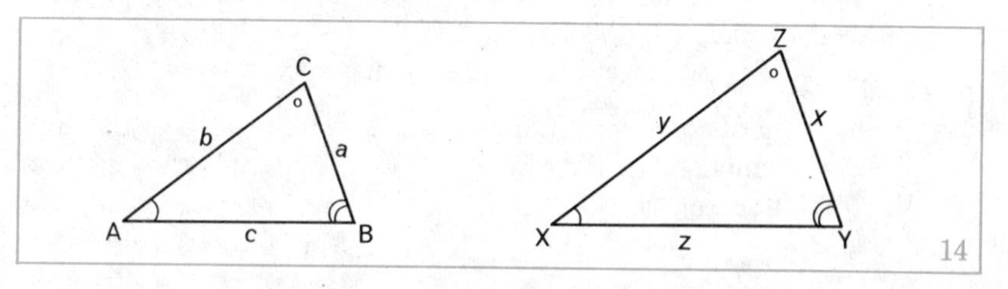

14

If in triangles ABC and XYZ we know that angle A = angle X, angle B = angle Y and angle C = angle Z (see Figure 14)

then
$$\frac{BC}{YZ} = \frac{CA}{ZX} = \frac{AB}{XY}.$$

If in triangles ABC and XYZ we know that $\frac{BC}{YZ} = \frac{CA}{ZX} = \frac{AB}{XY}$ *then* angle A = angle X, angle B = angle Y and angle C = angle Z.

Indicating the sides opposite the angles by the corresponding small letters, as shown in Figure 14, we have:

$$\left.\begin{array}{l}\angle A = \angle X\\ \angle B = \angle Y\\ \angle C = \angle Z\end{array}\right\} \Leftrightarrow \triangle\text{s ABC and XYZ are similar} \Leftrightarrow \frac{a}{x} = \frac{b}{y} = \frac{c}{z}.$$

Exercise 6

1 In Figure 15, name pairs of equal angles, and equal ratios of sides in the two triangles.

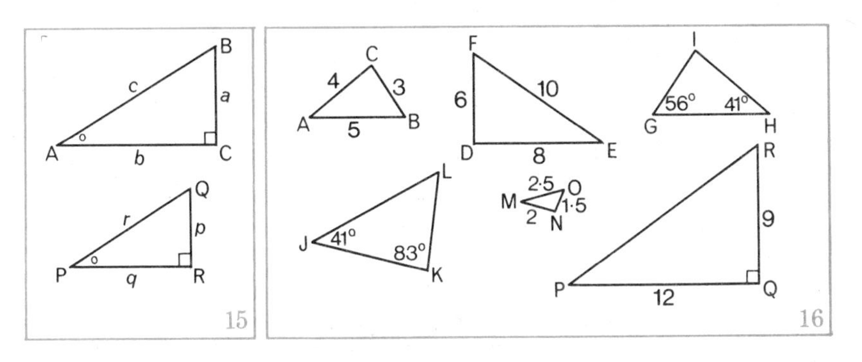

2 In Figure 16, list sets of similar triangles. In each case say if you know the triangles are similar because they are equiangular or because they have their sides in proportion.

3 In triangles ABC and PQR, $\angle A = 29°$, $\angle B = 114°$, $\angle P = 37°$ and $\angle Q = 29°$. Explain why the triangles are similar, and write down equal ratios of sides.

4 In triangles DEF and XYZ, DE = 12 cm, EF = 18 cm, FD = 27 cm, ZX = 24 cm, XY = 36 cm and YZ = 16 cm. Explain why the triangles are similar, and list pairs of equal angles.

5 Use the converse of Pythagoras' theorem to prove that triangles with sides of length 5, 12 and 13 cm, and 8, 15 and 17 cm are right-angled. Are the triangles similar? Give a reason for your answer.

6 Which of the following sets of side lengths give triangles which are similar to one or other of the given triangles in question **5**?

a 4, 7·5, 8·5 m *b* 15, 36, 39 km *c* 10, 24, 26 m
d 1·25, 3, 3·25 m *e* 1, 2·4, 2·6 cm *f* 88, 165, 187 mm

7 Figure 17 shows a method of finding the height of a tree (h m). The tree casts a shadow 25 m long on horizontal ground, and a vertical post 3 m high casts a shadow 4 m long. Identify two similar triangles, write down two equal ratios, and calculate the height of the tree to the nearest metre.

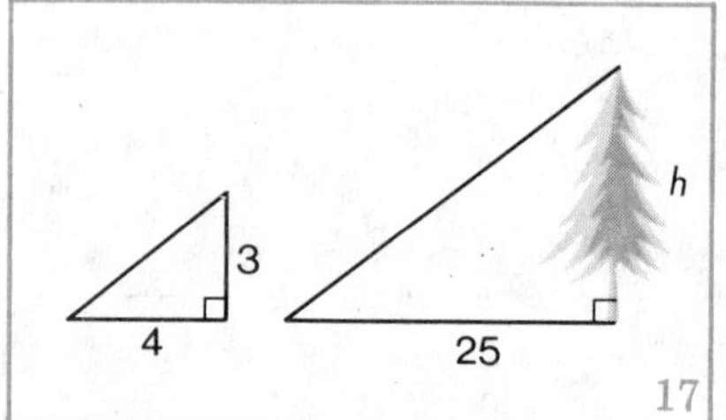

17

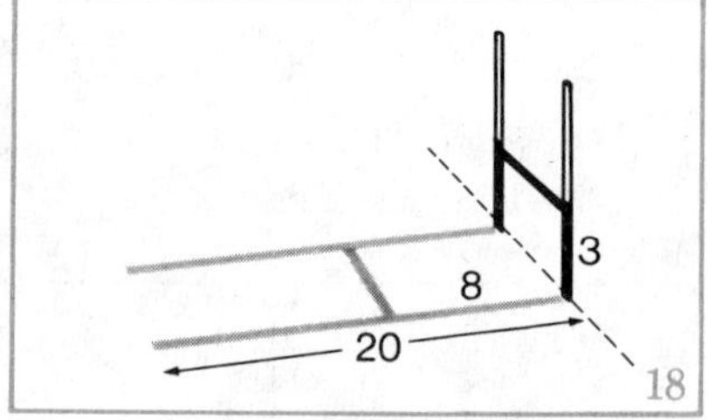

18

8 Figure 18 shows a rugby goal and its shadow. Each post casts a shadow 20 m long. The part of the post below the crossbar is 3 m high and casts a shadow 8 m long. Calculate the height of the post.

Exercise 7

1 Figure 19 shows pairs of similar cardboard triangles, placed in different positions. Copy each diagram, mark in pairs of equal angles and then write down equal ratios of sides in each case, using the given letters.

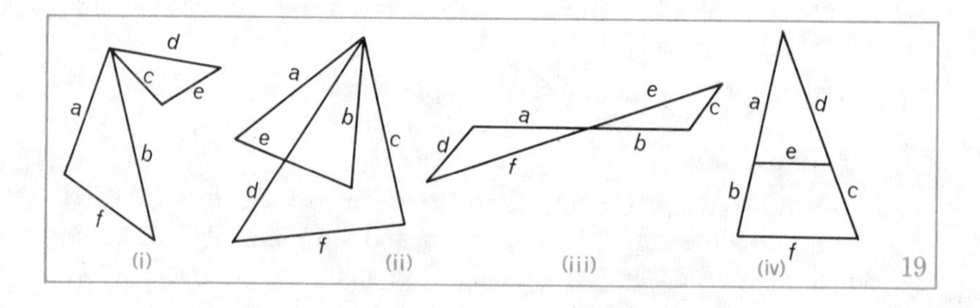

19

2 Figure 20 shows the side view of a camp stool, with the lengths marked in centimetres. Prove that the triangles are equiangular to each other. How far apart are the feet of the stool?

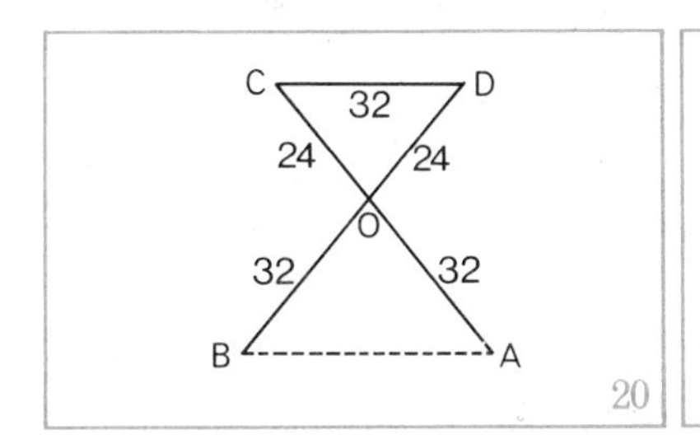

20

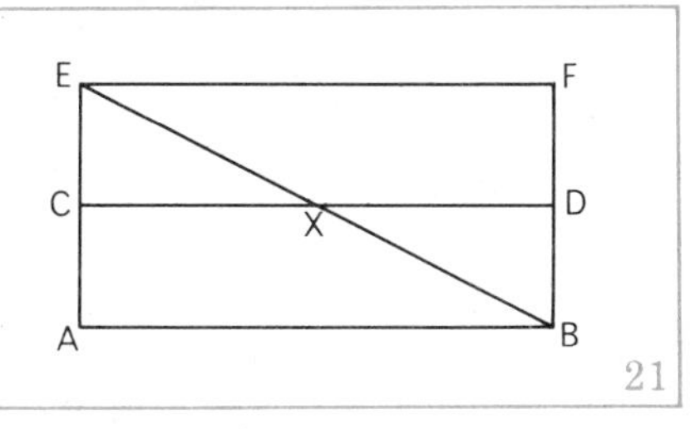

21

3 Figure 21 shows a 'barred gate' diagram. Name three triangles which are similar to triangle ECX. Copy and complete

a $\frac{EC}{EA} = \frac{EX}{\quad} = \frac{CX}{\quad}$ *b* $\frac{BD}{BF} = \frac{BX}{\quad} = \frac{DX}{\quad}$.

4 In Figure 22 the units are centimetres.

a Prove that triangles ABC and ADE are equiangular to each other.

b Write down three equal ratios of sides.

c Choosing two suitable ratios from ***b***, calculate the lengths of AD and BD.

d Denote the length of AC by x cm. Choosing two suitable ratios from ***b***, calculate x.

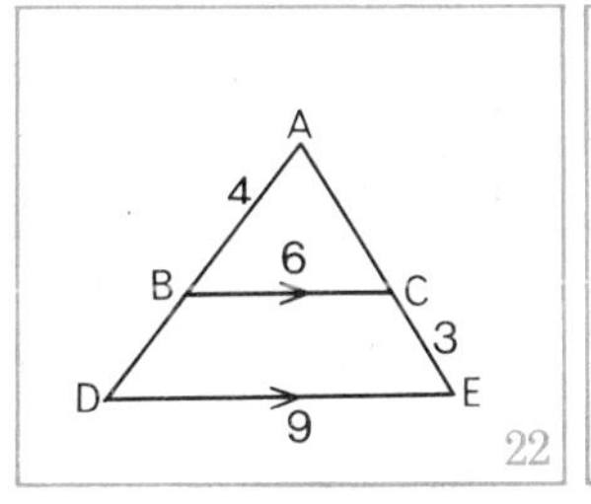

22

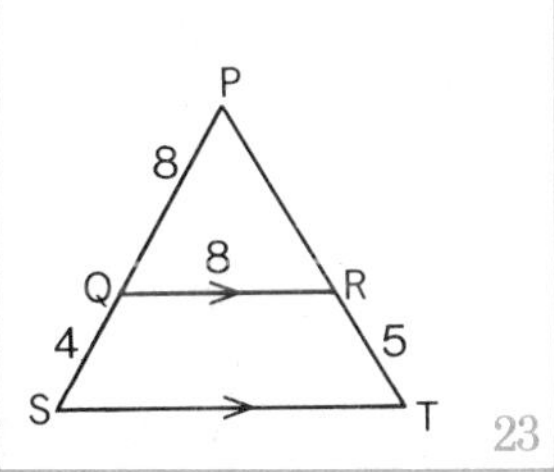

23

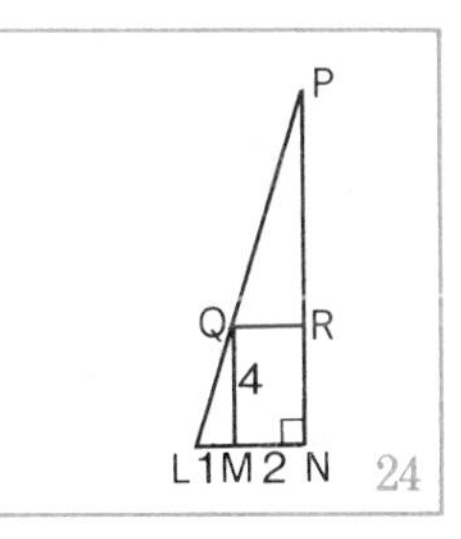

24

5 In Figure 23, calculate the lengths of ST and PR.

6 Figure 24 shows a ladder LP leaning against a wall NP, touching the corner of a rectangular shed at Q. The units are metres. Name three similar triangles in the diagram, and calculate how far up the wall the ladder reaches.

7 Calculate the values of a, b, c and d in Figure 25, all the lengths being in centimetres.

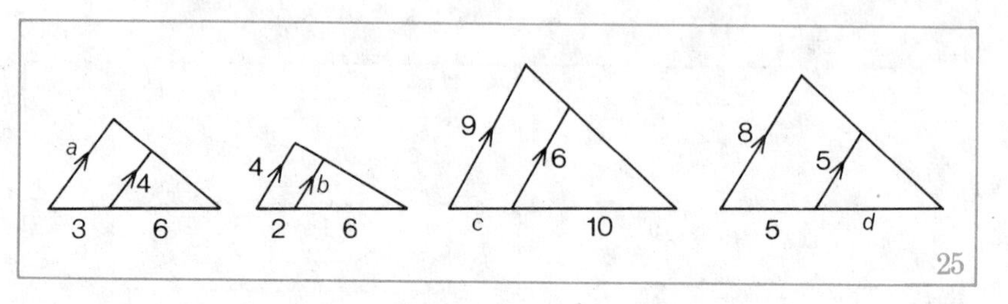

25

In Figure 26, the two triangles are equiangular, so we can write

$$\frac{a}{a+b} = \frac{c}{c+d}$$
$$\Leftrightarrow a(c+d) = c(a+b)$$
$$\Leftrightarrow ac+ad = ca+cb$$
$$\Leftrightarrow \quad ad = bc$$
$$\Leftrightarrow \quad \frac{a}{b} = \frac{c}{d}$$

This useful result shows that a line parallel to one side of a triangle cuts the other two sides in proportion.

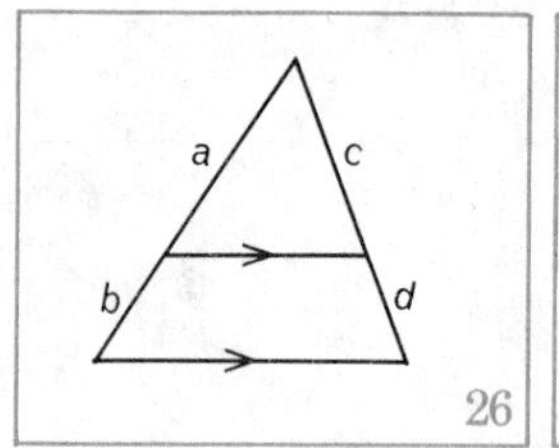

26

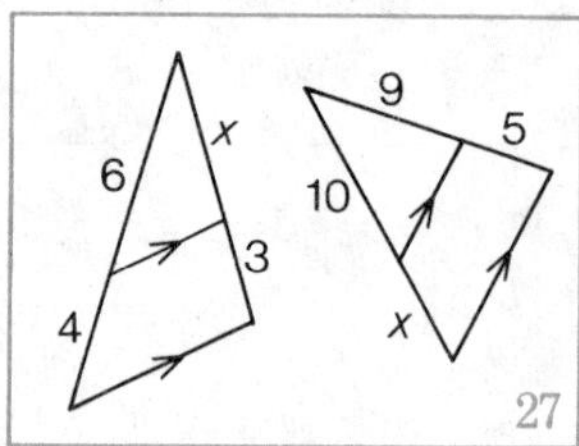

27

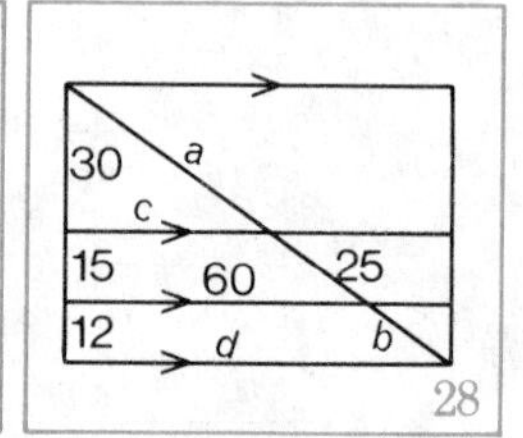

28

Exercise 7B

1 Calculate x in each diagram in Figure 27. The units are centimetres.

2 Calculate a, b, c and d in Figure 28. The units are centimetres.

3 The ratio of the lengths of corresponding sides in two similar triangles is 4:5. If the lengths of two corresponding sides differ by 2 cm, calculate their lengths.

4 In Figure 29, ABC is a triangle right-angled at A, and AD is perpendicular to BC.

a Prove that triangles DBA and DAC are equiangular.

b If BD = 9 cm and DC = 4 cm, use triangles DBA and DAC to calculate the length of AD.

5 *a* Prove that triangles ABC and DBA are equiangular in Figure 29.

b If BC = 16 cm and BA = 12 cm, use triangles ABC and DBA to calculate the length of BD.

c Deduce the lengths of DC, AD and AC.

6 In Figure 30, POR and QOS cut two parallel straight lines as shown.

a Prove that triangles POQ and ROS are similar.

b If PQ = 6 cm, SR = 9 cm and PR = 12 cm, calculate the lengths of PO and OR. (*Hint.* Let PO = x cm.)

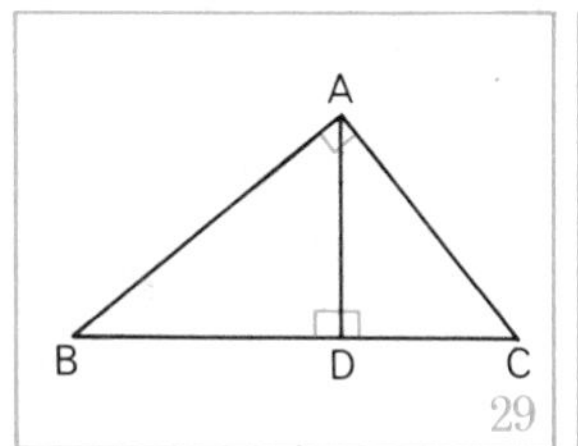

29

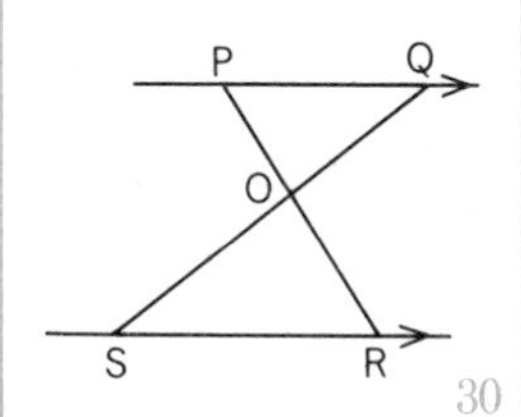

30

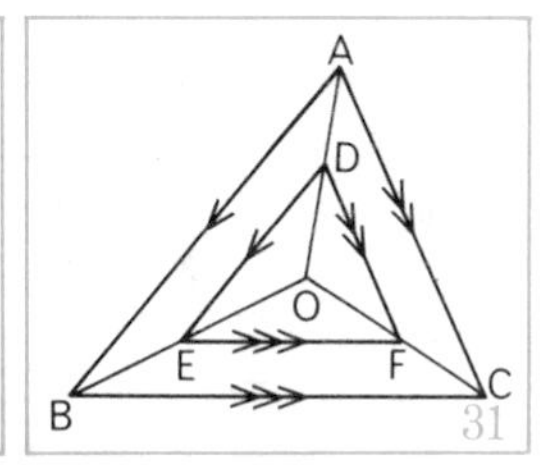

31

7 *a* In Figure 31, O is any point inside the triangle ABC. DE is parallel to AB. If lines are drawn through D and E parallel to AC and BC respectively, prove that these parallels meet OC in the same point. (Use the theorem on page 98).

b If F is this point, prove that triangles ABC and DEF are similar.

8 In Figure 29, BC = a units, CA = b units, AB = c units, BD = x units and DC = y units.

a Use similar triangles to show that $c^2 = ax$.

b Find a similar expression for b^2.

c Prove Pythagoras' theorem by adding these results.

9 In Figure 32 ABCD is a quadrilateral in which $\angle$ABC is the supplement of $\angle$ADC. AB and DC produced meet at E.

a Prove that triangles EAD and ECB are similar.

b If EA = 12 m, EB = 7·5 m and EC = 6 m, calculate the length of CD.

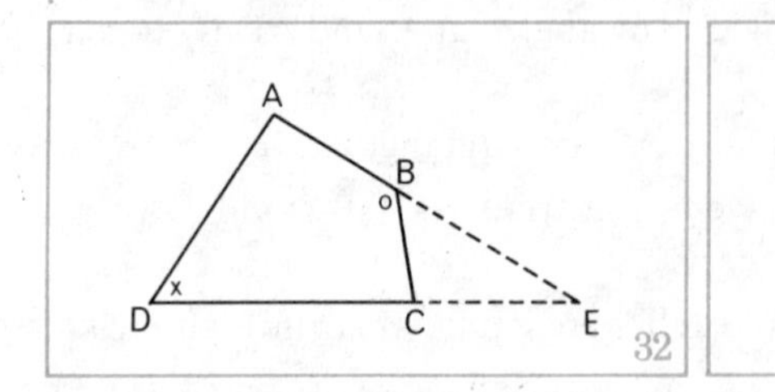

32

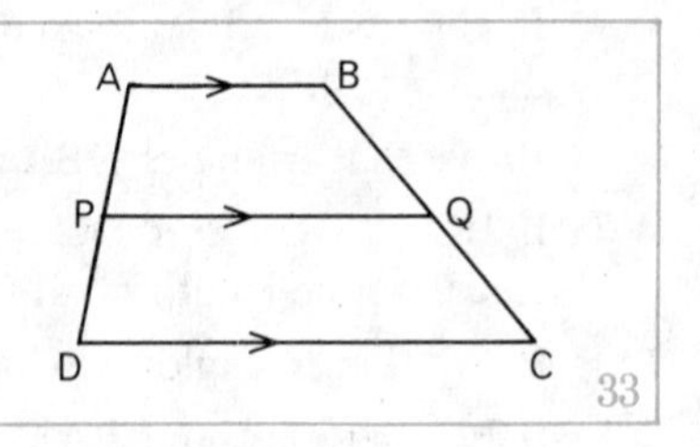

33

10a In Figure 33 AB, PQ and DC are parallel. Prove that $\frac{AP}{PD} = \frac{BQ}{QC}$.
(*Hint.* Join AC to cut PQ at R.)

b Suggest other constructions that would help you to prove this result.

c If AP = 5 cm, PD = 4 cm and BC = 13·5 cm, calculate the lengths of BQ and QC.

6 Gradient

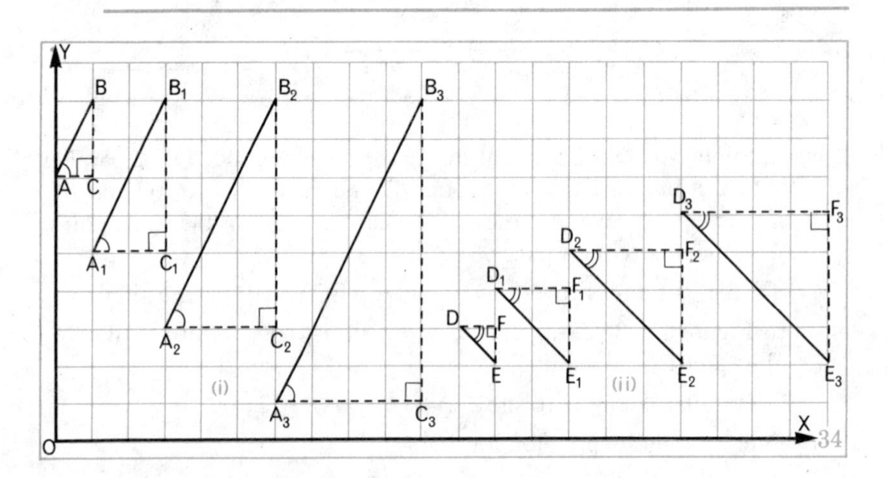

34

In Figure 34 (i), $AB \| A_1B_1 \| A_2B_2 \| A_3B_3$.

Since these lines are parallel they all make the same angle with the positive direction of the *x*-axis.

Triangles ABC, $A_1B_1C_1$, $A_2B_2C_2$ and $A_3B_3C_3$ are equiangular and have their sides in proportion.

So $\dfrac{CB}{AC} = \dfrac{C_1B_1}{A_1C_1} = \dfrac{C_2B_2}{A_2C_2} = \dfrac{C_3B_3}{A_3C_3}$.

We can describe CB as the y-component of the line segment AB, and AC as the x-component of the line segment AB.

We define the gradient of AB as $\dfrac{y\text{-component of AB}}{x\text{-component of AB}} = \dfrac{CB}{AC}$

The gradient of AB is often written m_{AB}.

In Figure 34 (i), the gradient of AB $= \frac{2}{1} = 2$,
the gradient of $A_1B_1 = \frac{4}{2} = 2$,
the gradient of $A_2B_2 = \frac{6}{3} = 2$ and
the gradient of $A_3B_3 = \frac{8}{4} = 2$.

Similarly in Figure 34 (ii) the gradients of DE, D_1E_1, D_2E_2 and D_3E_3 are all equal since $\dfrac{-1}{1} = \dfrac{-2}{2} = \dfrac{-3}{3} = \dfrac{-4}{4} = -1$.

Parallel lines have the same gradient.

Notes. (i) The gradient of AB is independent of the length of AB; it depends only on the angle which the line makes with the x-axis.
(ii) The gradient of BA is the same as the gradient of AB. For BA the x-component is -1 and the y-component is -2; hence the gradient is $\dfrac{-2}{-1}$, i.e. 2.
(iii) Lines which have a positive gradient slope upwards from left to right. Those which have a negative gradient slope downwards from left to right.
(iv) For a line parallel to the x-axis, the y-component is zero, and the gradient is of the form $\dfrac{0}{4}$, i.e. 0.
(v) For a line parallel to the y-axis, the x-component is zero, and the gradient is of the form $\dfrac{4}{0}$ which has no meaning. We cannot calculate the gradient of such a line.

Exercise 8

1 Calculate the gradient of each line shown in Figure 35.

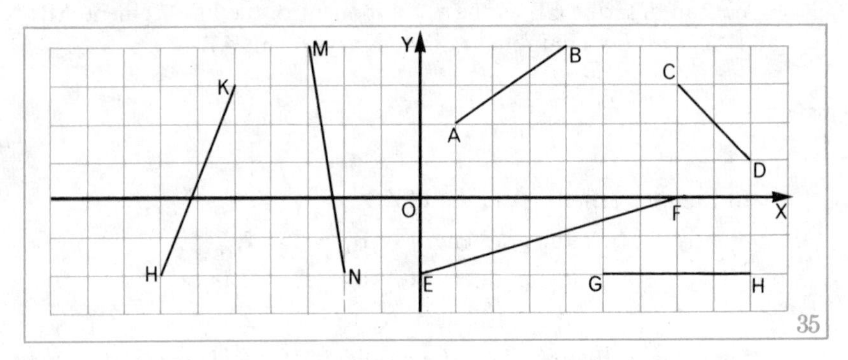

35

2 Plot the following pairs of points and find the gradient of the straight line through each pair (if possible).

a (5, 3), (6, 7)
b (5, − 3), (6, 7)
c (− 5, − 3), (− 10, − 3)
d (5, 3), (6, 0)
e (− 5, 3), (− 6, 0)
f (5, 3), (5, − 4)

3 For rectangular axes, what can you say about the slope of a line if the gradient is *a* positive *b* negative *c* zero?

4 a Draw a line through the origin whose gradient is 2. Write down an equation which is true for the coordinates (x, y) of every point on this line.

b Write down the coordinates of a number of points whose coordinates satisfy the equation $y = 3x$. Calculate the gradient of the line segments joining several pairs of these points.

5 a Draw a line through the origin whose gradient is $-\frac{1}{2}$. Write down an equation which is true for the coordinates (x, y) of every point on this line.

b Write down the coordinates of a number of points whose coordinates satisfy the equation $y = -2x$. Calculate the gradient of the line segments joining several pairs of these points.

6 a Check that the line joining A(0, 2) to (1, 5) has gradient 3. Work out the y-coordinate of the following points if the line joining A to each point is to have gradient 3.

(2,), (5,), (20,), (− 4,)

Write down an equation which is true for the coordinates (x, y) of every point on the line through (0, 2) of gradient 3.

b Write down the coordinates of a number of points whose coordinates satisfy the equation $y = \frac{1}{2}x+4$. Calculate the gradient of the line segments joining several pairs of these points.

7 a Check that the line joining A(0, 3) to $(-10, 7)$ has gradient $-\frac{2}{5}$. Work out the y-coordinates of the following points if the line joining A to each point has gradient $-\frac{2}{5}$.

(20,), (15,), (10,), (5,), (−5,)

b Write down the coordinates of a number of points whose coordinates satisfy the equation $y = -x+3$. Calculate the gradient of the line segments joining several pairs of these points.

Note. As we have seen in Book 3, the coordinates of every point on a straight line which is not parallel to the y-axis satisfy an equation of the form $y = mx+c$ where m is the gradient of every segment of the line, and $(0, c)$ is the point where the line meets the y-axis.

Every point whose coordinates satisfy an equation of this form lies on such a line.

8 Write down the equations of the lines through the origin with gradients stated. Draw all the lines on the same diagram, and label each with its equation.

a 4 *b* $\frac{2}{3}$ *c* $\frac{3}{2}$ *d* -4 *e* $-\frac{2}{3}$ *f* $-\frac{3}{2}$ *g* 0

9 Write down the equations of the lines through the point (0, 4) with the gradients given in question ***8***. Sketch and label the lines.

10 Write down the equations of the lines with gradient 2 which pass through the following points. Sketch and label the lines.

a (0, 4) *b* (0, 0) *c* (0, 1) *d* (0, −4)

11 Write down the equations of the lines of gradient -1 which pass through the points given in question ***10***. Sketch and label the lines.

12 For each of the lines whose equations are given here, state its gradient and the coordinates of the point where it cuts the y-axis. Sketch and label each line.

a $y = 2x+5$ *b* $y = -x+3$ *c* $y = -2x-4$

d $y = \frac{3}{4}x-3$ *e* $y-x = 2$ *f* $y-4 = 3x$

g $x+y = 4$ *h* $2x+y = 5$ *i* $5x-y+7 = 0$

j $2y = x+6$ *k* $3y+2x+6 = 0$ *l* $4x-3y-2 = 0$

Summary

1. *Scale drawings and scale models have the same shape*, but not the same *size*, as the original. Ratios of corresponding lengths are equal.

2. *Two shapes are similar if* :
(i) they are *equiangular* (pairs of angles taken in order are equal), *and*
(ii) pairs of *corresponding sides are in proportion.*
Conditions (i) and (ii) must be proved separately, as one does not imply the other.

3. In the case of *similar triangles* condition (i) $\Leftrightarrow$ condition (ii).

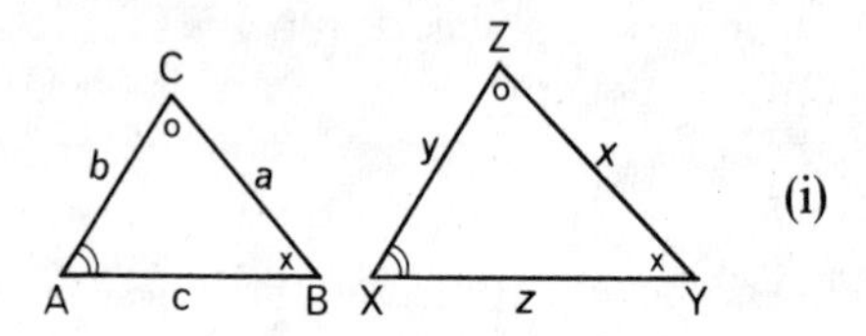

$$\left.\begin{array}{l}\angle A = \angle X \\ \angle B = \angle Y \\ \angle C = \angle Z\end{array}\right\} \Leftrightarrow \Delta\text{s ABC, XYZ are similar} \quad \Leftrightarrow \quad \frac{a}{x} = \frac{b}{y} = \frac{c}{z}.$$

4. From *similar triangles* PQR and PST,

$$\frac{a}{a+b} = \frac{c}{c+d} = \frac{e}{f}.\ \text{Also, if QR} \parallel \text{ST, } \frac{a}{b} = \frac{c}{d}.$$

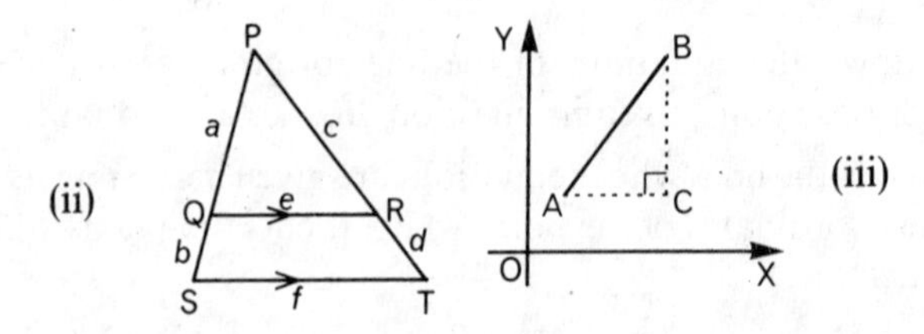

5. In Figure (iii) the *gradient* of AB $= \dfrac{y\text{-component of AB}}{x\text{-component of AB}} = \dfrac{\text{CB}}{\text{AC}}$

An Introduction to Vectors

1 *Revision: translation*

In the chapter on Translation in Book 4, we saw that translations had two important properties, magnitude and direction, which arise in many everyday activities.

Figure 1 shows a simplified map of Aberdeen (A), Banchory (B) and Stonehaven (S). We can motor as fast as we like from Aberdeen in the direction of Banchory but we shall not reach Stonehaven. Direction as well as speed is important.

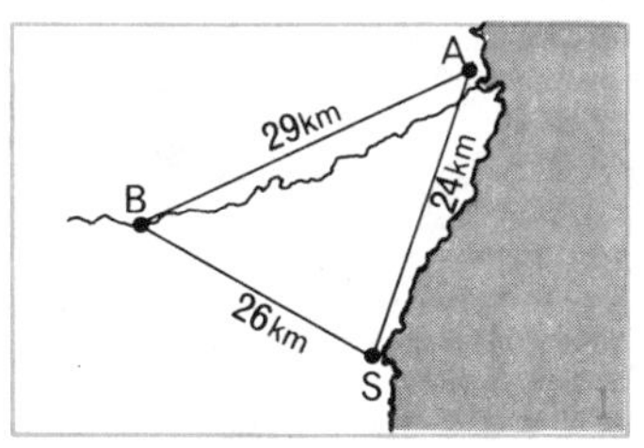

1

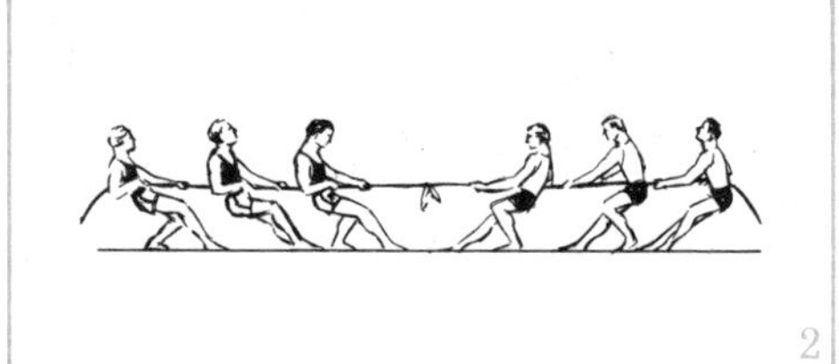

2

The strength with which we can pull will not achieve very much unless we pull in the right direction. In Figure 2, two tug-of-war teams may not get very far, however hard they pull. Forces are being exerted in opposite directions.

Before looking at the problem of magnitude and direction more generally, here are some reminders of topics from the Translation Chapter.

(i) Combination of displacements: $\overrightarrow{AB} \oplus \overrightarrow{BC} = \overrightarrow{AC}$

(ii) Translation as a mapping of all points in a plane through a given distance in a given direction

(iii) Representation of a translation by any one of a set of directed line segments, all with the same magnitude and direction

(iv) Component form of a translation, e.g. $\begin{pmatrix} a \\ b \end{pmatrix}$

(v) 'Addition' of translations in line segment or in component form.

Exercise 1

1 a In Figure 1 a car goes from A to B and then to S. Is it true to say that A is 55 km from S? Or 24 km? Could you make out a case for either answer?

b If you travelled 29 km in a straight line from A and then a further 26 km in a straight line, what would be your greatest and least possible distances from A?

c What is the locus of all points 29 km from A? What further information would you need to find the exact position of B?

2 A ship sails 12 km from a port P on a bearing 320° and then 17 km on a bearing 210°. Make an accurate scale drawing, and find the final distance and bearing of the ship from P.

3 For Figure 3 complete the following:

a $\overrightarrow{DA} \oplus \overrightarrow{AB} = \ldots$ *b* $\overrightarrow{BC} \oplus \ldots = \overrightarrow{BD}$ *c* $\overrightarrow{BD} \oplus \overrightarrow{DB} = \ldots$

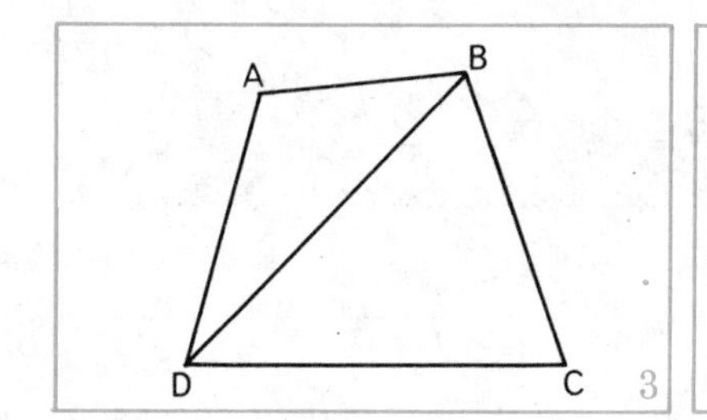

3

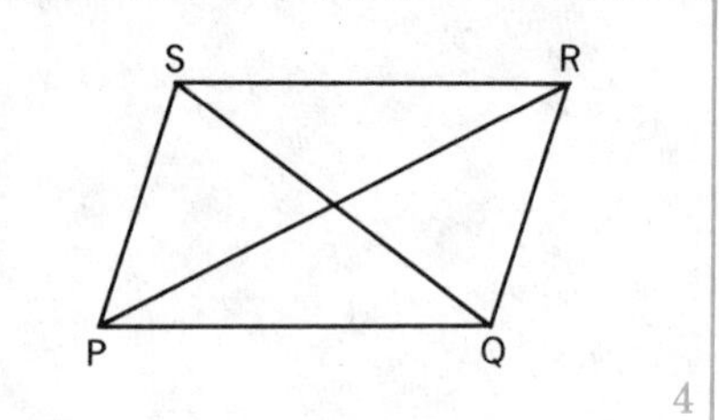

4

4 In Figure 4, PQRS is a parallelogram.

a Name a pair of directed line segments which represent the same translation.

b Name another pair of directed line segments which represent a translation different from the one chosen in ***a***.

c The translation represented by $\overrightarrow{PQ}$ is followed by the translation represented by *(1)* $\overrightarrow{QR}$ and *(2)* $\overrightarrow{PS}$. Name a directed line segment which represents the resulting translation in each case.

d Name a directed line segment which represents the translation represented by $\overrightarrow{QP}$ followed by the translation represented by $\overrightarrow{QR}$.

5 In Figure 5, $\overrightarrow{AB}$ represents the translation $\begin{pmatrix} 2 \\ -3 \end{pmatrix}$. Write in number pair form the translations represented by:

a $\overrightarrow{XY}$ *b* $\overrightarrow{RS}$ *c* $\overrightarrow{LM}$ *d* $\overrightarrow{PQ}$

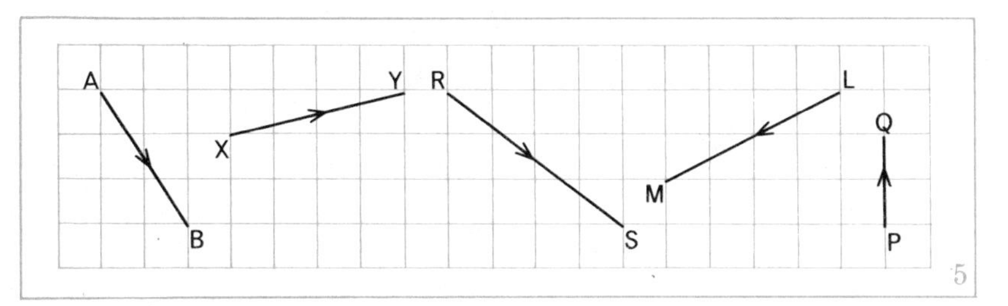

5

6 On squared paper draw directed line segments $\overrightarrow{AB}$, $\overrightarrow{PS}$ and $\overrightarrow{QR}$ representing the translations *a* $\begin{pmatrix}4\\2\end{pmatrix}$ *b* $\begin{pmatrix}-4\\-1\end{pmatrix}$ *c* $\begin{pmatrix}1\\-4\end{pmatrix}$.

7 a Using Figure 5, draw on squared paper representatives of the translations represented by:

(*1*) $\overrightarrow{AB} \oplus \overrightarrow{XY}$ (*2*) $\overrightarrow{XY} \oplus \overrightarrow{RS}$ (*3*) $\overrightarrow{PQ} \oplus \overrightarrow{LM}$.

b Write down each of these three translations in component form.

8 Fill in the missing components in the following equations:

a $\begin{pmatrix}4\\1\end{pmatrix} + \begin{pmatrix} \\ \end{pmatrix} = \begin{pmatrix}8\\5\end{pmatrix}$ *b* $\begin{pmatrix}4\\-3\end{pmatrix} + \begin{pmatrix}0\\0\end{pmatrix} = \begin{pmatrix} \\ \end{pmatrix}$

c $\begin{pmatrix}a\\-b\end{pmatrix} + \begin{pmatrix} \\ \end{pmatrix} = \begin{pmatrix}0\\0\end{pmatrix}$ *d* $\begin{pmatrix}2\\ \end{pmatrix} + \begin{pmatrix} \\ -3\end{pmatrix} = \begin{pmatrix}5\\4\end{pmatrix}$

9 In Figure 5, use Pythagoras' theorem to calculate the magnitudes of the translations represented by $\overrightarrow{AB}$, $\overrightarrow{XY}$, $\overrightarrow{RS}$ and $\overrightarrow{LM}$. (Leave in the square roots where necessary).

---

2 *Vectors and their addition*

In order to deal with quantities like displacement, translation, velocity and force, which have in common the properties of *magnitude* and *direction*, we now study the mathematics of *vectors*.

In Figure 6 the directed line segments all have the same magnitude and direction. The set of all the directed line segments in the plane with this magnitude and direction is called a (geometrical) *vector*;

the magnitude is called the *magnitude* or *length of the vector*, and the direction is called the *direction of the vector*. Each of the directed line segments is called a *representative* of the vector.

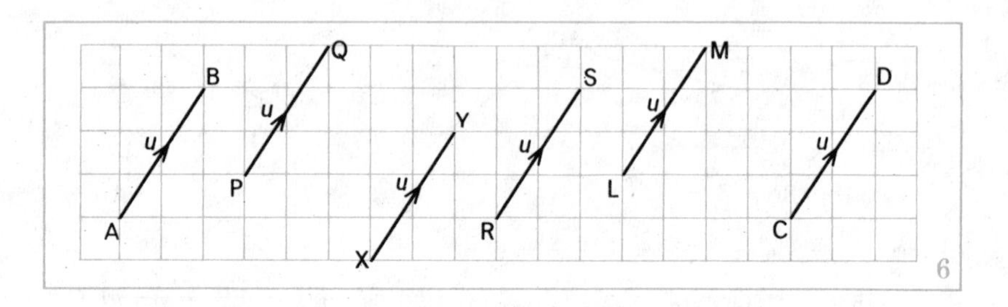

6

We often denote a vector by a single small letter in bold italic type, e.g. ***a***, ***b***, ***u***, ***v***. As it is not easy to copy this in your books, it is usual to write a vector as a small letter underlined thus, $\underline{u}$ or $\underset{\sim}{u}$, read 'vector ***u***'. We often label a directed line segment by the vector it represents (see Figure 6).

Given a vector ***u*** there is a unique translation of the plane represented by the same directed line segments as ***u***. For this reason a geometrical vector is often defined as a translation. Because of this close connection between vectors and translations it is natural to 'add' two vectors ***u*** and ***v*** in the same way as we 'added' two translations, and to denote the sum by $\boldsymbol{u} \oplus \boldsymbol{v}$.

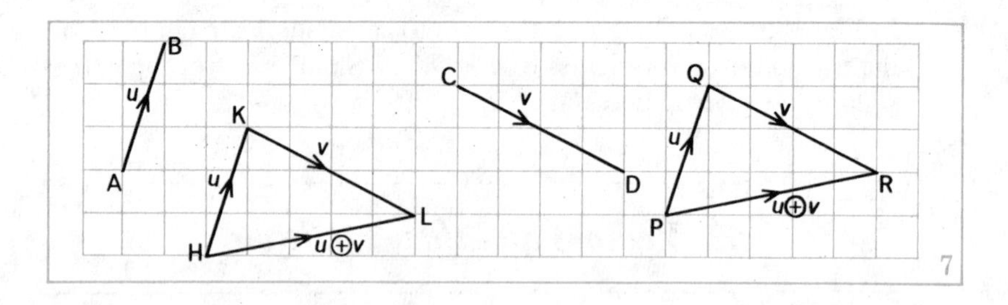

7

In Figure 7, $\overrightarrow{AB}$, $\overrightarrow{HK}$ and $\overrightarrow{PQ}$ all represent the vector ***u***, and $\overrightarrow{CD}$, $\overrightarrow{KL}$ and $\overrightarrow{QR}$ all represent the vector ***v***. From the diagram we see that each of $\overrightarrow{HL}$ and $\overrightarrow{PR}$ represents $\boldsymbol{u} \oplus \boldsymbol{v}$. Any triangle congruent to triangle HKL with its sides parallel to those of triangle HKL would give the required result.

Exercise 2

1 Copy Figure 8 on squared paper, where $\overrightarrow{AB}$ represents vector $\boldsymbol{u}$. Draw three more representatives of $\boldsymbol{u}$.

2 In Figure 9 the directed line segments represent vectors $\boldsymbol{r}$, $\boldsymbol{s}$ and $\boldsymbol{t}$ as shown. Draw these on squared paper, and obtain representatives of:

a $\boldsymbol{r} \oplus \boldsymbol{s}$ *b* $\boldsymbol{s} \oplus \boldsymbol{r}$ *c* $\boldsymbol{s} \oplus \boldsymbol{t}$ *d* $\boldsymbol{t} \oplus \boldsymbol{s}$

Which law do your answers illustrate?

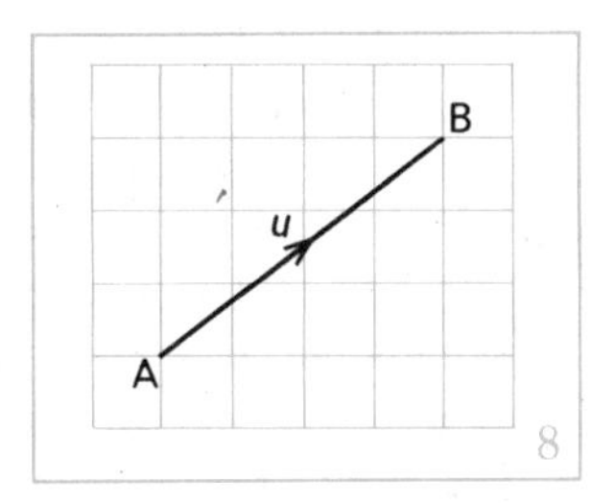

8

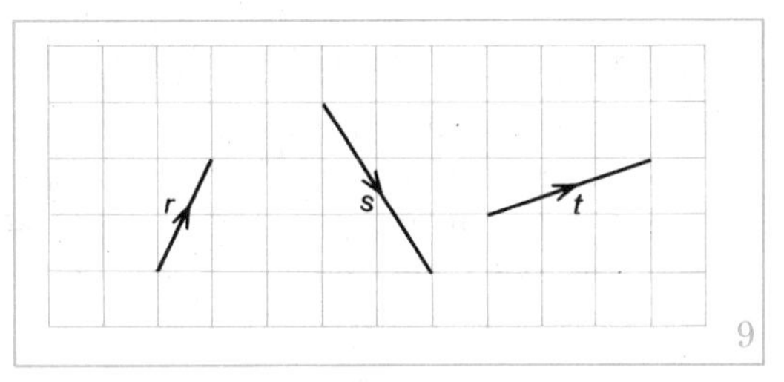

9

3 Use squared paper to obtain representatives of:

a $\boldsymbol{r} \oplus \boldsymbol{s}$ *b* $(\boldsymbol{r} \oplus \boldsymbol{s}) \oplus \boldsymbol{t}$ *c* $\boldsymbol{s} \oplus \boldsymbol{t}$ *d* $\boldsymbol{r} \oplus (\boldsymbol{s} \oplus \boldsymbol{t})$.

Which law do your answers illustrate?

4 Questions **2** and **3** suggest that the operation $\oplus$ (followed by) is commutative and associative. Verify this by drawing representatives of vectors of your own choice on squared paper.

5 In Figure 10, the vectors $\boldsymbol{u}$, $\boldsymbol{v}$ and $\boldsymbol{w}$ are represented by $\overrightarrow{AB}$, $\overrightarrow{BC}$ and $\overrightarrow{CD}$.

a Sketch the figure. Draw and name directed line segments which represent $\boldsymbol{u} \oplus \boldsymbol{v}$ and $(\boldsymbol{u} \oplus \boldsymbol{v}) \oplus \boldsymbol{w}$.

b Draw and name directed line segments which represent $\boldsymbol{v} \oplus \boldsymbol{w}$ and $\boldsymbol{u} \oplus (\boldsymbol{v} \oplus \boldsymbol{w})$.

c Why is it possible to write $\boldsymbol{u} \oplus \boldsymbol{v} \oplus \boldsymbol{w}$ without brackets?

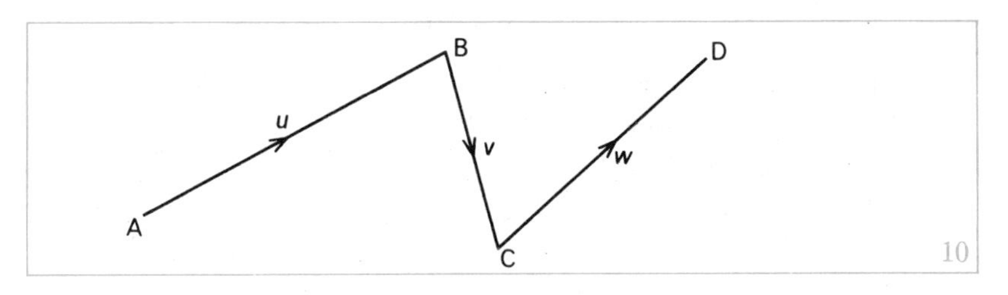

10

6 Here is an example of vectors outside geometry.

An aircraft heads due north from a point A. It can fly 240 km/h in still air. A strong west wind carries it 100 km/h towards the east.

a Make a scale drawing to show where you think the aircraft will be after 6 minutes, 12 minutes, . . . , 1 hour.

b What is the actual speed and direction of the aircraft?

c Verify that you can find the answer by 'adding' two vectors with the magnitudes and directions of the two velocities.

The triangle and parallelogram rules for adding vectors

We see from the above Exercise that the operation denoted by $\oplus$ obeys the commutative and associative laws in the same way as the operation denoted by $+$. It seems reasonable to replace $\oplus$ by $+$, and to talk of *adding* vectors. In some situations, however, it will be necessary to make a clear distinction in our minds between vector addition and numerical addition.

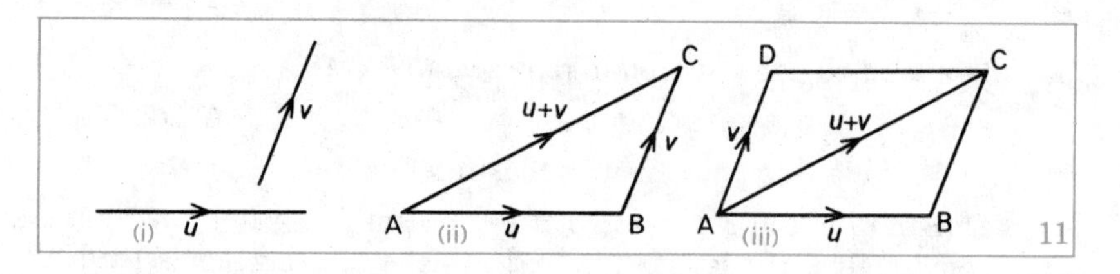

11

To add two vectors $\boldsymbol{u}$ and $\boldsymbol{v}$ represented by the directed line segments shown in Figure 11 (i), we place the segments 'nose-to-tail' as shown in (ii). *The third side* $\overrightarrow{AC}$ *of* $\triangle$ABC represents $\boldsymbol{u}+\boldsymbol{v}$.

Figure 11 (iii) illustrates a situation in which we often meet representatives of vectors, both of which start from the same point. In this case *the diagonal* $\overrightarrow{AC}$ *of the parallelogram* ABCD represents $\boldsymbol{u}+\boldsymbol{v}$; this agrees with Figure 11 (ii) since $\overrightarrow{BC}$ also represents $\boldsymbol{v}$.

Exercise 3

1 *a* Copy on squared paper the directed line segments $\overrightarrow{AB}$ and $\overrightarrow{XY}$ of Figure 12 to represent vectors $\boldsymbol{u}$ and $\boldsymbol{v}$ respectively.

b Draw a triangle PQR in which $\overrightarrow{PQ}$ represents $\boldsymbol{u}$ and $\overrightarrow{QR}$ represents $\boldsymbol{v}$, to illustrate the addition of these vectors.

c Draw a triangle HKL in which $\vec{HK}$ represents $\boldsymbol{v}$ and $\vec{KL}$ represents $\boldsymbol{u}$. Name the directed line segment which represents $\boldsymbol{v}+\boldsymbol{u}$.

d Draw a parallelogram EFGH in which $\vec{FG}$ represents $\boldsymbol{u}$ and $\vec{FE}$ represents $\boldsymbol{v}$. Name the directed line segment which represents $\boldsymbol{u}+\boldsymbol{v}$.

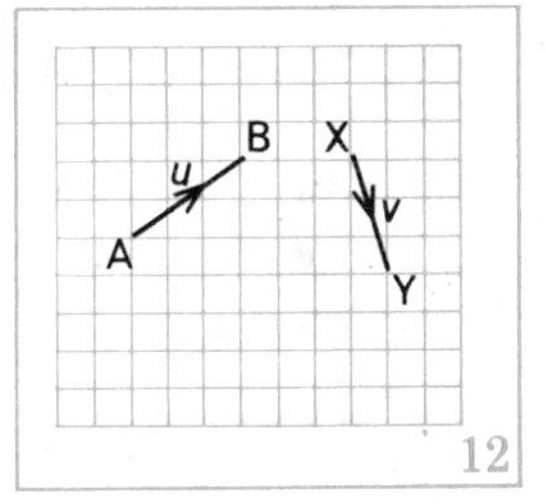

12

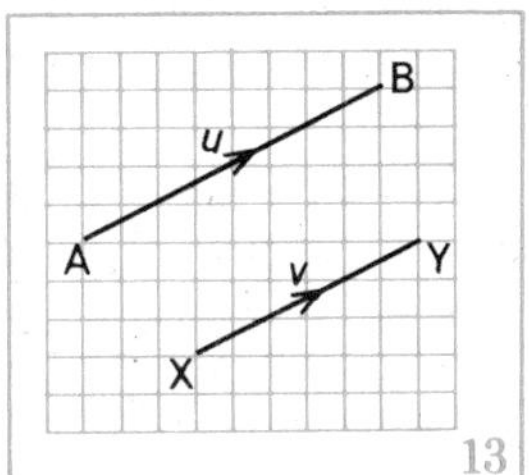

13

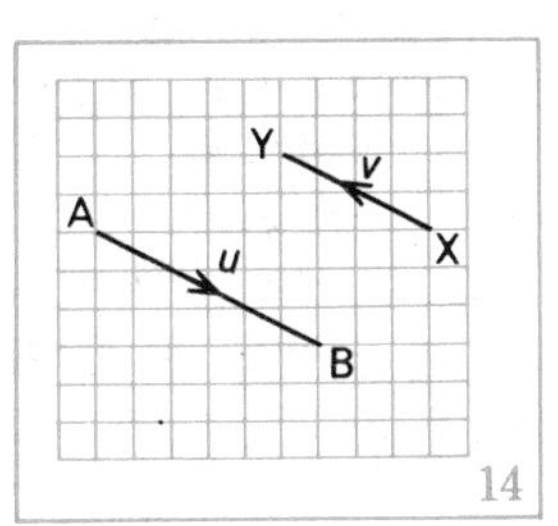

14

2 In Figure 13, directed line segments $\vec{AB}$ and $\vec{XY}$, representing vectors $\boldsymbol{u}$ and $\boldsymbol{v}$, are parallel and have the same direction.

a On squared paper add the vectors by a nose-to-tail arrangement. Do you get a triangle?

b Check if it is still true that $\boldsymbol{u}+\boldsymbol{v} = \boldsymbol{v}+\boldsymbol{u}$.

3 In Figure 14, $\boldsymbol{u}$ and $\boldsymbol{v}$ are represented by $\vec{AB}$ and $\vec{XY}$, which are parallel but have opposite directions.

a Is it still true that $\boldsymbol{u}+\boldsymbol{v} = \boldsymbol{v}+\boldsymbol{u}$?

b What interesting special case can occur?

4 In Figure 15 the various line segments are taken to be representatives of vectors. Find in the figure directed line segments equal to the following:

a $\vec{AE}+\vec{EC}$ *b* $\vec{DB}+\vec{BE}$ *c* $\vec{AD}+\vec{DB}+\vec{BC}$

d $\vec{CB}+\vec{BE}+\vec{EA}+\vec{AD}$

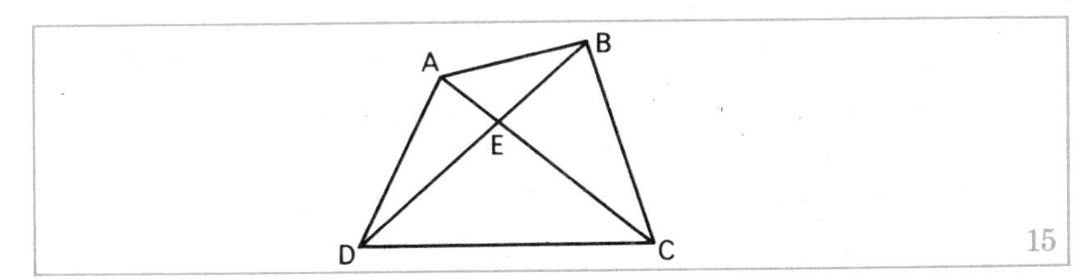

15

5 Using Figure 15, copy and complete the following:

a $\vec{AE}+\ldots = \vec{AB}$ *b* $\vec{AD}+\ldots+\vec{EC} = \vec{AC}$ *c* $\vec{DE}+\ldots = \vec{DB}$

d $\vec{EB}+\ldots = \vec{ED}$ *e* $\ldots+\vec{DA} = \vec{CA}$

3 Vectors in number pair form

As vectors can be represented in the same way as translations by directed line segments, we can represent vectors by the same number pair notation.

In Figure 16, $\overrightarrow{AB}$, $\overrightarrow{HK}$ and $\overrightarrow{PQ}$ are all representatives of the same vector $\boldsymbol{u}$. The corresponding translation maps points 'one to the right and three up', and as in the Translation chapter (Book 4, page 98), can be denoted by $\begin{pmatrix}1\\3\end{pmatrix}$. It follows that the vector $\boldsymbol{u}$ is completely determined by $\begin{pmatrix}1\\3\end{pmatrix}$, and we write $\boldsymbol{u} = \begin{pmatrix}1\\3\end{pmatrix}$. This is sometimes called a *column vector*, and 1 and 3 are the *components* of $\boldsymbol{u}$. Similarly $\overrightarrow{CD}$, $\overrightarrow{KL}$ and $\overrightarrow{QR}$ each represent the vector $\begin{pmatrix}4\\-2\end{pmatrix}$. When we alter the components we get a new magnitude, or a new direction, or both.

There is a one-to-one correspondence between vectors and ordered number pairs.

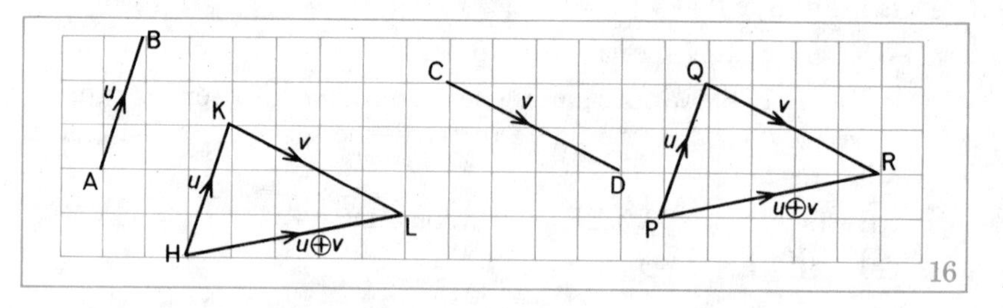

16

If we consider the addition law in triangle HKL we see that $\boldsymbol{u}+\boldsymbol{v} = \begin{pmatrix}1\\3\end{pmatrix} + \begin{pmatrix}4\\-2\end{pmatrix} = \begin{pmatrix}5\\1\end{pmatrix}$ by adding components, just as in the case of translations in number pair form.

Exercise 4

1 a In Figure 16 express in number pair form the vectors represented by $\overrightarrow{AB}$, $\overrightarrow{CD}$, $\overrightarrow{DC}$, $\overrightarrow{BA}$, $\overrightarrow{HL}$, $\overrightarrow{PR}$ and $\overrightarrow{LH}$ respectively.

b State how we can get the components of $\overrightarrow{PR}$ from those of $\overrightarrow{PQ}$ and $\overrightarrow{QR}$.

2 On squared paper draw representatives of the vectors:

a $\begin{pmatrix}4\\5\end{pmatrix}$ *b* $\begin{pmatrix}-2\\1\end{pmatrix}$ *c* $\begin{pmatrix}-2\\-3\end{pmatrix}$ *d* $\begin{pmatrix}0\\4\end{pmatrix}$ *e* $\begin{pmatrix}-2\\0\end{pmatrix}$

3 a If $\boldsymbol{p} = \begin{pmatrix}3\\-1\end{pmatrix}$ and $\boldsymbol{q} = \begin{pmatrix}-2\\5\end{pmatrix}$, draw representatives on squared paper of $\boldsymbol{p}$, $\boldsymbol{q}$ and $\boldsymbol{p}+\boldsymbol{q}$.

b What are the components of $\boldsymbol{p}+\boldsymbol{q}$ from your diagram?

c Check that you get the same answer by adding components.

4 Repeat question *3* for $\boldsymbol{p} = \begin{pmatrix}-1\\2\end{pmatrix}$ and $\boldsymbol{q} = \begin{pmatrix}-2\\-3\end{pmatrix}$.

5 Write $\boldsymbol{u}+\boldsymbol{v}$ and $\boldsymbol{v}+\boldsymbol{u}$ in component form when $\boldsymbol{u}$ and $\boldsymbol{v}$ are:

a $\begin{pmatrix}2\\-3\end{pmatrix}$ and $\begin{pmatrix}-1\\4\end{pmatrix}$ *b* $\begin{pmatrix}6\\8\end{pmatrix}$ and $\begin{pmatrix}-3\\-10\end{pmatrix}$

c $\begin{pmatrix}-1\\2\end{pmatrix}$ and $\begin{pmatrix}-5\\-7\end{pmatrix}$

Which law do your answers illustrate?

6 If $\boldsymbol{u} = \begin{pmatrix}3\\4\end{pmatrix}$, $\boldsymbol{v} = \begin{pmatrix}-1\\3\end{pmatrix}$ and $\boldsymbol{w} = \begin{pmatrix}-2\\-1\end{pmatrix}$ express in components:

a $\boldsymbol{u}+\boldsymbol{v}$ *b* $(\boldsymbol{u}+\boldsymbol{v})+\boldsymbol{w}$ *c* $\boldsymbol{v}+\boldsymbol{w}$ *d* $\boldsymbol{u}+(\boldsymbol{v}+\boldsymbol{w})$

7 Repeat question *6* if $\boldsymbol{u} = \begin{pmatrix}-1\\9\end{pmatrix}$, $\boldsymbol{v} = \begin{pmatrix}5\\4\end{pmatrix}$ and $\boldsymbol{w} = \begin{pmatrix}-3\\2\end{pmatrix}$

Which law do the answers to questions *6* and *7* illustrate?

8 If $\boldsymbol{u} = \begin{pmatrix}a\\b\end{pmatrix}$ and $\boldsymbol{u} = \begin{pmatrix}c\\d\end{pmatrix}$, what can we say about a, b, c and d?

4 *Zero vector. Negative of a vector*

Exercise 5

1 Copy the directed line segments $\overrightarrow{AB}$ and $\overrightarrow{XY}$ of Figure 17 on squared paper. Try to find a directed line segment which represents their sum. What can you say about its length and direction?

2 On plain paper draw a pair of directed line segments $\overrightarrow{CD}$ and $\overrightarrow{EF}$ which are equal in length but opposite in direction. What answer do you get when you add them by the nose-to-tail rule?

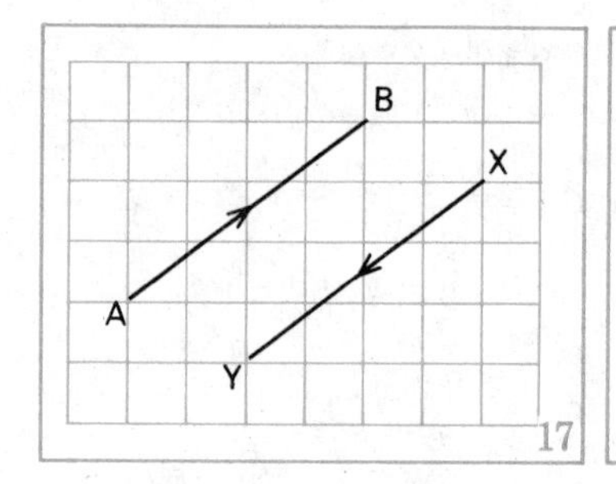

17

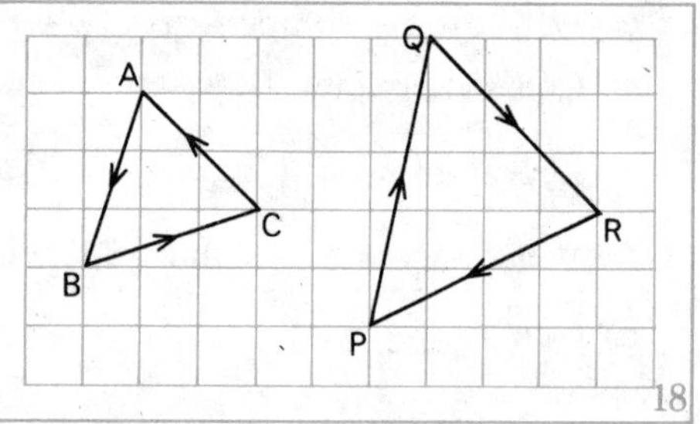

18

3 In Figure 18, simplify *a* $\overrightarrow{AB}+\overrightarrow{BA}$ *b* $\overrightarrow{AB}+\overrightarrow{BC}+\overrightarrow{CA}$ *c* $\overrightarrow{PQ}+\overrightarrow{QR}+\overrightarrow{RP}$

4 Simplify *a* $\begin{pmatrix}3\\4\end{pmatrix}+\begin{pmatrix}-3\\-4\end{pmatrix}$ *b* $\begin{pmatrix}-2\\1\end{pmatrix}+\begin{pmatrix}2\\-1\end{pmatrix}$ *c* $\begin{pmatrix}5\\-3\end{pmatrix}+\begin{pmatrix}-5\\3\end{pmatrix}$

In questions *1–3* we find that the answer is a line segment representing a vector which has no magnitude and no definite direction.

We write the zero line segment, or the zero vector, $\boldsymbol{0}$. In number pair notation $\boldsymbol{0}=\begin{pmatrix}0\\0\end{pmatrix}$.

Since $\overrightarrow{AB}+\overrightarrow{BA}=\boldsymbol{0}$ we can say that $\overrightarrow{BA}$ is the additive inverse or negative of $\overrightarrow{AB}$, i.e. $\overrightarrow{BA}=-\overrightarrow{AB}$.

Similarly $\begin{pmatrix}3\\4\end{pmatrix}+\begin{pmatrix}-3\\-4\end{pmatrix}=\begin{pmatrix}0\\0\end{pmatrix}$,

and hence $\begin{pmatrix}-3\\-4\end{pmatrix}$ is the negative of $\begin{pmatrix}3\\4\end{pmatrix}$.

Zero vector. Negative of a vector

If $\boldsymbol{u}+\boldsymbol{v} = \boldsymbol{0}$ then $\boldsymbol{v}$ is the negative of $\boldsymbol{u}$ and is naturally written $-\boldsymbol{u}$.
$-\boldsymbol{u}$ is the vector consisting of the set of all directed line segments in the direction opposite to that of the line segments of $\boldsymbol{u}$, but with the same magnitude.

5 ABCD is a quadrilateral with its diagonals intersecting at E, as in Figure 15. Copy and complete the following:

a $\overrightarrow{DB}+\ldots = \boldsymbol{0}$ *b* $\overrightarrow{DC}+\ldots+\overrightarrow{ED} = \boldsymbol{0}$

c $\overrightarrow{AE}+\overrightarrow{ED}+\ldots = \overrightarrow{AD}$

d $\overrightarrow{AB}+\overrightarrow{BE}+\overrightarrow{EC}+\ldots+\overrightarrow{CD}+\overrightarrow{DA} = \boldsymbol{0}$

6 Simplify *a* $\boldsymbol{u}+\boldsymbol{0}$ *b* $\boldsymbol{0}+\boldsymbol{v}$ *c* $\boldsymbol{u}+\boldsymbol{0}+\boldsymbol{v}$

7 Simplify *a* $\begin{pmatrix}2\\-3\end{pmatrix}+\begin{pmatrix}0\\0\end{pmatrix}$ *b* $\begin{pmatrix}0\\0\end{pmatrix}+\begin{pmatrix}-4\\5\end{pmatrix}$ *c* $\begin{pmatrix}2\\-3\end{pmatrix}+\begin{pmatrix}0\\0\end{pmatrix}+\begin{pmatrix}-4\\-5\end{pmatrix}$

8 Express the directed line segments $-\overrightarrow{BA}$, $-\overrightarrow{CD}$ and $-\overrightarrow{RQ}$ without using negative signs.

9 Copy Figure 15 in your notebooks and simplify:

a $\overrightarrow{DE}+(-\overrightarrow{BE})$ *b* $\overrightarrow{AC}+(-\overrightarrow{BC})$ *c* $\overrightarrow{CD}+\overrightarrow{BA}+(-\overrightarrow{BD})$

10 On squared paper draw representatives of the vectors $\boldsymbol{p}$ and $\boldsymbol{q}$ with components $\begin{pmatrix}3\\4\end{pmatrix}$ and $\begin{pmatrix}-2\\3\end{pmatrix}$ respectively. Draw representatives of:

a $-\boldsymbol{p}$ *b* $-\boldsymbol{q}$ *c* $(-\boldsymbol{p})+(-\boldsymbol{q})$ *d* $\boldsymbol{p}+\boldsymbol{q}$ *e* $-(\boldsymbol{p}+\boldsymbol{q})$

What can you say about $(-\boldsymbol{p})+(-\boldsymbol{q})$ and $-(\boldsymbol{p}+\boldsymbol{q})$?

11 Write down the negatives of the vectors:

a $\begin{pmatrix}5\\4\end{pmatrix}$ *b* $\begin{pmatrix}-3\\5\end{pmatrix}$ *c* $\begin{pmatrix}4\\-7\end{pmatrix}$ *d* $\begin{pmatrix}0\\0\end{pmatrix}$ *e* $\begin{pmatrix}a\\b\end{pmatrix}$

12 Copy and complete:

a $\begin{pmatrix}3\\-4\end{pmatrix}+\begin{pmatrix}\ \\\ \end{pmatrix}=\begin{pmatrix}0\\0\end{pmatrix}$ *b* $\begin{pmatrix}4\\\ \end{pmatrix}+\begin{pmatrix}\ \\-7\end{pmatrix}=\begin{pmatrix}0\\0\end{pmatrix}$

c $\begin{pmatrix}a\\-b\end{pmatrix}+\begin{pmatrix}\ \\\ \end{pmatrix}=\begin{pmatrix}0\\0\end{pmatrix}$

5 Subtraction of vectors

For real numbers a and b we defined subtraction by $a-b = a+(-b)$. Similarly we define subtraction of vectors by $\boldsymbol{u}-\boldsymbol{v} = \boldsymbol{u}+(-\boldsymbol{v})$.

To subtract a vector, add its negative.

Useful results

In Figure 19 (i), $\overrightarrow{AC}-\overrightarrow{AB}$
$= \overrightarrow{AC}+\overrightarrow{BA}$ (adding the negative)
$= \overrightarrow{BA}+\overrightarrow{AC}$
$= \overrightarrow{BC}$

Similarly in (ii), $\overrightarrow{PQ} = \overrightarrow{OQ}-\overrightarrow{OP}$. (Think carefully about the order of the letters in $\overrightarrow{BC}$ and $\overrightarrow{PQ}$.)

In components, $\begin{pmatrix} 2 \\ -3 \end{pmatrix} - \begin{pmatrix} -5 \\ 1 \end{pmatrix} = \begin{pmatrix} 2 \\ -3 \end{pmatrix} + \begin{pmatrix} 5 \\ -1 \end{pmatrix} = \begin{pmatrix} 7 \\ -4 \end{pmatrix}$

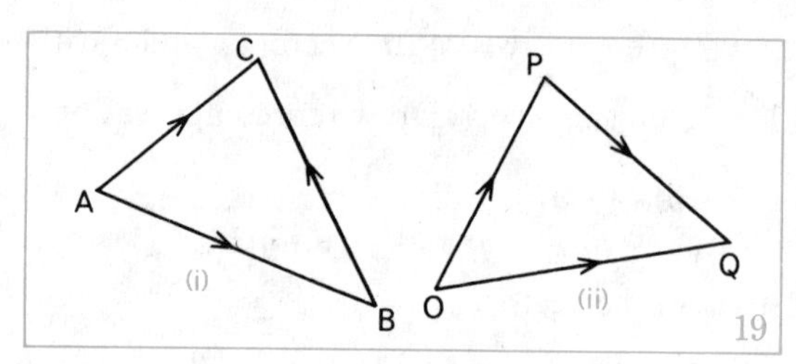

19

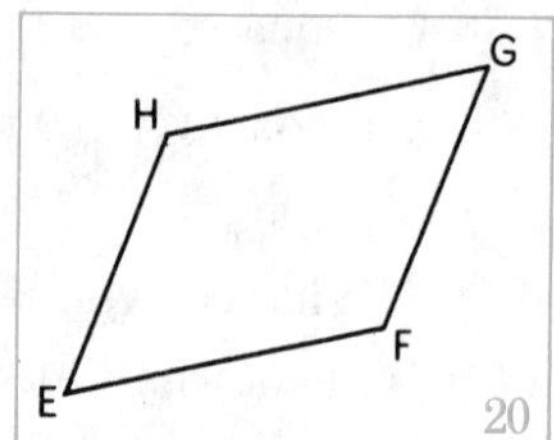

20

Exercise 6

1 In Figure 19 (i), simplify:

a $\overrightarrow{AC}-\overrightarrow{AB}$ *b* $\overrightarrow{BA}-\overrightarrow{BC}$ *c* $\overrightarrow{BC}-\overrightarrow{BA}$
d $\overrightarrow{CA}-\overrightarrow{CB}$ *e* $\overrightarrow{CB}-\overrightarrow{CA}$

If P, S and L are any three points, simplify $\overrightarrow{PS}-\overrightarrow{PL}$.

2 In Figure 20, EFGH is a parallelogram. Simplify:

a $\overrightarrow{EF}-\overrightarrow{EH}$ *b* $\overrightarrow{EH}-\overrightarrow{EF}$ *c* $\overrightarrow{FG}-\overrightarrow{FE}$ *d* $\overrightarrow{GE}-\overrightarrow{GH}$

3 Simplify *a* $\begin{pmatrix} 3 \\ 9 \end{pmatrix} - \begin{pmatrix} 2 \\ 5 \end{pmatrix}$ *b* $\begin{pmatrix} -2 \\ 4 \end{pmatrix} - \begin{pmatrix} 3 \\ 7 \end{pmatrix}$ *c* $\begin{pmatrix} a \\ b \end{pmatrix} - \begin{pmatrix} c \\ d \end{pmatrix}$

4 In each of the following find $\boldsymbol{x}$ in component form.

a $x+\begin{pmatrix}2\\3\end{pmatrix}=\begin{pmatrix}5\\7\end{pmatrix}$ *b* $x-\begin{pmatrix}2\\3\end{pmatrix}=\begin{pmatrix}5\\7\end{pmatrix}$

5 In Figure 21, the directed line segments represent vectors as shown. Simplify:

a $\boldsymbol{b}-\boldsymbol{u}$ *b* $\boldsymbol{w}-\boldsymbol{a}$ *c* $\boldsymbol{v}+\boldsymbol{u}-\boldsymbol{b}$ *d* $\boldsymbol{b}+\boldsymbol{a}-\boldsymbol{v}-\boldsymbol{w}$

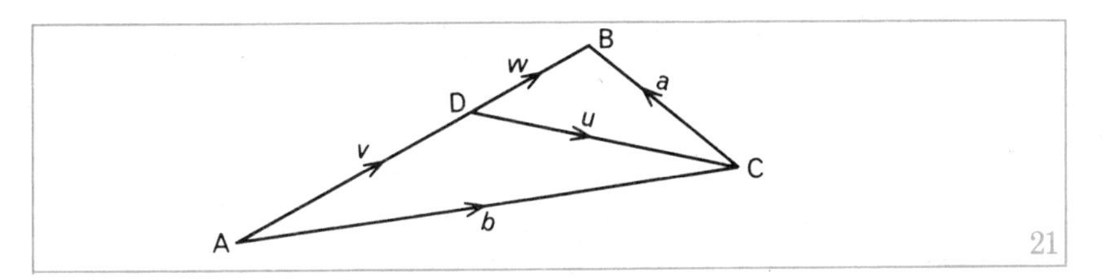

21

6 In Figure 21, find $\boldsymbol{x}$ in its simplest form from the equations:

a $\boldsymbol{x}+\boldsymbol{a}=\boldsymbol{w}$ *b* $\boldsymbol{b}+\boldsymbol{x}=\boldsymbol{v}+\boldsymbol{u}+\boldsymbol{a}$

c $\boldsymbol{v}=\boldsymbol{b}-\boldsymbol{x}$ *d* $\boldsymbol{x}-\boldsymbol{w}=\boldsymbol{v}$

7 Make a sketch of Figure 21 and on it construct a representative of $\boldsymbol{b}-\boldsymbol{a}$. (Remember $\boldsymbol{b}-\boldsymbol{a}=\boldsymbol{b}+(-\boldsymbol{a})$.)

8 Simplify:

a $-\begin{pmatrix}-3\\-2\end{pmatrix}$ *b* $\begin{pmatrix}3\\4\end{pmatrix}-\begin{pmatrix}-3\\-4\end{pmatrix}$ *c* $\begin{pmatrix}2\\-1\end{pmatrix}-\begin{pmatrix}1\\3\end{pmatrix}+\begin{pmatrix}4\\-2\end{pmatrix}$

9 If $\boldsymbol{u}=\begin{pmatrix}9\\-3\end{pmatrix}$, $\boldsymbol{v}=\begin{pmatrix}-3\\2\end{pmatrix}$ and $\boldsymbol{w}=\begin{pmatrix}4\\6\end{pmatrix}$, express in component form:

a $-\boldsymbol{u}+\boldsymbol{v}+\boldsymbol{w}$ *b* $\boldsymbol{u}-\boldsymbol{v}+\boldsymbol{w}$ *c* $\boldsymbol{u}+\boldsymbol{v}-\boldsymbol{w}$

6 Multiplication of a vector by a number

In Figure 22 (i) the directed line segment $\overrightarrow{AB}$ represents a vector $\boldsymbol{u}$. It follows that each of the equal line segments $\overrightarrow{BC}$, $\overrightarrow{CD}$, etc., represents the vector $\boldsymbol{u}$.

Since $\overrightarrow{AC}=\overrightarrow{AB}+\overrightarrow{BC}$, $\overrightarrow{AC}$ represents $\boldsymbol{u}+\boldsymbol{u}$, which we naturally write $2\boldsymbol{u}$.

Similarly $\overrightarrow{AD}$ represents $3\boldsymbol{u}$, and so on.

In Figure 22 (ii) three equal vectors $\boldsymbol{x}$ when added together make up the vector $\boldsymbol{u}$. Since $3\boldsymbol{x} = \boldsymbol{u}$, we can write $\boldsymbol{x} = \frac{1}{3}\boldsymbol{u}$.

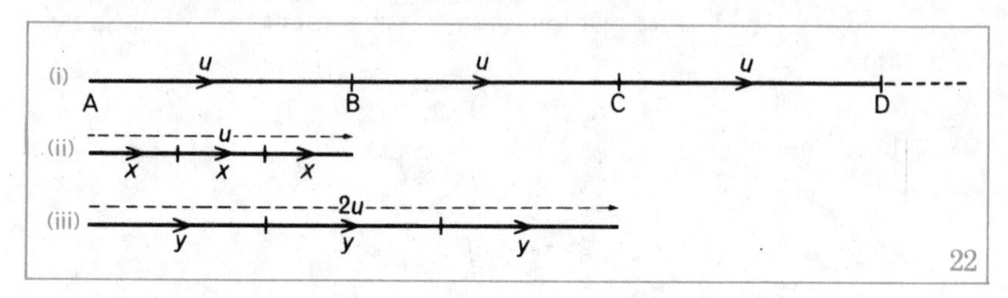

22

Similarly in Figure 22 (iii), three equal vectors $\boldsymbol{y}$ when added together make up $2\boldsymbol{u}$. Since $3\boldsymbol{y} = 2\boldsymbol{u}$, we can write $\boldsymbol{y} = \frac{2}{3}\boldsymbol{u}$.

Note that we always put the multiplier in front. We write $2\boldsymbol{u}$, not $\boldsymbol{u}2$; and $\frac{1}{7}\boldsymbol{u}$, not $\frac{\boldsymbol{u}}{7}$.

In Figure 23, $\overrightarrow{AB}$ represents vector $\boldsymbol{u}$, which can be written $\begin{pmatrix}4\\2\end{pmatrix}$ in component form. $\overrightarrow{CD} = 2\overrightarrow{AB}$ and hence represents $2\boldsymbol{u}$, which is $\begin{pmatrix}8\\4\end{pmatrix}$ in component form.

We have $2\boldsymbol{u} = 2\begin{pmatrix}4\\2\end{pmatrix} = \begin{pmatrix}2\times4\\2\times2\end{pmatrix} = \begin{pmatrix}8\\4\end{pmatrix}$

Similarly $\overrightarrow{EF}$ represents $\frac{1}{2}\boldsymbol{u}$ and we have

$\frac{1}{2}\boldsymbol{u} = \frac{1}{2}\begin{pmatrix}4\\2\end{pmatrix} = \begin{pmatrix}\frac{1}{2}\times4\\\frac{1}{2}\times2\end{pmatrix} = \begin{pmatrix}2\\1\end{pmatrix}$

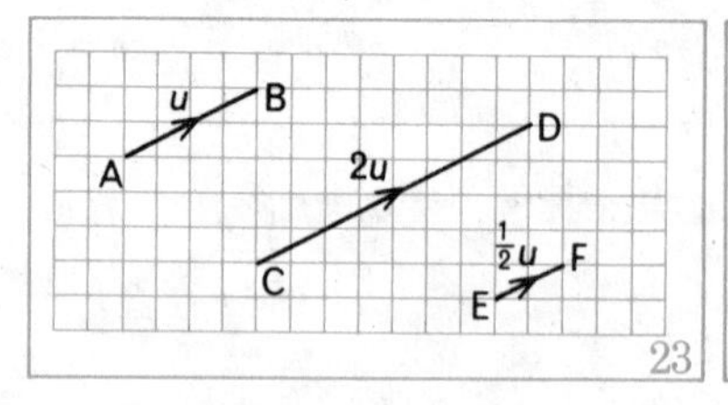

23

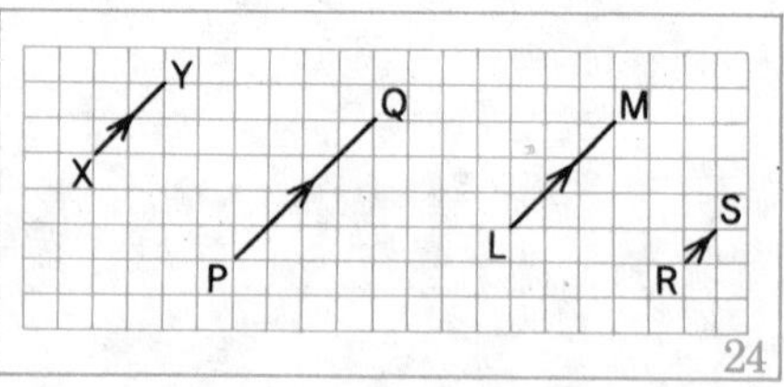

24

Exercise 7

1 In Figure 24, $\overrightarrow{XY}$ is a representative of $\boldsymbol{p}$. Which vectors are represented by *a* $\overrightarrow{PQ}$ *b* $\overrightarrow{LM}$ *c* $\overrightarrow{RS}$?

2 *a* On squared paper draw a directed line segment $\overrightarrow{AB}$ representing the vector $\boldsymbol{u} = \begin{pmatrix} 2 \\ -1 \end{pmatrix}$.

b Draw directed line segments to represent $2\boldsymbol{u}$ and $3\boldsymbol{u}$.

c Write down the components of $2\boldsymbol{u}$ and $3\boldsymbol{u}$ from your diagram.

d State how these components can be found from the components of $\boldsymbol{u}$.

3 *a* Draw directed line segments on plain paper to represent vectors $\boldsymbol{u}$, $2\boldsymbol{u}$ and $3\boldsymbol{u}$.

b Show by a sketch that $2\boldsymbol{u}+3\boldsymbol{u} = (2+3)\boldsymbol{u} = 5\boldsymbol{u}$.

c Show by a sketch that $4\boldsymbol{u}+5\boldsymbol{u} = (4+5)\boldsymbol{u} = 9\boldsymbol{u}$.

4 If $\boldsymbol{p} = \begin{pmatrix} 2 \\ 1 \end{pmatrix}$, draw directed line segments on squared paper to represent:

a $\boldsymbol{p}$ *b* $2\boldsymbol{p}$ *c* $-(2\boldsymbol{p})$ *d* $-\boldsymbol{p}$ *e* $2(-\boldsymbol{p})$.

What can you state about $-(2\boldsymbol{p})$ and $2(-\boldsymbol{p})$?

5 On squared paper draw line segments to represent a vector $\boldsymbol{t}$ of your own choice. Then draw representatives of:

a $-2\boldsymbol{t}$ *b* $-3\boldsymbol{t}$ *c* $\frac{1}{2}\boldsymbol{t}$

6 Illustrate by sketches *a* $2(3\boldsymbol{u}) = 6\boldsymbol{u}$ *b* $2(-3\boldsymbol{u}) = -6\boldsymbol{u}$

In question ***4*** we saw that $2(-\boldsymbol{p}) = -(2\boldsymbol{p})$, and hence we may write $-2\boldsymbol{p}$ for either of them.

In the above examples we have been illustrating the following general laws. Note how closely the laws of vector algebra agree with the laws of ordinary algebra.

(i) $k\boldsymbol{u}$, where k is a number, represents a vector with magnitude k times that of $\boldsymbol{u}$, with the same direction as $\boldsymbol{u}$ when k is positive, and the opposite direction when k is negative. When $k = 0$, we take for granted that $0\boldsymbol{u} = \boldsymbol{0}$.

(ii) $k\boldsymbol{u}+m\boldsymbol{u} = (k+m)\boldsymbol{u}$.

(iii) $k(-\boldsymbol{u}) = -(k\boldsymbol{u}) = -k\boldsymbol{u}$.

(iv) $k(m\boldsymbol{u}) = (km)\boldsymbol{u}$.

We have seen that $2\boldsymbol{u}+3\boldsymbol{u} = (2+3)\boldsymbol{u}$. Can we show that there is a second distributive law in vector algebra? For example, is it true that $2\boldsymbol{u}+2\boldsymbol{v} = 2(\boldsymbol{u}+\boldsymbol{v})$?

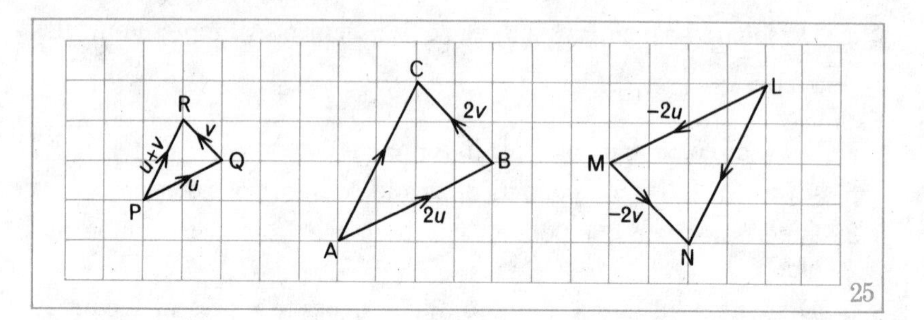

In Figure 25, $\overrightarrow{PQ}$, $\overrightarrow{QR}$ and $\overrightarrow{PR}$ represent $\boldsymbol{u}$, $\boldsymbol{v}$ and $\boldsymbol{u}+\boldsymbol{v}$ respectively. Since $\overrightarrow{AC} = 2\overrightarrow{PR}$, $\overrightarrow{AC}$ represents $2(\boldsymbol{u}+\boldsymbol{v})$.

But $\overrightarrow{AC} = \overrightarrow{AB}+\overrightarrow{BC}$ which represents $2\boldsymbol{u}+2\boldsymbol{v}$.

Hence $2\boldsymbol{u}+2\boldsymbol{v} = 2(\boldsymbol{u}+\boldsymbol{v})$.

Using triangle LMN we can show that $\overrightarrow{LN}$ represents $-2\boldsymbol{u}-2\boldsymbol{v}$ or $-2(\boldsymbol{u}+\boldsymbol{v})$.

By using ratios of sides in similar triangles we can show that $k\boldsymbol{u}+k\boldsymbol{v} = k(\boldsymbol{u}+\boldsymbol{v})$ where k is any number.

Exercise 7B

1 Using line segments representing vectors $\boldsymbol{u}$ and $\boldsymbol{v}$ as in Figure 25, construct representatives of:

a $\boldsymbol{u}+\boldsymbol{v}$ *b* $3\boldsymbol{u}+3\boldsymbol{v}$ *c* $3(\boldsymbol{u}+\boldsymbol{v})$

What can you say about the last two answers?

2 Use the distributive law $k(\boldsymbol{u}+\boldsymbol{v}) = k\boldsymbol{u}+k\boldsymbol{v}$, and any other laws required, to simplify:

a $2(\boldsymbol{u}+\boldsymbol{v})+(3\boldsymbol{u}-\boldsymbol{v})$ *b* $\boldsymbol{u}-\frac{1}{2}(\boldsymbol{v}+\boldsymbol{u})$

c $\frac{5}{6}(\boldsymbol{u}-\boldsymbol{v})+\frac{1}{6}\boldsymbol{v}$ *d* $2(4\boldsymbol{u}+\frac{1}{2}\boldsymbol{v})-12(\frac{2}{3}\boldsymbol{u}-\frac{1}{4}\boldsymbol{v})$

3 If $\boldsymbol{u} = \begin{pmatrix}1\\2\end{pmatrix}$ and $\boldsymbol{v} = \begin{pmatrix}3\\-1\end{pmatrix}$, verify in number pair form that:

a $3\boldsymbol{u}+3\boldsymbol{v} = 3(\boldsymbol{u}+\boldsymbol{v})$ *b* $5\boldsymbol{u}-5\boldsymbol{v} = 5(\boldsymbol{u}-\boldsymbol{v})$

4 For the vectors $\boldsymbol{u}$ and $\boldsymbol{v}$ of question 3, check in number pair form that:

a $2\boldsymbol{u}+3\boldsymbol{u} = 5\boldsymbol{u}$ *b* $7\boldsymbol{v}-3\boldsymbol{v} = 4\boldsymbol{v}$

5 If $\boldsymbol{p} = \begin{pmatrix}3\\1\end{pmatrix}$ and $\boldsymbol{q} = \begin{pmatrix}1\\-5\end{pmatrix}$, simplify in component form:

a $3(\boldsymbol{p}+\boldsymbol{q})$ *b* $\frac{1}{2}(\boldsymbol{p}-\boldsymbol{q})$ *c* $2(\boldsymbol{p}-2\boldsymbol{q})+3\boldsymbol{p}$

6 Solve each of the following vector equations for $\boldsymbol{x}$:

a $3(\boldsymbol{x}+\boldsymbol{u}) = 5\boldsymbol{v}$ *b* $2\boldsymbol{x}+5\boldsymbol{u} = 7\boldsymbol{v}+5\boldsymbol{x}$

7 Solve for $\boldsymbol{x}$ in component form each of the vector equations:

a $2\left[\boldsymbol{x}+\begin{pmatrix}2\\3\end{pmatrix}\right] = 5\begin{pmatrix}2\\2\end{pmatrix}$ *b* $3\boldsymbol{x}+4\begin{pmatrix}1\\2\end{pmatrix} = 3\begin{pmatrix}4\\-2\end{pmatrix}+\boldsymbol{x}$

8 Representatives of the vectors $\boldsymbol{u}$, $\boldsymbol{v}$, $\boldsymbol{w}$ and $\boldsymbol{p}$ are shown in Figure 26. On squared paper construct representatives of the following, and state the answers in component form:

a $\boldsymbol{u}+\boldsymbol{w}$ *b* $2(\boldsymbol{u}+\boldsymbol{w})$ *c* $-3(\boldsymbol{v}+\boldsymbol{w})$ *d* $\frac{1}{2}(\boldsymbol{u}+\boldsymbol{p})$

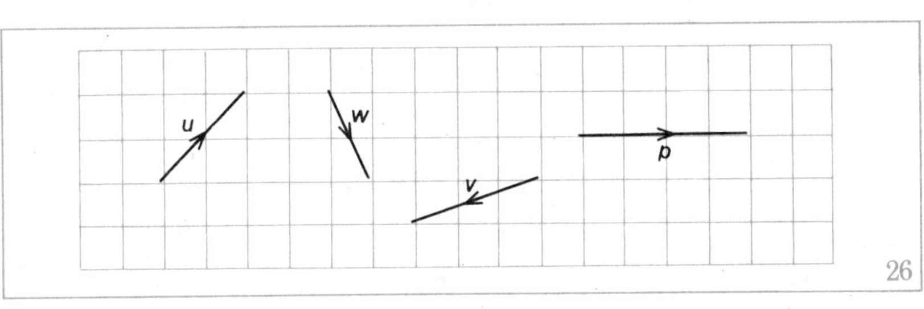

26

9 *a* Given that $\boldsymbol{v} = k\boldsymbol{u}$ (where k is a positive number and $\boldsymbol{u} \neq \boldsymbol{0}$), what must be true about $\boldsymbol{u}$ and $\boldsymbol{v}$ with respect to magnitude and direction?

b Repeat the question when k is negative.

10*a* What can you deduce, given that $3\boldsymbol{u} = 2\boldsymbol{u}$?

b What can you deduce, given that $3\boldsymbol{u} = 2\boldsymbol{v}$, $(\boldsymbol{u} \neq \boldsymbol{0})$?

c What can you deduce, given that $a\boldsymbol{u} = b\boldsymbol{v}$, where $\boldsymbol{u}$ and $\boldsymbol{v}$ are not in the same or opposite directions and are not zero vectors.

7 *Position vectors*

In Figure 27, O is the origin and P is the point (3, 6). The vector represented by $\overrightarrow{OP}$ is called the *position vector* of P. The position vector of P is denoted by $\boldsymbol{p}$, and $\boldsymbol{p} = \begin{pmatrix}3\\6\end{pmatrix}$.

The coordinates of a point are the components of its position vector.

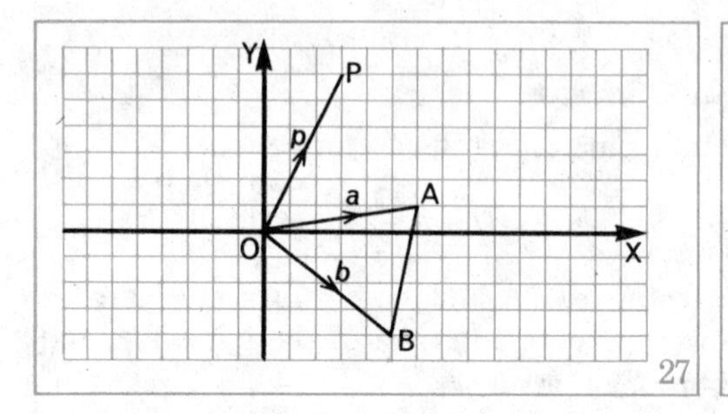

27

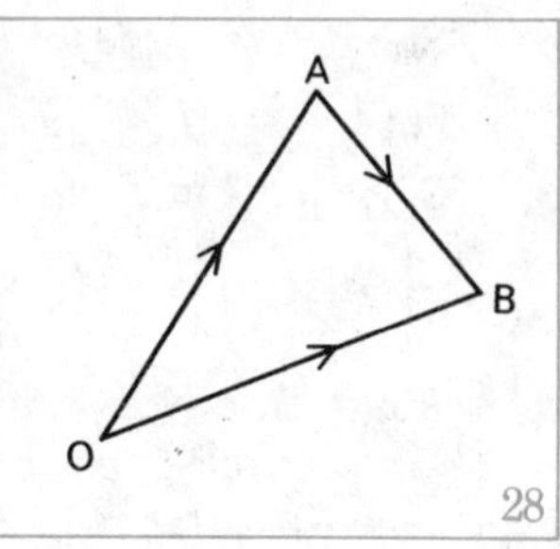

28

Similarly, the position vector of A (6, 1) is $\boldsymbol{a} = \begin{pmatrix} 6 \\ 1 \end{pmatrix}$

and the position vector of B (5, −4) is $\boldsymbol{b} = \begin{pmatrix} 5 \\ -4 \end{pmatrix}$.

If a and b are the position vectors of points A and B, then $\overrightarrow{AB}$ represents b − a

From Section 5, $\overrightarrow{AB} = \overrightarrow{OB} - \overrightarrow{OA}$ for any chosen origin (see Figure 28).

Hence $\overrightarrow{AB}$ represents $\boldsymbol{b} - \boldsymbol{a}$.

If $\boldsymbol{a} = \boldsymbol{b}$, A and B must be different names for the same point.

Example. A is the point (6, 1) and B is (5, −4). Find the components of the vector represented by $\overrightarrow{AB}$ (see Figure 27).

$$\overrightarrow{AB} \text{ represents } \boldsymbol{b} - \boldsymbol{a}$$
$$= \begin{pmatrix} 5 \\ -4 \end{pmatrix} - \begin{pmatrix} 6 \\ 1 \end{pmatrix}$$
$$= \begin{pmatrix} -1 \\ -5 \end{pmatrix}$$

Note. In Exercises 8, 8B and 9, O is understood to be the origin.

Exercise 8

1 a P is the point (2, 3) and Q is (7, 5). Write down the position vectors $\boldsymbol{p}$ and $\boldsymbol{q}$ in component form.

b Calculate the components of the vector represented by $\overrightarrow{PQ}$.

2 Repeat question ***1*** for the points P(0, −1) and Q(4, 2).

3 A is the point (4, 0), B is (6, 2), C is (−2, 1). Write down the component form of their position vectors $\boldsymbol{a}$, $\boldsymbol{b}$ and $\boldsymbol{c}$. Calculate the components of the vectors represented by $\overrightarrow{AB}$, $\overrightarrow{BC}$ and $\overrightarrow{CA}$.

4 In Figure 29, PQRS is a parallelogram.

a Explain why $\overrightarrow{PQ} = \overrightarrow{SR}$.

b Use the result in ***a*** to find the coordinates of Q, given that P is the point (1, 1), Q is (x, y), R is (7, 6) and S is (3, 4).

5 Repeat question ***4*** for P (0, 2), Q (x, y), R (10, 8), S (3, 7).

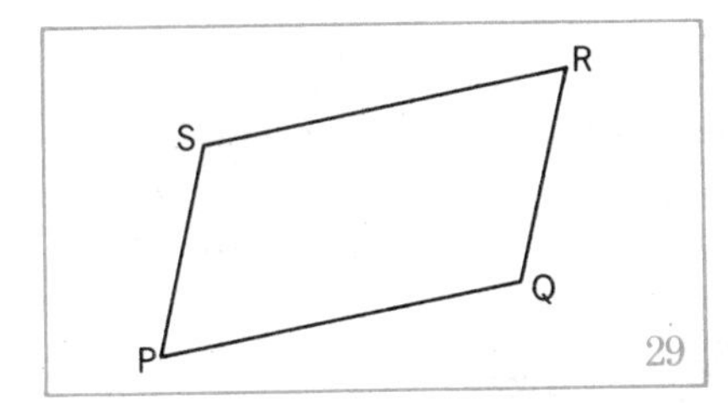

29

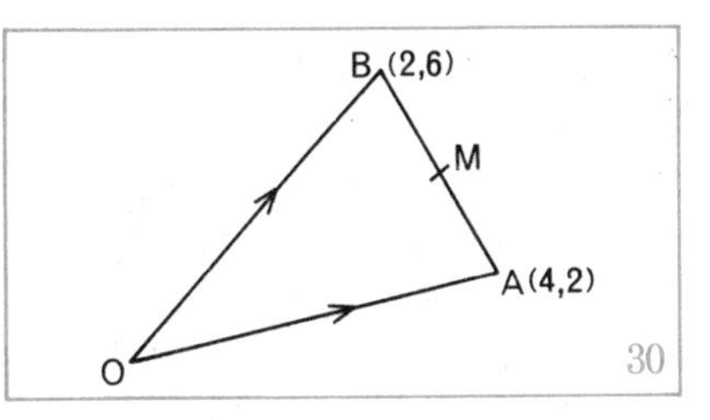

30

6 In Figure 30, A is the point (4, 2) and B is (2, 6). M is the midpoint of AB.

a What are the components of $\boldsymbol{a}$ and $\boldsymbol{b}$?

b What are the components of the vectors represented by $\overrightarrow{AB}$ and $\overrightarrow{BA}$?

c What are the components of the vector represented by $\overrightarrow{BM}$? (Remember that $\overrightarrow{BM} = \frac{1}{2}\overrightarrow{BA}$.)

d Use vector addition to find $\boldsymbol{m}$ (i.e. the vector represented by $\overrightarrow{OM}$).

e Check the last answer by thinking of $\overrightarrow{OA} + \overrightarrow{AM}$.

f Write down the coordinates of M.

7 Repeat question ***6*** when A and B are respectively:

a (7, 6) and (9, −4) *b* (−2, 8) and (−4, −7)

8 Verify that the midpoints obtained in questions ***6*** and ***7*** can be found by taking the average of the x-coordinates and the average of the y-coordinates of A and B.

9 a $\boldsymbol{p}$ and $\boldsymbol{q}$ are the position vectors of P and Q. If P is the point (3, −4) and $\boldsymbol{p} = \boldsymbol{q}$, what are the coordinates of Q?

b If $\boldsymbol{s} = \begin{pmatrix} 2 \\ -5 \end{pmatrix}$ and $\boldsymbol{t} = \begin{pmatrix} 2 \\ -5 \end{pmatrix}$, what can you state about the points S and T?

10 A is the point (2, 3), B is (7, 5) and C is $(-4, -2)$. Find the coordinates of:

a P, if $\overrightarrow{OP} = \overrightarrow{AB}$ *b* Q, if $\overrightarrow{OQ} = \overrightarrow{BC}$ *c* R if $\overrightarrow{BR} = \overrightarrow{CA}$

11a PQRS is a parallelogram with vertices P (0, 1), Q (6, 2), R (8, 5) S (2, 4). If the diagonals intersect at T, calculate the coordinates of T by using *(1)* $\overrightarrow{OP}+\overrightarrow{PT}$ *(2)* $\overrightarrow{OR}+\overrightarrow{RT}$

b What is the connection between the coordinates of T and those of Q and S?

The midpoint of a line and the centroid of a triangle

Here we use position vectors of points, but dispense with coordinates.

Example 1. From Figure 31, in which M is the midpoint of AB, $\boldsymbol{m}$, $\boldsymbol{a}$ and $\boldsymbol{b}$ are the position vectors of M, A and B with respect to an origin O. Show that $\boldsymbol{m} = \frac{1}{2}(\boldsymbol{a}+\boldsymbol{b})$.

$\overrightarrow{OM}$ represents $\boldsymbol{m}$, $\overrightarrow{OA}$ represents $\boldsymbol{a}$ and $\overrightarrow{OB}$ represents $\boldsymbol{b}$.

$$\begin{aligned} \overrightarrow{OM} &= \overrightarrow{OA}+\overrightarrow{AM} \\ &= \overrightarrow{OA}+\tfrac{1}{2}\overrightarrow{AB} \\ \Leftrightarrow \boldsymbol{m} &= \boldsymbol{a}+\tfrac{1}{2}(\boldsymbol{b}-\boldsymbol{a}) \\ &= \boldsymbol{a}+\tfrac{1}{2}\boldsymbol{b}-\tfrac{1}{2}\boldsymbol{a} \\ &= \tfrac{1}{2}(\boldsymbol{a}+\boldsymbol{b}) \end{aligned}$$

or

$$\begin{aligned} \overrightarrow{AM} &= \overrightarrow{MB} \\ \Leftrightarrow \boldsymbol{m}-\boldsymbol{a} &= \boldsymbol{b}-\boldsymbol{m} \\ \Leftrightarrow 2\boldsymbol{m} &= \boldsymbol{b}+\boldsymbol{a} \\ \Leftrightarrow \boldsymbol{m} &= \tfrac{1}{2}(\boldsymbol{a}+\boldsymbol{b}) \end{aligned}$$

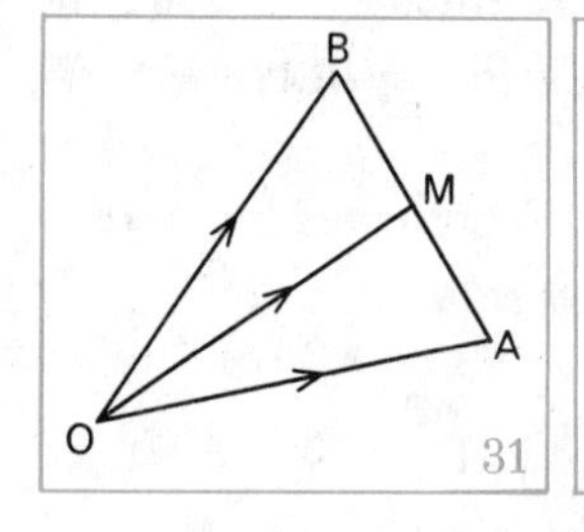

31

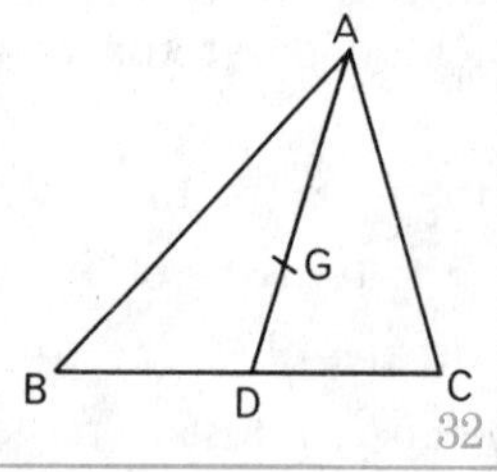

32

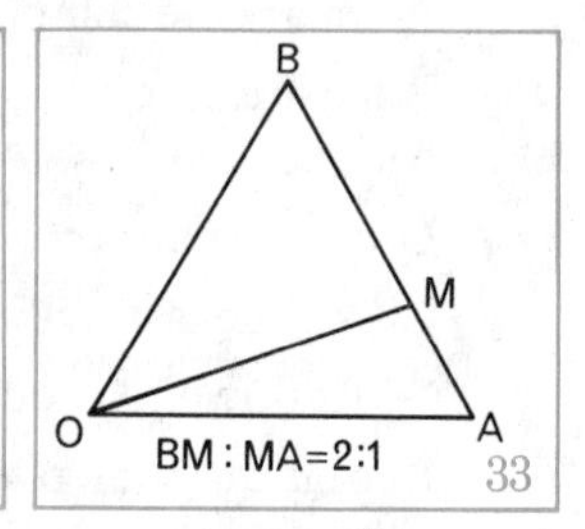

33

Example 2. In Figure 32, D is the midpoint of BC, so that AD is a median in △ABC. G divides AD in the ratio 2:1. Show that $\boldsymbol{g} = \frac{1}{3}(\boldsymbol{a}+\boldsymbol{b}+\boldsymbol{c})$.

With respect to an arbitrary origin O, $\overrightarrow{OG}$, $\overrightarrow{OA}$, $\overrightarrow{OB}$ and $\overrightarrow{OC}$ represent $\boldsymbol{g}$, $\boldsymbol{a}$, $\boldsymbol{b}$ and $\boldsymbol{c}$, the position vectors of G, A, B and C respectively.

$$\begin{aligned} \overrightarrow{OG} &= \overrightarrow{OA}+\overrightarrow{AG} \\ &= \overrightarrow{OA}+\tfrac{2}{3}\overrightarrow{AD} \end{aligned}$$

$$\Leftrightarrow g = a+\tfrac{2}{3}(d-a)$$
$$= a+\tfrac{2}{3}[\tfrac{1}{2}(b+c)-a)]$$
$$= a+\tfrac{1}{3}b+\tfrac{1}{3}c-\tfrac{2}{3}a$$
$$= \tfrac{1}{3}(a+b+c)$$

In the same way it can be shown that, if H divides the median BE in the ratio 2:1 and K divides the median CF in the ratio 2:1, then $\boldsymbol{h} = \frac{1}{3}(\boldsymbol{a}+\boldsymbol{b}+\boldsymbol{c}) = \boldsymbol{k}$.

Hence G, H and K are different names for the same point.

From these two examples some results are worth remembering.

(i) The position vector of the midpoint of AB is $\frac{1}{2}(\boldsymbol{a}+\boldsymbol{b})$.

(ii) The medians of a triangle are concurrent, i.e. they pass through the same point. The point of concurrence is called the *centroid* of the triangle. The position vector of the centroid of $\triangle$ABC is $\frac{1}{3}(\boldsymbol{a}+\boldsymbol{b}+\boldsymbol{c})$.

(iii) The centroid divides each median in the ratio 2:1.

Exercise 8B

1 In Figure 33, BA is divided at M in the ratio 2:1. Hence $\overrightarrow{BM} = 2\overrightarrow{MA}$. Copy and complete:

$$\overrightarrow{BM} = 2\overrightarrow{MA}$$
$$\Leftrightarrow \quad m-b = 2(\ldots - m)$$
$$\Leftrightarrow \quad m-b = \ldots - 2m$$
$$\Leftrightarrow m+\ldots = 2a+\ldots$$
$$\Leftrightarrow \quad 3m = \ldots + \ldots$$
$$\Leftrightarrow \quad m = \tfrac{1}{3}(2a+b)$$

2 In Figure 33, A is the point (7, 2) and B is (4, 5). Use the result of question ***1*** to calculate the components of $\boldsymbol{m}$, and write down the coordinates of M.

3 a Use the method of question ***1*** to show that $\boldsymbol{m} = \frac{1}{4}(3\boldsymbol{a}+\boldsymbol{b})$ if BA is divided at M in the ratio 3:1 (or $\overrightarrow{BM} = 3\overrightarrow{MA}$).

b If A is the point (2, 0), and B is (6, 4), calculate the coordinates of M.

4 a If BA is divided at M in the ratio 3:2 ($2\overrightarrow{BM} = 3\overrightarrow{MA}$), show that $\boldsymbol{m} = \frac{1}{5}(3\boldsymbol{a}+2\boldsymbol{b})$.

b If A is the point (9, 2) and B is the point (4, −8), calculate the coordinates of M.

5 a If A is (a, b) and B is (c, d), use the result of Worked Example 1 on page 124 to show that the midpoint of AB is $(\frac{1}{2}(a+c), \frac{1}{2}(b+d))$.

b Write down the coordinates of the midpoints of the lines joining:

(*1*) (4, 2) and (8, 6) (*2*) (−2, −4) and (4, −4) (*3*) (6, 0) and (0, 8).

6 Prove the result, $\boldsymbol{g} = \frac{1}{3}(\boldsymbol{a}+\boldsymbol{b}+\boldsymbol{c})$ of Worked Example 2 on page 124, by starting with $\boldsymbol{g}-\boldsymbol{a} = 2(\boldsymbol{d}-\boldsymbol{g})$.

7 a A, B and C are the points (4, 3), (5, 1) and (−3, 5) respectively. Write down $\boldsymbol{a}$, $\boldsymbol{b}$ and $\boldsymbol{c}$ in component form.

b Calculate the components of the position vector of M, the midpoint of BC.

c Write down the coordinates of M.

d Calculate the components of the position vector of G, the centroid of triangle ABC.

e Write down the coordinates of G.

8 In triangle ABC, D and E are the midpoints of AB and AC.

a Express $\boldsymbol{d}$ and $\boldsymbol{e}$ in terms of $\boldsymbol{a}$, $\boldsymbol{b}$ and $\boldsymbol{c}$.

b Which vectors do $\overrightarrow{BC}$ and $\overrightarrow{DE}$ represent?

c What can you state about $\overrightarrow{BC}$ and $\overrightarrow{DE}$?

9 In Figure 34, ABCD is a quadrilateral with P, Q, R and S the midpoints of AB, BC, CD and DA respectively.

a Express $\boldsymbol{p}$, $\boldsymbol{q}$, $\boldsymbol{r}$ and $\boldsymbol{s}$ in terms of $\boldsymbol{a}$, $\boldsymbol{b}$, $\boldsymbol{c}$ and $\boldsymbol{d}$.

b Which vectors do $\overrightarrow{PQ}$ and $\overrightarrow{SR}$ represent?

c What kind of figure is PQRS?

d State a theorem which is true about every quadrilateral.

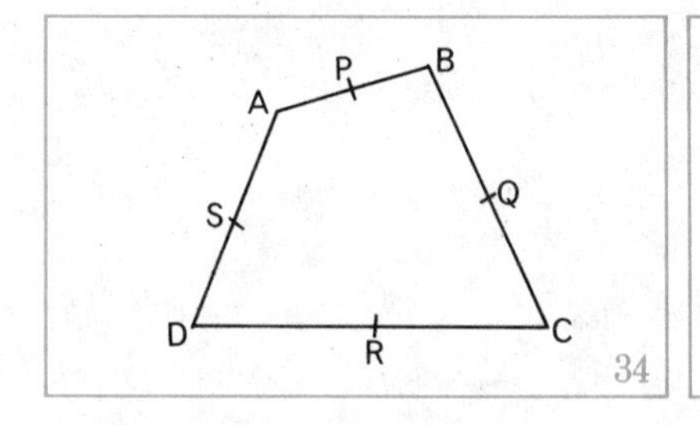

34

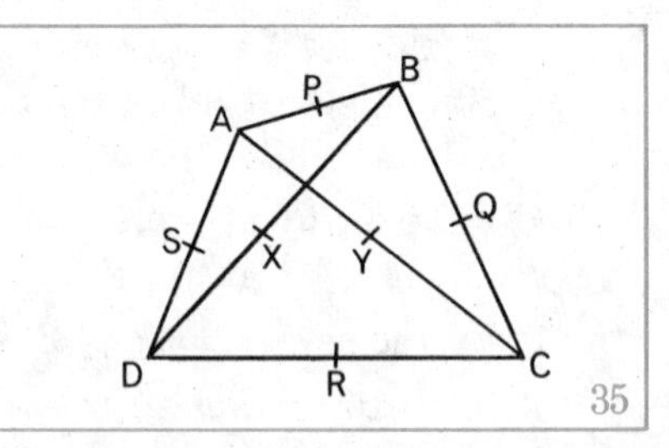

35

10 Figure 35 shows a quadrilateral as in Figure 34, with the midpoints of its sides marked. X and Y are the midpoints of the diagonals BD and AC.

a Write down the position vectors $\boldsymbol{p}$ and $\boldsymbol{r}$ in terms of $\boldsymbol{a}$, $\boldsymbol{b}$, $\boldsymbol{c}$ and $\boldsymbol{d}$.

b What is the position vector of the midpoint of PR?

c Calculate the position vectors of the midpoints of QS and XY.

d What theorem can you state about the quadrilateral ABCD?

11 In Figure 35, show that $\vec{AB}+\vec{AD}+\vec{CB}+\vec{CD} = 4\vec{YX}$.

8 *Magnitude (or length) of a vector*

In Figure 36 (i), $\vec{AB}$ represents a vector $\boldsymbol{u}$. AC and CB are parallel to the usual rectangular axes. $\boldsymbol{u} = \begin{pmatrix} 5 \\ 12 \end{pmatrix}$.

From Pythagoras' theorem the magnitude, or length, of $\vec{AB}$ $= \sqrt{(5^2+12^2)} = 13$.

Hence the magnitude of $\boldsymbol{u}$ is 13.

The magnitude of $\vec{AB}$ is usually written $|\vec{AB}|$, and the magnitude of $\boldsymbol{u}$ is written $|\boldsymbol{u}|$. We say that $\vec{AB}$ *represents* $\boldsymbol{u}$, and that $|\vec{AB}| = |\boldsymbol{u}|$. ('The length of segment AB is equal to the magnitude of vector $\boldsymbol{u}$.')

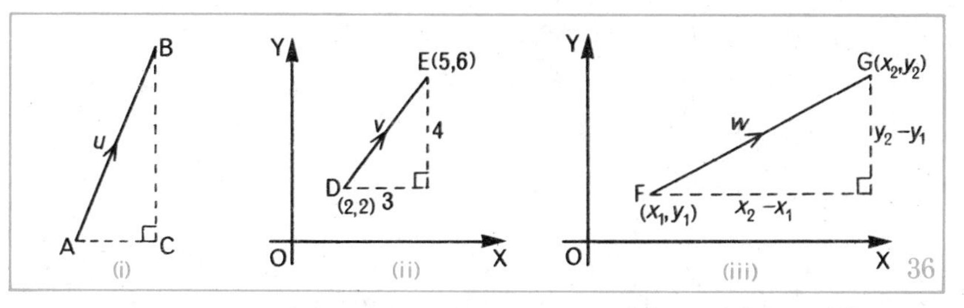

36

In (ii), $\vec{DE} = \vec{OE} - \vec{OD}$, so $\vec{DE}$ represents $\begin{pmatrix} 5 \\ 6 \end{pmatrix} - \begin{pmatrix} 2 \\ 2 \end{pmatrix}$, i.e. $\begin{pmatrix} 3 \\ 4 \end{pmatrix}$.

If $\vec{DE}$ represents $\boldsymbol{v}$, then $|\boldsymbol{v}| = |\vec{DE}| = \sqrt{(3^2+4^2)} = 5$.

In (iii), $\boldsymbol{w} = \begin{pmatrix} x_2 \\ y_2 \end{pmatrix} - \begin{pmatrix} x_1 \\ y_1 \end{pmatrix} = \begin{pmatrix} x_2 - x_1 \\ y_2 - y_1 \end{pmatrix}$,

and $|\boldsymbol{w}| = |\vec{FG}| = \sqrt{[(x_2-x_1)^2+(y_2-y_1)^2]}$.

Note the close connection between the formula for $|\boldsymbol{w}|$ and the Distance Formula in Book 4 Geometry, Chapter 2. Both express Pythagoras' theorem in terms of coordinates.

The magnitude of a vector may be zero, but is never negative. $\boldsymbol{u}$ and $-\boldsymbol{u}$ have opposite directions, but $|\boldsymbol{u}| = |-\boldsymbol{u}|$.

Exercise 9

1 On squared paper draw representatives of the following vectors, and calculate their magnitudes:

a $\begin{pmatrix}6\\8\end{pmatrix}$ *b* $\begin{pmatrix}8\\6\end{pmatrix}$ *c* $\begin{pmatrix}-8\\6\end{pmatrix}$ *d* $\begin{pmatrix}-8\\-6\end{pmatrix}$

e $\begin{pmatrix}6\\-8\end{pmatrix}$ *f* $\begin{pmatrix}8\\-6\end{pmatrix}$ *g* $\begin{pmatrix}-6\\8\end{pmatrix}$ *h* $\begin{pmatrix}-6\\-8\end{pmatrix}$

How many different magnitudes are there in this list?
How many different vectors are there in this list?

2 Write down the lengths of each of the following vectors, leaving in square roots where necessary:

a $\begin{pmatrix}-2\\-1\end{pmatrix}$ *b* $\begin{pmatrix}1\\-8\end{pmatrix}$ *c* $\begin{pmatrix}-4\\7\end{pmatrix}$ *d* $\begin{pmatrix}-12\\5\end{pmatrix}$

3 State whether the following are true or false:

a If $|\boldsymbol{u}| = |\boldsymbol{v}|$, then $\boldsymbol{u} = \boldsymbol{v}$ *b* If $\boldsymbol{u} = \boldsymbol{v}$, then $|\boldsymbol{u}| = |\boldsymbol{v}|$

4 Calculate the magnitude of the vector represented by $\overrightarrow{PQ}$ if P and Q are the points:

a (5, 0), (10, 4) *b* (7, 4), (1, 12) *c* $(-1, -1), (-5, -6)$

d $(4, -1), (-3, -4)$ *e* $(a, 4a), (-2a, 8a)$

f $(-2m, 5m), (-4m, -2m)$

5 $\overrightarrow{AB}$ and $\overrightarrow{BC}$ represent vectors $\boldsymbol{u}$ and $\boldsymbol{v}$ such that $|\boldsymbol{u}|+|\boldsymbol{v}| = |\boldsymbol{u}+\boldsymbol{v}|$, and $\boldsymbol{u} \neq 0$.

a What can you say about A, B and C?

b Why is it correct to say that in this case $\boldsymbol{v} = k\boldsymbol{u}$?

6 If ABC is a triangle in which $\overrightarrow{AB}$ and $\overrightarrow{BC}$ represent vectors $\boldsymbol{u}$ and $\boldsymbol{v}$, why is it true to say that $|\boldsymbol{u}|+|\boldsymbol{v}| > |\boldsymbol{u}+\boldsymbol{v}|$? (This is often called the *triangle inequality*.)

Summary

1 *The set of all directed line segments with the same magnitude and direction is a geometrical vector.*
The directed line segments are *representatives* of the vector.
The vector may be written $\boldsymbol{u}$ or, in components, $\begin{pmatrix} a \\ b \end{pmatrix}$.

2 *Addition.* Vectors may be added by the triangle or parallelogram rules.

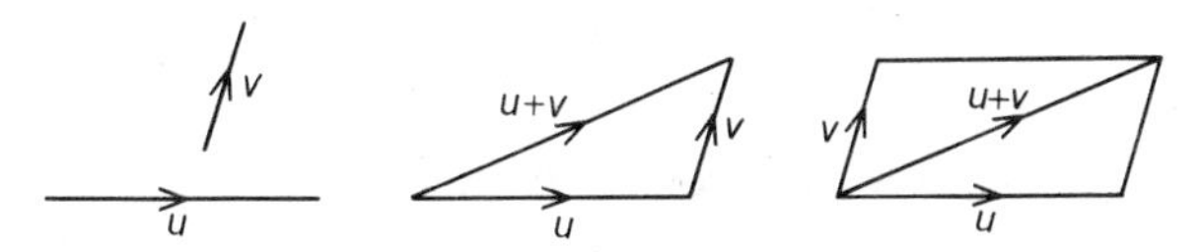

$$\boldsymbol{u}+\boldsymbol{v} = \boldsymbol{v}+\boldsymbol{u} \text{ (commutative law)}$$
$$(\boldsymbol{u}+\boldsymbol{v})+\boldsymbol{w} = \boldsymbol{u}+(\boldsymbol{v}+\boldsymbol{w}) \text{ (associative law)}$$
$$\boldsymbol{u}+\boldsymbol{0} = \boldsymbol{u} = \boldsymbol{0}+\boldsymbol{u} \text{ (identity element)}$$
$$\boldsymbol{u}+(-\boldsymbol{u}) = \boldsymbol{0} = (-\boldsymbol{u})+\boldsymbol{u} \text{ (additive inverse)}$$

All of the above can be expressed in component form.

3 *Subtraction.* Vectors are subtracted by the rule

$$\boldsymbol{u}-\boldsymbol{v} = \boldsymbol{u}+(-\boldsymbol{v}); \begin{pmatrix} a \\ b \end{pmatrix} - \begin{pmatrix} c \\ d \end{pmatrix} = \begin{pmatrix} a \\ b \end{pmatrix} + \begin{pmatrix} -c \\ -d \end{pmatrix} = \begin{pmatrix} a-c \\ b-d \end{pmatrix}$$

4 *Multiplication by a number.* If k is positive, $k\boldsymbol{u}$ has the same direction as $\boldsymbol{u}$ and k times the magnitude.
If k is negative, $k\boldsymbol{u}$ has the opposite direction to $\boldsymbol{u}$ and k times the magnitude.
If k is zero, $0\boldsymbol{u} = \boldsymbol{0}$. $k\begin{pmatrix} a \\ b \end{pmatrix} = \begin{pmatrix} ka \\ kb \end{pmatrix}$

5 *Distributive Laws.*

$$k\boldsymbol{u}+k\boldsymbol{v} = k(\boldsymbol{u}+\boldsymbol{v})$$
$$k\boldsymbol{u}+m\boldsymbol{u} = (k+m)\boldsymbol{u}$$

6 *Position vectors.* If O is the origin and P is the point (h, k), the vector represented by $\overrightarrow{OP}$ is the *position vector* of P. The position vector is denoted by $\boldsymbol{p}$, and $\boldsymbol{p} = \begin{pmatrix} h \\ k \end{pmatrix}$.

The coordinates of a point are the components of its position vector.

$\overrightarrow{PQ} = \overrightarrow{OQ} - \overrightarrow{OP}$ and represents $\boldsymbol{q} - \boldsymbol{p}$.

If M is the midpoint of PQ, $\boldsymbol{m} = \frac{1}{2}(\boldsymbol{p} + \boldsymbol{q})$.

If G is the centroid of $\triangle$ABC, $\boldsymbol{g} = \frac{1}{3}(\boldsymbol{a} + \boldsymbol{b} + \boldsymbol{c})$.

7 *Magnitude of a vector.* If $\overrightarrow{AB}$ represents the vector $\boldsymbol{u} = \begin{pmatrix} a \\ b \end{pmatrix}$, then

$$|\boldsymbol{u}| = |\overrightarrow{AB}| = \sqrt{(a^2 + b^2)}.$$

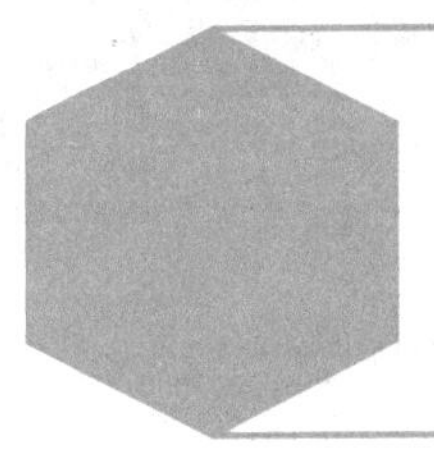

Revision Exercises

Revision Exercise on Chapter 1
Similar Shapes

Revision Exercise 1

1 On a map the scale used is such that 1 cm on the map represents a distance of 1 km on the earth's surface. What would be the length of a line on the map which represents a distance of 8·5 km on the earth's surface?

2 A model railway is made to a scale of 1 cm representing 50 cm. If the width of a British Rail track is 145 cm, what should be the track width on the model?

If the length of a model engine on the same scale is 18 cm, calculate the length of the real engine in metres.

3 A photograph measures 40 mm by 30 mm. If the longer side of an enlargement measures 160 mm, what is the length of the shorter side?

If the enlargement is displayed on a rectangular cardboard mount with a uniform border 40 mm wide all round, explain why the enlargement and the mount are not similar rectangles.

4 Figure 1 shows a plan of a rugby pitch drawn to a scale of 1 cm representing 30 m. What are the actual dimensions of the pitch?

How far is each of the dotted lines from the centre line on the actual pitch?

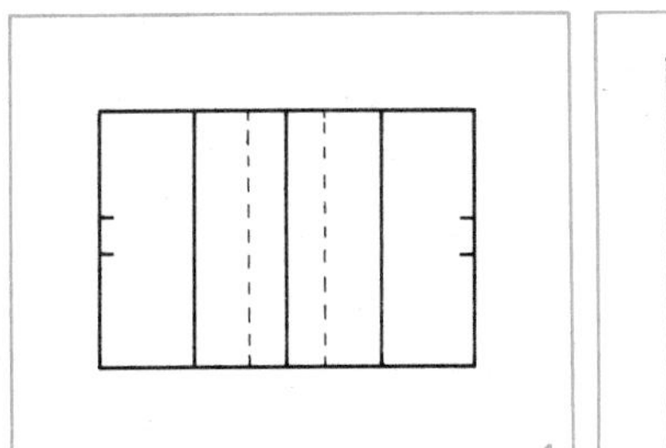

1

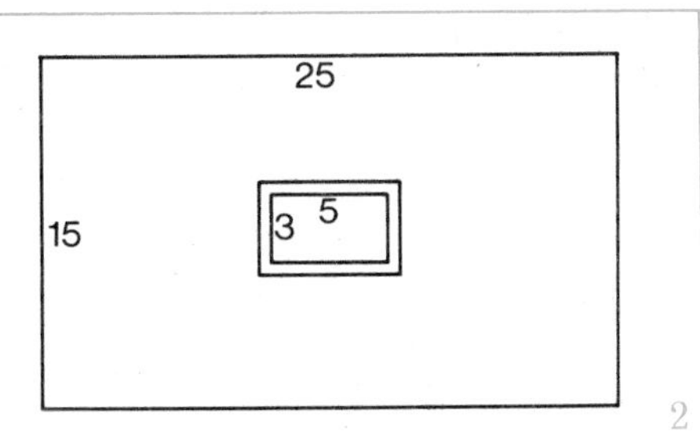

2

5 a Figure 2 is a plan of a large rectangular garden measuring 25 m by 15 m. To what scale is the plan drawn?

b A central flower bed measures 5 m by 3 m and is surrounded by a path 1 m wide. Are any or all of the rectangles similar? Give reasons for your answers.

6 In triangles ABC and DEF, angle A = 69°, angle B = 53°, angle E = 53° and angle F = 58°. Prove that the triangles are similar. Name pairs of corresponding sides.

7 In triangles ABC and PQR, AB = 8 cm, BC = 11 cm, CA = 16 cm, PQ = 48 cm, QR = 33 cm and RP = 24 cm. Prove that the triangles are similar. Name pairs of equal angles.

8 In Figure 3, there are two triangles similar to triangle ABC. Name them and state why they are similar. Name three sets of angles that must be equal.

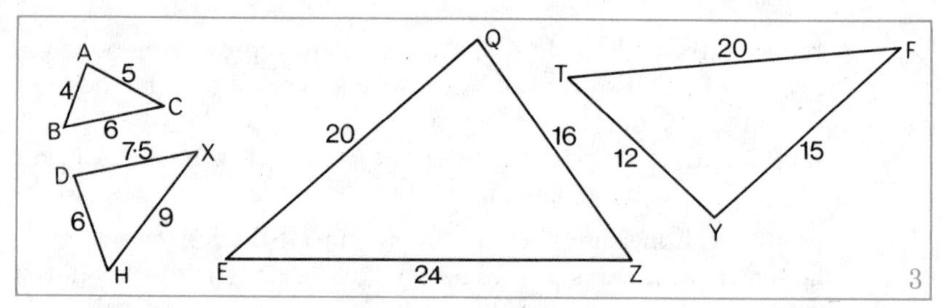

9 a In Figure 4 (i) prove that triangles ABC and ADE are equiangular. Are they similar? Give a reason for your answer.

b In Figure 4 (ii) prove that quadrilaterals ABCD and AXYD are equiangular. Are they similar? Give a reason for your answer.

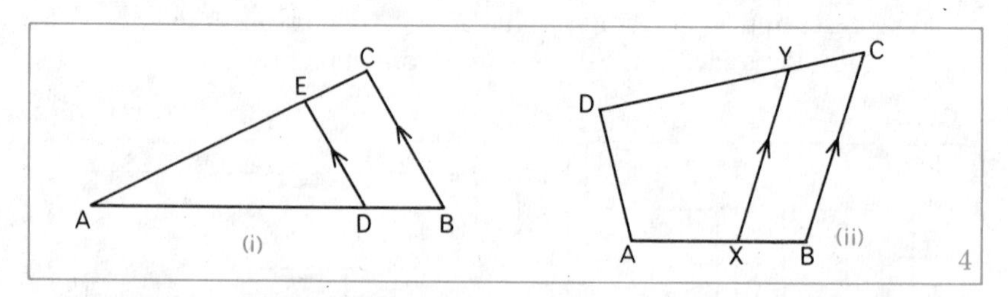

10a Sketch two quadrilaterals which are equiangular but not similar.

b Sketch two quadrilaterals which have their sides, taken in order, in proportion, but are not similar.

11 State which of the following are true and which are false.

a If two triangles are equiangular their corresponding sides must be in proportion.

b If two quadrilaterals are equiangular their corresponding sides must be in proportion.

c If two triangles are similar, they must be congruent.

d If two triangles are congruent, they must be similar.

e A scale drawing of a shape must be similar to the shape.

f All circles are similar to one another.

g If the sides of a rectangle are each reduced by the same length, the new rectangle is similar to the original rectangle.

12 Two similar rectangles have their sides in the ratio 5:4. If a diagonal of the larger rectangle measures 17·5 cm, calculate the length of a diagonal of the smaller.

13 Triangles ABC and DEF have angle A = angle D and angle B = angle E.

a Why must angle C = angle F?

b Name equal ratios of sides in the two triangles.

c If AX and DY are altitudes of the triangles ABC and DEF, prove that triangles ABX and DEY are equiangular.

d Show that $\frac{AX}{DY} = \frac{BC}{EF}$. (Use the results of ***b*** and ***c***.)

14 PQR and XYZ are similar triangles with P, Q and R corresponding to X, Y and Z respectively. PS and XW bisect angles P and X, and meet QR and YZ in S and W respectively.

Prove that triangles PQS and XYW are equiangular and show that $\frac{PS}{XW} = \frac{QR}{YZ}$.

15 In Figure 5, triangles ABC and DEF are similar with equal angles marked. Calculate the values of *c* and *d*.

16 In the similar quadrilaterals ABCD and PQRS in Figure 6, A, B, C and D correspond to P, Q, R and S respectively. Calculate the values of *x*, *a* and *b*.

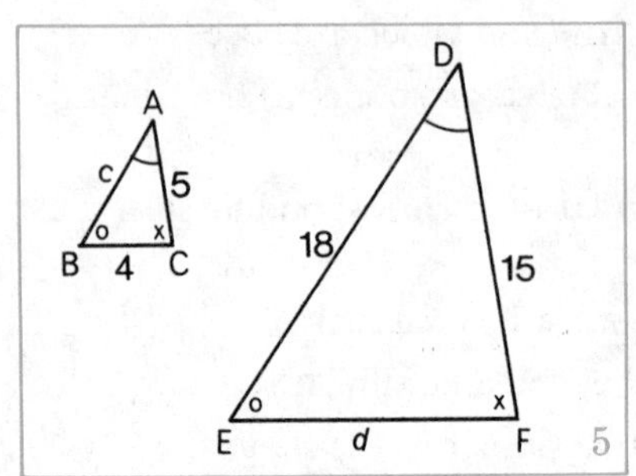

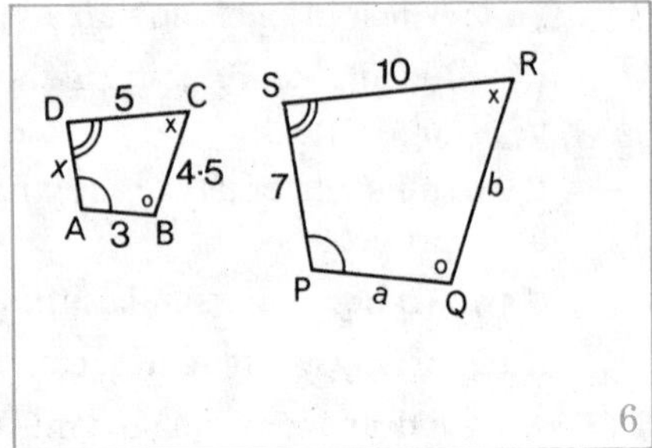

17 Figure 7 shows five squares with sides 1, 2, 3, 4 and 5 units long. State the ratios of the areas of squares whose sides are in the ratio:

a 2:1 *b* 3:1 *c* 1:4 *d* 2:3
e 4:3 *f* 2:5 *g* $m:1$ *h* $m:n$

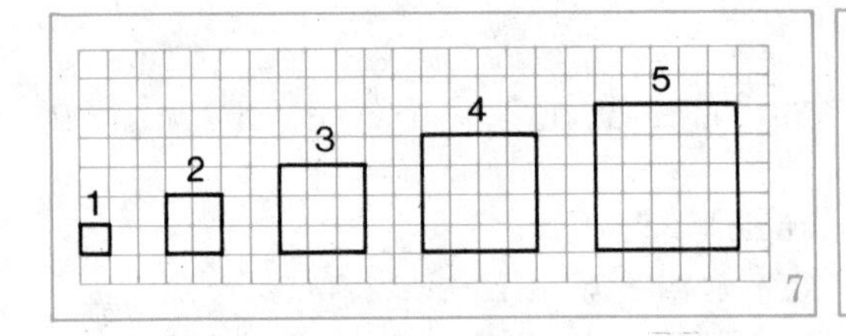

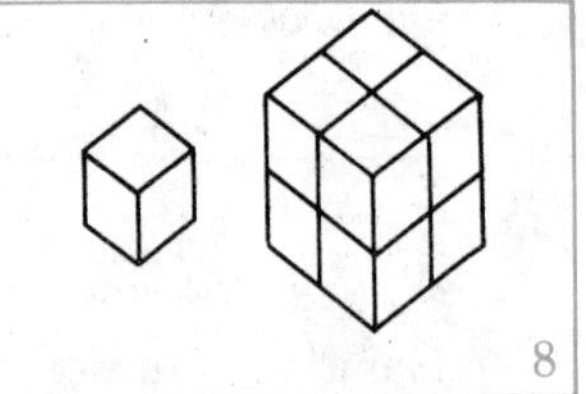

18 Figure 8 shows a sketch of a 1-cm cube, and a 2-cm cube built up from 1-cm cubes. Make a corresponding sketch for a 3-cm cube.

What is the ratio of the volumes of two cubes whose edges are in the ratio:

a 2:1 *b* 3:1 *c* 3:2 *d* 2:3?

19 Two rectangular boxes have the lengths of their corresponding edges in the ratio 5:4. If the smaller box holds $\frac{1}{2}$ kg of sweets, show that the larger box holds very nearly 1 kg.

20 A new car is designed and a model is made from the same materials by reducing all lengths to the ratio 1:10.

a If the overall length of the car is 5 m, what is the length of the model?

b What is the ratio of the areas of the windscreens in the model and in the car?

c What is the ratio of the total volumes of the model and the car?

d If the capacity of the fuel tank in the car is 50 litres, what is the capacity of the fuel tank in the model?

21 In Figure 9 a man 1·8 m tall stands 2·4 m from a lamppost. He observes that his shadow is 1·8 m long. Calculate the height, h metres, of the lamp above the pavement.

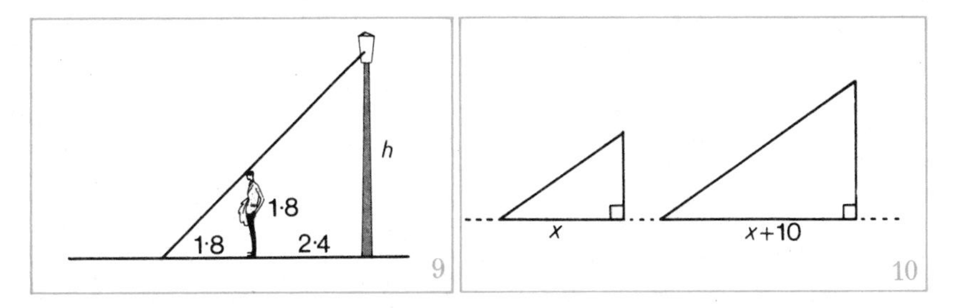

22 In Figure 10 two upright poles cast shadows of x metres and $(x+10)$ metres respectively. If the length of the shorter pole is $\frac{3}{5}$ of the length of the longer, calculate the value of x.

23 In Figure 11, calculate the values of x, y, p and q.

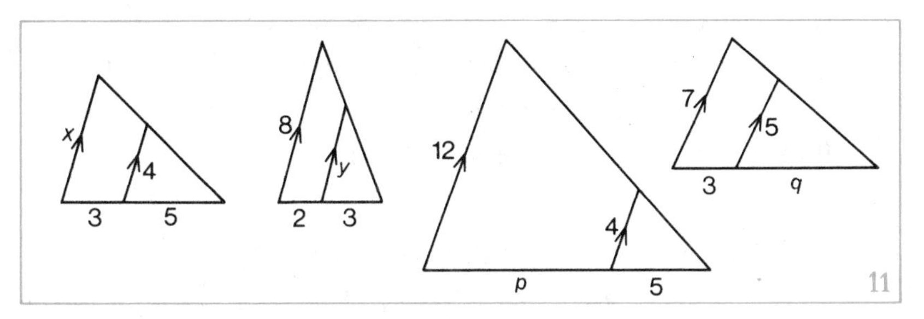

24 State the gradients and the coordinates of the point of intersection with the y-axis of the lines with equations:

a $y = \frac{1}{2}x+2$ *b* $y = -2x-5$ *c* $y+3x = 4$

d $3y = x-6$ *e* $4y+x = 8$ *f* $2x+3y = 5$

25 Calculate, where possible, the gradient of the lines joining the pairs of points;

a (2, 3), (9, 4) *b* (6, −7), (−8, 3) *c* (−1, −3), (−4, −7)

d (4, 6), (4, −5) *e* (−7, 5), (3, 5) *f* (−14, −14), (1, 1)

26 Write down the equations of the lines which have the given gradients and pass through the given points:

a $\frac{1}{2}$, (0, 3) *b* −4, (0, 1) *c* $-\frac{1}{3}$, (0, −4)

d 0, (0, 4) *e* 3, (0, −1) *f* k, (0, p)

27 In Figure 12, RS is parallel to BC. AR = 12 units and RB = 8 units. Calculate the ratio in which SB divides RC.

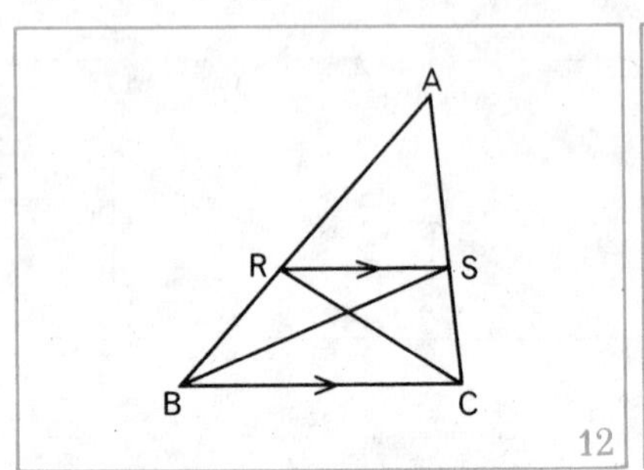

12

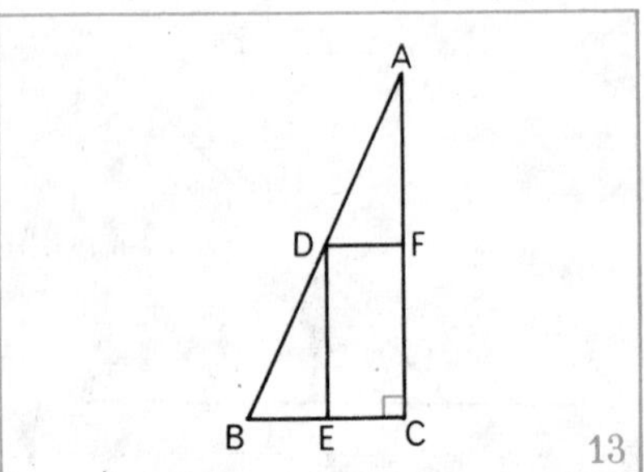

13

28 Figure 13 shows a ladder leaning against a wall so that its upper end is 7·2 m above the ground and its lower end is 3 m out from the foot of the wall. The ladder just touches at D the front upper edge of a cupboard whose cross-section is shown as the rectangle DECF.

a State why triangles ABC, ADF and DBE are all similar.

b If DF is 1·5 m, show that the height of the cupboard is 3·6 m.

c If DF is 0·9 m, calculate the height DE.

d If the height DE is 1·8 m, calculate the breadth DF.

Revision Exercises on Chapter 2
Introduction to Vectors

Revision Exercise 2

1 State which of the following are true and which are false:

a If the distance from A to B is 10 km and the distance from B to C is 8 km, then the distance from A to C must be 18 km.

b If two adjacent sides of a parallelogram are representatives of vectors $\boldsymbol{u}$ and $\boldsymbol{v}$, the diagonals represent vectors $\boldsymbol{u}+\boldsymbol{v}$ and $\boldsymbol{u}-\boldsymbol{v}$.

c If a translation maps A and B onto C and D respectively, then $\overrightarrow{AC}$ and $\overrightarrow{BD}$ represent the same vector.

d In triangle ABC, $\overrightarrow{AB}+\overrightarrow{BC}+\overrightarrow{AC}$ represents the zero vector.

2 For Figure 14 copy and complete the following:

a $\vec{AC}+\vec{CD} = \ldots$ *b* $\vec{AC}+\vec{CB}+\vec{BG} = \ldots$ *c* $\vec{BG}+\vec{AB}+\ldots = \vec{AD}$

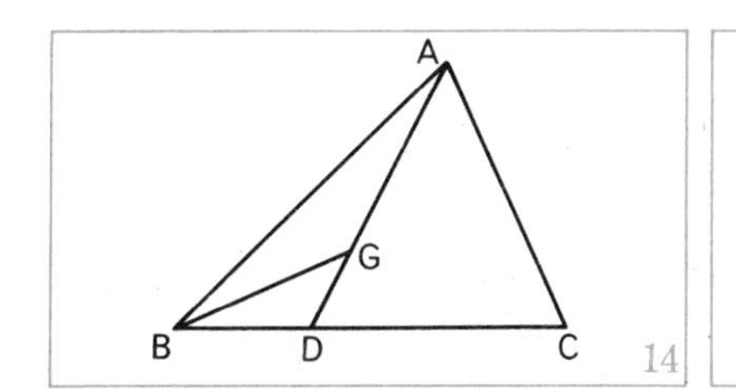

14

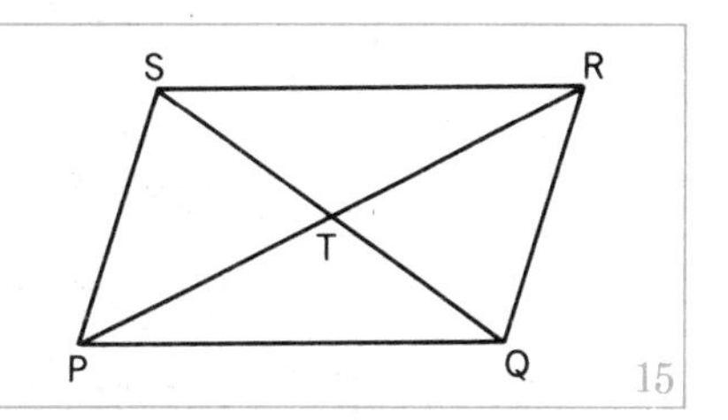

15

3 For Figure 15, where PQRS is a parallelogram, copy and complete the following:

a $\vec{PS}+\frac{1}{2}\vec{SQ} = \ldots$ *b* $\frac{1}{2}\vec{SQ}+\frac{1}{2}\vec{PR} = \ldots$ *c* $\vec{PQ}+\vec{RT}+\vec{QR} = \ldots$

4 Using the vectors $\boldsymbol{u}$ and $\boldsymbol{v}$ which are represented in Figure 16, construct on squared paper representatives of:

a $\boldsymbol{u}+\boldsymbol{v}$ *b* $2(\boldsymbol{u}+\boldsymbol{v})$ *c* $\boldsymbol{u}-\boldsymbol{v}$ *d* $\boldsymbol{u}+2\boldsymbol{v}$

e $-(\boldsymbol{u}+\boldsymbol{v})$ *f* $\frac{1}{2}\boldsymbol{u}+\boldsymbol{v}$.

Express your answers also in component form.

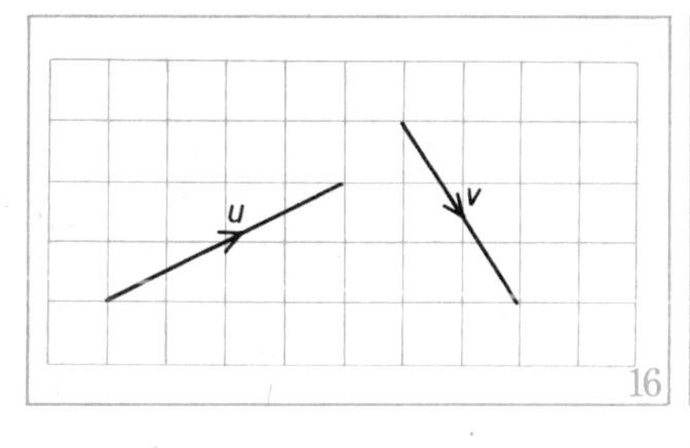

16

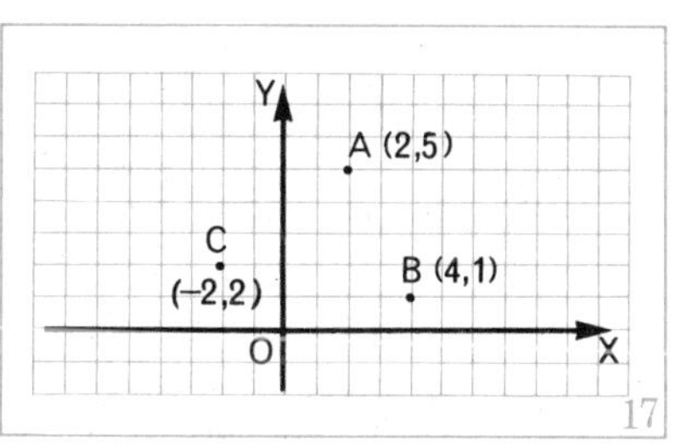

17

5 In Figure 17, A is the point (2, 5), B is (4, 1) and C (−2, 2). Express in number pair form the vectors represented by:

a $\vec{AB}$ *b* $\vec{BC}$ *c* $\vec{CA}$ *d* $-\vec{AB}$ *e* $\vec{AB}+\vec{BC}$ *f* $\vec{CA}-\vec{CB}$

6 Figure 18 shows a tiling of congruent right-angled triangles. Using only the letters in the figure, name the directed line segments which could represent the same vectors as:

a $2\vec{AB}$ *b* $3\vec{AB}$ *c* $4\vec{AB}$ *d* $\vec{BA}$

7 In Figure 18, find in terms of $\overrightarrow{AB}$ the vector sums represented by the following:

a $\overrightarrow{NP}+2\overrightarrow{PR}$ *b* $\overrightarrow{PQ}+3\overrightarrow{MH}$ *c* $\overrightarrow{RM}+\overrightarrow{MF}$ *d* $2\overrightarrow{QM}+\overrightarrow{NP}$

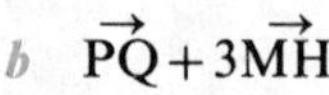

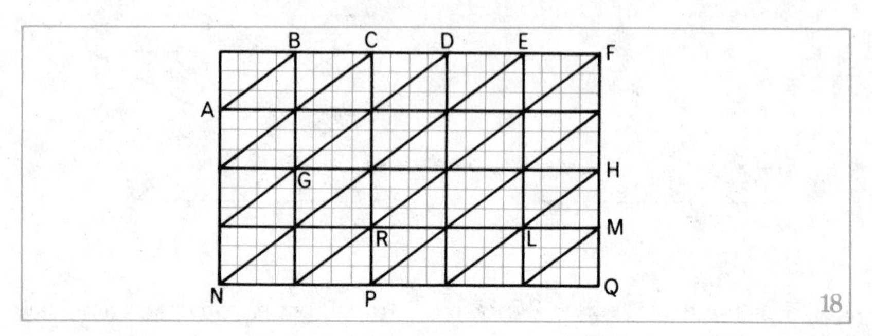

18

8 If, in Figure 18, $\overrightarrow{AB}$ represents the vector $\begin{pmatrix}4\\3\end{pmatrix}$, write in component form the vectors represented by:

a $\overrightarrow{LH}$ *b* $\overrightarrow{GD}$ *c* $\overrightarrow{RF}$ *d* $\overrightarrow{EN}$ *e* $\overrightarrow{LM}$ *f* $\overrightarrow{QH}$

9 On squared paper draw representatives of the vectors:

a $\begin{pmatrix}3\\5\end{pmatrix}$ *b* $\begin{pmatrix}-4\\6\end{pmatrix}$ *c* $\begin{pmatrix}7\\-5\end{pmatrix}$ *d* $\begin{pmatrix}-4\\-3\end{pmatrix}$

10 Draw three directed line segments to represent vectors $\boldsymbol{u}$, $\boldsymbol{v}$ and $\boldsymbol{w}$. Verify by drawing that:

a $\boldsymbol{u}+\boldsymbol{v} = \boldsymbol{v}+\boldsymbol{u}$ *b* $(\boldsymbol{u}+\boldsymbol{v})+\boldsymbol{w} = \boldsymbol{u}+(\boldsymbol{v}+\boldsymbol{w})$

11 Check that the commutative and associative laws, illustrated in question ***10***, hold for the addition of the vectors $\boldsymbol{u} = \begin{pmatrix}2\\3\end{pmatrix}$, $\boldsymbol{v} = \begin{pmatrix}-4\\1\end{pmatrix}$ and $\boldsymbol{w} = \begin{pmatrix}6\\-4\end{pmatrix}$.

12 Write down the negatives of:

a $\begin{pmatrix}2\\3\end{pmatrix}$ *b* $\begin{pmatrix}-4\\1\end{pmatrix}$ *c* $\begin{pmatrix}0\\0\end{pmatrix}$ *d* $\begin{pmatrix}h\\-k\end{pmatrix}$ *e* $\begin{pmatrix}p-q\\p+q\end{pmatrix}$

13 Simplify the following:

a $\begin{pmatrix}2\\-1\end{pmatrix}+\begin{pmatrix}-3\\4\end{pmatrix}$ *b* $\begin{pmatrix}6\\-2\end{pmatrix}-\begin{pmatrix}3\\-4\end{pmatrix}$ *c* $\begin{pmatrix}4\\1\end{pmatrix}+\begin{pmatrix}3\\-2\end{pmatrix}-\begin{pmatrix}2\\6\end{pmatrix}$

14 Copy and complete:

a $\begin{pmatrix}3\\-4\end{pmatrix}+\begin{pmatrix}\ \\\ \end{pmatrix}=\begin{pmatrix}4\\1\end{pmatrix}$ *b* $\begin{pmatrix}4\\-3\end{pmatrix}+\begin{pmatrix}3\\1\end{pmatrix}+\begin{pmatrix}\ \\\ \end{pmatrix}=\begin{pmatrix}0\\0\end{pmatrix}$

c $\begin{pmatrix}1\\1\end{pmatrix}+\begin{pmatrix}2\\2\end{pmatrix}-\begin{pmatrix}\ \\ \ \end{pmatrix}=\begin{pmatrix}0\\0\end{pmatrix}$ *d* $\begin{pmatrix}p\\ \ \end{pmatrix}+\begin{pmatrix}\ \\q\end{pmatrix}=\begin{pmatrix}0\\0\end{pmatrix}$

15 A is the point (3, 1), B is (5, 2). If $\overrightarrow{AB}$ and $\overrightarrow{CD}$ represent the same vector, write down the coordinates of D when C is:

a (1, 5) *b* (3, 9) *c* (−4, 2) *d* (2, −4).

If D is the point (−3, −2) what are the coordinates of C?

16 P is the point (8, 6), Q is (2, −2) and R is (0, 10).

a Write down the position vectors $\boldsymbol{p}$, $\boldsymbol{q}$, $\boldsymbol{r}$ in component form.

b If K, M, N are the midpoints of PQ, QR and RP respectively, calculate the components of $\boldsymbol{k}$, $\boldsymbol{m}$, $\boldsymbol{n}$. Hence write down the coordinates of K, M and N.

17 A is the point (4, 1), B is (0, 4), C is (−1, −1). Calculate the magnitudes of the vectors represented by $\overrightarrow{AB}$, $\overrightarrow{BC}$ and $\overrightarrow{CA}$.

Revision Exercise 2B

1 Simplify *a* $3\begin{pmatrix}2\\-1\end{pmatrix}-2\begin{pmatrix}3\\-2\end{pmatrix}$ *b* $2\begin{pmatrix}4\\-5\end{pmatrix}+3\begin{pmatrix}0\\0\end{pmatrix}$

c $4(\boldsymbol{u}+\boldsymbol{v})-3(\boldsymbol{u}-\boldsymbol{v})$ *d* $2(\boldsymbol{u}-\boldsymbol{v})-(\boldsymbol{u}+\boldsymbol{v})$

2 Given that $\boldsymbol{u}=\begin{pmatrix}2\\-3\end{pmatrix}$ and $\boldsymbol{v}=\begin{pmatrix}-1\\2\end{pmatrix}$, find in component form:

a $\boldsymbol{u}+\boldsymbol{v}$ *b* $3\boldsymbol{u}+2\boldsymbol{v}$ *c* $4\boldsymbol{u}-5\boldsymbol{v}$

3 Solve for $\boldsymbol{x}$ the vector equations:

a $2\boldsymbol{x}+\begin{pmatrix}4\\-1\end{pmatrix}=\begin{pmatrix}5\\9\end{pmatrix}+\boldsymbol{x}$ *b* $4\boldsymbol{x}+\begin{pmatrix}3\\1\end{pmatrix}+\begin{pmatrix}-7\\3\end{pmatrix}=\begin{pmatrix}0\\0\end{pmatrix}$

c $3(\boldsymbol{x}+\boldsymbol{a})=6\boldsymbol{a}$ *d* $6(\boldsymbol{x}-\boldsymbol{a})=4(\boldsymbol{x}+\boldsymbol{b})$

4 $\boldsymbol{u}=\begin{pmatrix}5\\-5\end{pmatrix}$ and $\boldsymbol{v}=\begin{pmatrix}-7\\1\end{pmatrix}$. Explain why $\boldsymbol{u}\neq\boldsymbol{v}$ but $|\boldsymbol{u}|=|\boldsymbol{v}|$.

5 Solve the following equations for x and y:

a $3\begin{pmatrix}x\\y\end{pmatrix}=\begin{pmatrix}6\\-9\end{pmatrix}$ *b* $\begin{pmatrix}x\\y\end{pmatrix}+\begin{pmatrix}2\\-1\end{pmatrix}=\begin{pmatrix}5\\-3\end{pmatrix}$

c $\frac{1}{2}\begin{pmatrix}x\\y\end{pmatrix}+\begin{pmatrix}-1\\3\end{pmatrix}=\begin{pmatrix}-1\\7\end{pmatrix}$

6 A is the point (2, 1) and B is (8, −3). Write down the position vectors $\boldsymbol{a}$ and $\boldsymbol{b}$ in component form. Use vectors to find the coordinates of:

a M, if $\overrightarrow{AM} = \overrightarrow{MB}$ *b* H, if $\overrightarrow{AH} = \frac{1}{4}\overrightarrow{AB}$ *c* K if $\overrightarrow{AK} = 3\overrightarrow{KB}$

7 *a* In Figure 19, $\overrightarrow{AA'} = \frac{1}{2}\overrightarrow{OA}$ and $\overrightarrow{BB'} = \frac{1}{2}\overrightarrow{OB}$. Write down one equation of the same kind connecting:

(*1*) $\overrightarrow{OA'}$ and $\overrightarrow{OA}$ (*2*) $\overrightarrow{OB'}$ and $\overrightarrow{OB}$ (*3*) $\overrightarrow{A'B'}$ and $\overrightarrow{AB}$.

b If BB′, OA and A′B′ are 3 cm, 4 cm and 5 cm long respectively, calculate the lengths of OB, AA′ and AB.

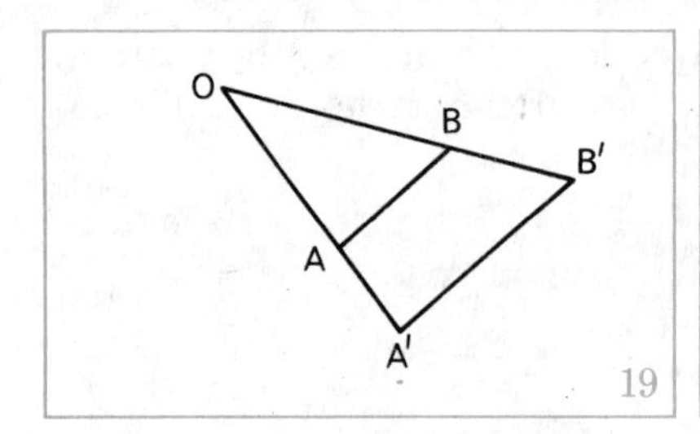

19

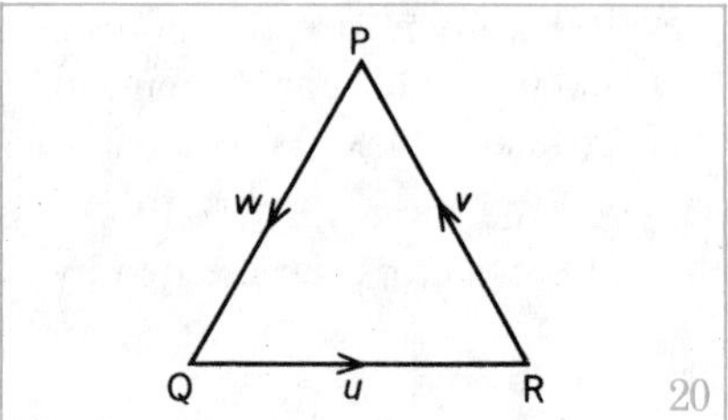

20

8 In Figure 20, △PQR is equilateral. $\overrightarrow{QR}$, $\overrightarrow{RP}$ and $\overrightarrow{PQ}$ represent the vectors $\boldsymbol{u}$, $\boldsymbol{v}$ and $\boldsymbol{w}$ respectively. State which of the following are true and which are false:

a $\boldsymbol{u}+\boldsymbol{v} = \boldsymbol{w}$ *b* $|\boldsymbol{u}+\boldsymbol{v}| = |\boldsymbol{w}|$ *c* $|\boldsymbol{u}|+|\boldsymbol{v}| = |\boldsymbol{w}|$

d $\boldsymbol{u}+\boldsymbol{v}+\boldsymbol{w} = 3\boldsymbol{u}$ *e* $|\boldsymbol{u}|+|\boldsymbol{v}|+|\boldsymbol{w}| = 3|\boldsymbol{u}|$ *f* $\boldsymbol{u} = -\boldsymbol{v}-\boldsymbol{w}$

9 ABCD is a quadrilateral with P, Q, R and S the midpoints of AB, BC, CD and DA respectively.

a With respect to an origin O show that $\boldsymbol{p} = \frac{1}{2}(\boldsymbol{a}+\boldsymbol{b})$ and that $\overrightarrow{PQ}$ represents $\frac{1}{2}(\boldsymbol{c}-\boldsymbol{a})$.

b Find a similar expression for the vector represented by $\overrightarrow{SR}$.

c State any geometrical relations between $\overrightarrow{PQ}$, $\overrightarrow{SR}$ and $\overrightarrow{AC}$.

d What kind of quadrilateral is PQRS?

e If AB and CD cross each other, what difference would there be in your first four answers?

f If A, B, C and D were not all in the same plane, would there be any difference in your first four answers?

10 ABCD is a parallelogram with A (1, 1), B (7, 3) and C (10, 7). M is the midpoint of AB, and DM cuts AC at P.

a State the coordinates of D.

b Express $\boldsymbol{m}$ in component form.

c Which vector, in component form, does $\overrightarrow{MD}$ represent?

d Given that $\overrightarrow{MP} = \frac{1}{3}\overrightarrow{MD}$, find the components of $\boldsymbol{p}$ and state the coordinates of P.

e What are the components of the vector represented by $\overrightarrow{AP}$?

f What can you state about AP and AC?

Arithmetic

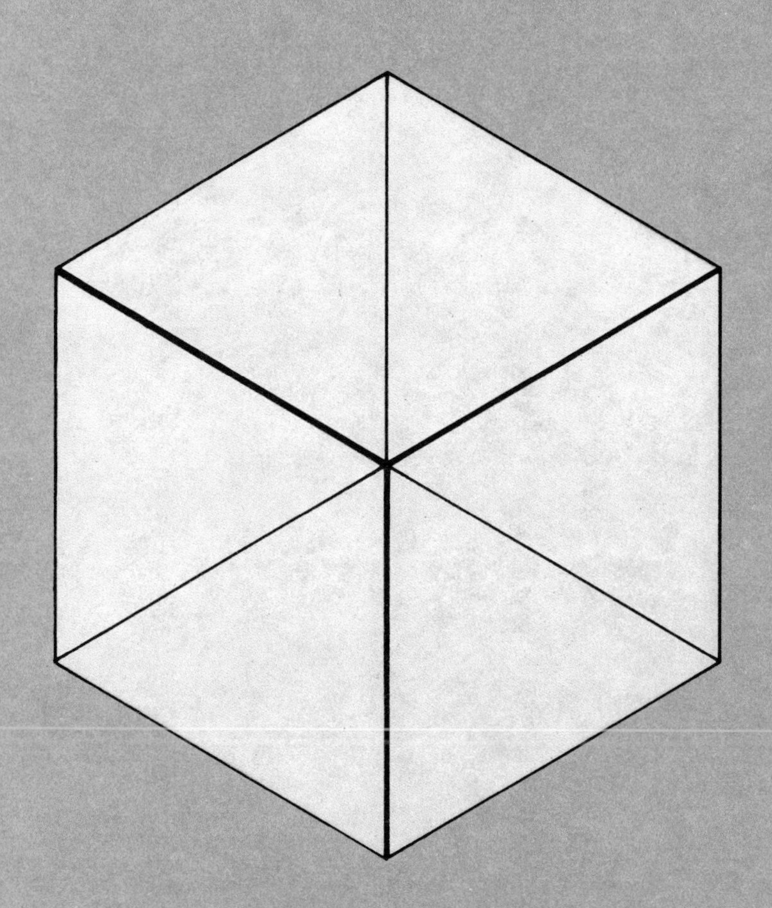

Arithmetic

Logarithms and Calculating Machines

1 The meaning of a logarithm

We have seen that the following are true for the number systems we have studied in earlier books:

	Addition	*Multiplication*
Commutative law	$a+b = b+a$	$a \times b = b \times a$
Associative law	$(a+b)+c = a+(b+c)$	$(a \times b) \times c = a \times (b \times c)$
Identity elements	$a+0 = a = 0+a$	$a \times 1 = a = 1 \times a$

As a result, we were able in the chapter on the Slide Rule in Book 4 to set up a correspondence between
the elements of set $M = \{1, 2, 4, 8, 16, \ldots\}$ under multiplication,
and
the elements of set $A = \{0, 1, 2, 3, 4, \ldots\}$ under addition, thus:

Multiplication set M	1	(2)	4	8	16	...
	↕	↕	↕	↕	↕	
Addition set A	0	1	2	3	4	...

We chose the identity element for addition, 0, as the first element of A, and the identity element for multiplication, 1, as the first element of M. We then chose 2, shown in brackets, as the second element of M, and this gave the rest of the elements in M: 4, 8, 16, ..., each term being two times the preceding term.

The correspondence between set M and set A can be illustrated by a graph, as shown in Figure 1. The points (1, 0), (2, 1), (4, 2), (8, 3) and (16, 4) were plotted, and then joined by a smooth curve. From the graph we can *estimate* corresponding elements in the two sets. For example, (3, 1·6) and (12, 3·6) are points on the curve, so 3 in set M corresponds to 1·6 in set A, and 12 in M corresponds to 3·6 in A.

We define the *logarithm* of a number in the multiplication set M to be the corresponding number in the addition set A.

For example, the logarithm of 8 is 3, written log 8 = 3. The number whose logarithm is 1 is called the *base* of the system. Here

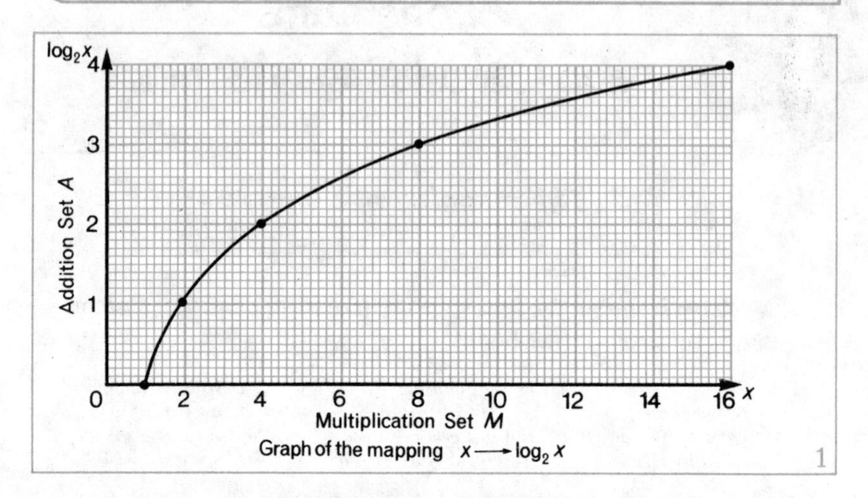

Graph of the mapping $x \longrightarrow \log_2 x$ 1

$\log 2 = 1$, so 2 is the base, and we write $\log_2 8 = 3$ ('log 8 to the base 2 is 3'). Also, $\log_2 15 = 3{\cdot}9$, and so on.

We can sum up as follows:

Corresponding elements Multiplication set M		Addition set A	*Point on graph*	*Logarithm*
1	$\longleftrightarrow$	0	(1, 0)	$\log_2 1 = 0$
2	$\longleftrightarrow$	1	(2, 1)	$\log_2 2 = 1$
4	$\longleftrightarrow$	2	(4, 2)	$\log_2 4 = 2$
8	$\longleftrightarrow$	3	(8, 3)	$\log_2 8 = 3$
...		...	...	... ...

Exercise 1

1 a Copy and complete this table:

Multiplication set M	1	2	4	.	.	.	.
Addition set A	0	1	2	3	4	5	6

b Copy and complete: $\log_2 4 = \ldots$, $\log_2 8 = \ldots$, $\log_2 16 = \ldots$, $\log_2 64 = \ldots$

2 Use the graph in Figure 1 to write down the numbers (to 1 decimal place where necessary) in the addition set which correspond to the following numbers in the multiplication set:

a 1 *b* 6 *c* 8 *d* 10 *e* 15

3 Which numbers in M correspond to the following numbers in A?

a 2 *b* 4 *c* 1·5 *d* 3·5 *e* 2·7

4 Express your answers to questions ***2*** and ***3*** in the form of logarithms, e.g. $\log_2 1 = 0$.

5 From the graph find approximations for:

a $\log_2 12$ *b* $\log_2 5$ *c* $\log_2 3$ *d* $\log_2 14$

6 From the graph write down numbers whose logarithms to base 2 are:

a 3 *b* 2·5 *c* 0·8 *d* 1·6

2 *Logarithms to base 10*

We can choose any base we please by choosing the number in the multiplication set which corresponds to 1 in the addition set. Since our calculations are normally based on the decimal system, 10 is an obvious choice. Then we have:

Multiplication set	1	10	100	1000	10 000	100 000	...
	↕	↕	↕	↕	↕	↕	
Addition set	0	1	2	3	4	5	...

We can now write $\log_{10}10 = 1$, $\log_{10}1000 = 3$, etc.

We can obtain approximations for the logarithms of many numbers from a graph. Figure 2 shows the graph of the mapping $x \to \log_{10}x$, and gives the logarithms to base 10 of numbers from 1 to 16.

For example, $\log_{10}5 = 0{\cdot}7$, approximately.

Note. (i) $\log_{10}1 = 0$

(ii) $\log_{10}10 = 1$

(iii) The logarithm to base 10 of every number between 1 and 10 has a value between 0 and 1.

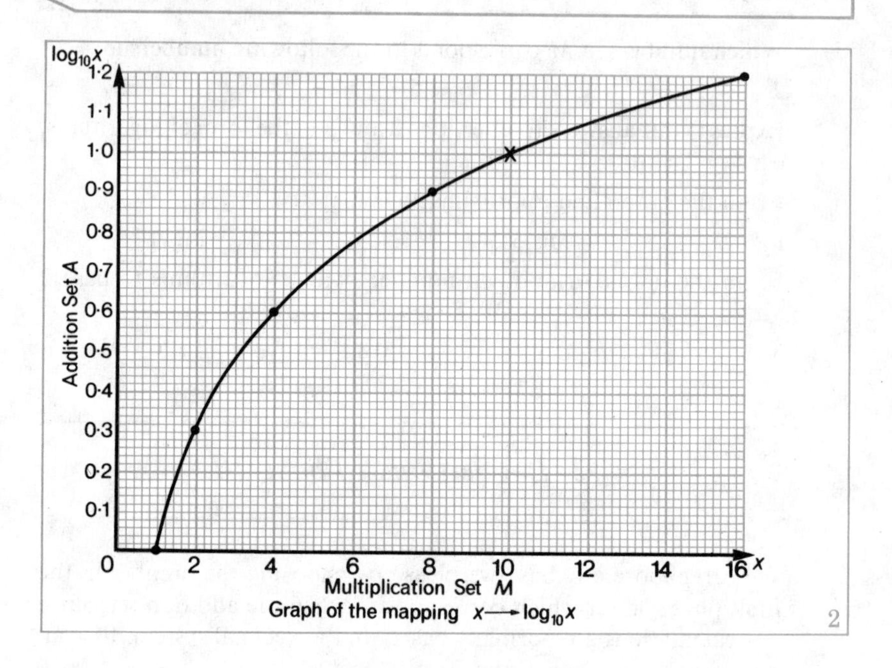

Graph of the mapping $x \longrightarrow \log_{10}x$ 2

Exercise 2

Note. In this Exercise and for the rest of the chapter we will take 10 as the base of the logarithms, i.e. log a means $\log_{10}a$.

1 Use the graph in Figure 2 to write down approximations, to 1 decimal place, for:

a log 2 *b* log 6·4 *c* log 8 *d* log 12·4

2 Use the graph to write down the numbers whose logarithms are:

a 0·2 *b* 0·4 *c* 0·6 *d* 0·8

Logarithms can be calculated to any required degree of accuracy, and the decimal parts are given in books of tables.

Example. Find log 3·14 from tables of logarithms.
From tables, the decimal part of log 3·14 is ·497.
Since the logarithm of every number between 1 and 10 has a value between 0 and 1, log 3·14 = 0·497.

3 Use *tables of logarithms* to write down the values of:

a log 3·1 *b* log 5·5 *c* log 8 *d* log 8·2

e log 8·23   *f* log 6·6   *g* log 7·49   *h* log 2·08

To find the number corresponding to a given logarithm we use tables of *antilogarithms.*

Example. Find the number whose logarithm is 0·789.
From antilogarithm tables, the number contains the digits 615.
Since the number must lie between 1 and 10, antilog 0·789 = 6·15.

4 Use *tables of antilogarithms* to find the numbers corresponding to these logarithms:

a 0·25   *b* 0·38   *c* 0·409   *d* 0·782

e 0·011   *f* 0·40   *g* 0·45   *h* 0·456

5 Copy and complete:

a

numbers	logarithms
7·65	
7·6	
7	
1·02	
3·33	

b

numbers	logarithms
	0·4
	0·46
	0·468
	0·901
	0·005

3 Using logarithms to multiply and divide numbers

From the mappings and definitions of Sections 1 and 2 we have:

Multiplication set M		*Addition set A*
1	$\longleftrightarrow$	0, or log 1
10	$\longleftrightarrow$	1, or log 10
100	$\longleftrightarrow$	2, or log 100
1000	$\longleftrightarrow$	3, or log 1000
...		...
a	$\longleftrightarrow$	log a
b	$\longleftrightarrow$	log b

From the correspondence between the elements of M under multiplication and the elements of A under addition,

$10 \times 10 \leftrightarrow 1+1$

i.e. $100 \leftrightarrow 2$, which agrees with the corresponding entry.

Also $100 \times 10 \leftrightarrow 2+1$

i.e. $1000 \leftrightarrow 3$, which also agrees with the corresponding entry.

In each case the *product* (or quotient) of two elements of M corresponds to the *sum* (or difference) of the corresponding elements of A.

$$\begin{aligned} \text{Again,}\quad 10\times 10 &\leftrightarrow \log 10+\log 10 \\ 100\times 10 &\leftrightarrow \log 100+\log 10 \\ a\times b &\leftrightarrow \log a+\log b \\ a\div b &\leftrightarrow \log a-\log b \end{aligned}$$

Multiplication of numbers corresponds to addition of their logarithms; division of numbers corresponds to subtraction of their logarithms.

$$a\times b \leftrightarrow \log a+\log b \qquad a\div b \leftrightarrow \log a-\log b$$

So we use logarithms to reduce the labour involved in multiplying and dividing numbers, by adding and subtracting their logarithms instead.

Example 1. Calculate $1{\cdot}23\times 3{\cdot}48$

$1{\cdot}23\times 3{\cdot}48 \leftrightarrow \log 1{\cdot}23+\log 3{\cdot}48$
i.e. $0{\cdot}090+0{\cdot}542$
i.e. $0{\cdot}632$

From tables, $0{\cdot}632 \leftrightarrow 4{\cdot}29$, so
$1{\cdot}23\times 3{\cdot}48 = 4{\cdot}29$, to 3 significant figures.

nos		logs
1·23	→	0·090
3·48	→	0·542
4·29	←	0·632

Note. It is convenient to put the working in a table, as shown.

Example 2. Calculate $8{\cdot}93\div 3{\cdot}29$

$8{\cdot}93\div 3{\cdot}29 \leftrightarrow \log 8{\cdot}93-\log 3{\cdot}29$
From the table, $8{\cdot}93\div 3{\cdot}29 = 2{\cdot}72$

nos	logs
8·93	0·951
3·29	0·517
2·72	0·434

Exercise 3

Use logarithms to calculate the products and quotients in questions ***1–11***.

1 $1{\cdot}48\times 4{\cdot}15$ *2* $1{\cdot}98\times 3{\cdot}6$ *3* $2{\cdot}7\times 3{\cdot}25$

4 $3{\cdot}72\div 1{\cdot}8$ *5* $9{\cdot}76\div 6{\cdot}74$ *6* $8{\cdot}73\div 8{\cdot}73$

7 $\dfrac{5{\cdot}6}{3{\cdot}9}$ *8* $\dfrac{2{\cdot}7\times 8{\cdot}8}{5{\cdot}3}$ *9* $\dfrac{6{\cdot}97\times 8{\cdot}37}{9{\cdot}12}$

10 $2{\cdot}19\times 1{\cdot}87\times 1{\cdot}34$ *11* $3{\cdot}14\times 1{\cdot}09\times 1{\cdot}09\times 2{\cdot}58$

12 Calculate the area of a rectangle of length 4·3 mm and breadth 1·9 mm.

13 Calculate the volume of a cuboid with length, breadth and height 2·15 cm, 1·39 cm and 2·06 cm respectively.

14 Calculate the area of a circle with radius 1·76 m.

4 *Logarithms of numbers greater than 10*

Every number can be written in *standard form*, $a \times 10^n$, where $1 \leqslant a < 10$ and n is an integer.

For example, $4850 = 4{\cdot}85 \times 10^3$

Then $4850 = 4{\cdot}85 \times 10^3 \leftrightarrow \log 4{\cdot}85 + \log 1000$

i.e. $0{\cdot}686 + 3$

So $4850 \leftrightarrow 3{\cdot}686$

and $\log 4850 = 3{\cdot}686$

For a number expressed in standard form, $a \times 10^n$, the integral part (characteristic) of its logarithm is n, and the decimal part (mantissa) of its logarithm is $\log a$.

$4850 = 4{\cdot}85 \times 10^3 \leftrightarrow \log 4{\cdot}85 + \log 10^3 = 0{\cdot}686 + 3$

log 4850 = 3 ·686

Integral part (characteristic) from 10^3, or by inspection

Decimal part (mantissa) = log 4·85, from tables

3

Example 1. Find: ***a*** log 31·6 ***b*** log 4 560 000

a $\log 31{\cdot}6 = \log (3{\cdot}16 \times 10^1) = 1{\cdot}500$.

b $\log 4\,560\,000 = \log (4{\cdot}56 \times 10^6) = 6{\cdot}659$.

Note. A quick way to find the index of the power of 10, and hence the integral part of the logarithm, is to find how many places the decimal point in the number is from its 'standard position' (after the first digit).

For example, in $3\overline{1}\cdot6$, 1 place, so log 31·6 = 1·...

and in $4\overline{560\,000}$, 6 places, so log 4 560 000 = 6·...

Example 2. Find the number whose logarithm is: ***a*** 2·871 ***b*** 5·432.

a From tables, antilog 0·871 = 7·43.
So the number is $7{\cdot}43 \times 10^2 = 743$.
b The number is $2{\cdot}70 \times 10^5 = 270\,000$.

Example 3. Calculate $\dfrac{48\,700}{3{\cdot}14 \times 567}$

$$\frac{48\,700}{3{\cdot}14 \times 567}$$
$= 2{\cdot}74 \times 10^1$
= 27·4, to 3 significant figures.

nos	logs
48 700	4·688
	3·251 ←
27·4	1·437
3·14	0·497
567	2·754
	3·251

Exercise 4

1 By expressing the following in standard form ($a \times 10^n$), or otherwise, find the characteristics of the logarithms of:

a 973 *b* 48·6 *c* 5000 *d* 4·96 *e* 5 270 000
f 15 *g* 48 000 *h* 600 *i* 2 *j* 93 000 000

2 Copy and complete the following tables:

a

numbers	logarithms
247	
3600	
1·5	
42·1	
1000	

b

numbers	logarithms
	1·370
	0·8
	2·06
	4·123
	5·000

Use logarithms to calculate the products and quotients in questions ***3–12***.

3 $4{\cdot}96 \times 18{\cdot}3$

4 $4{\cdot}23 \times 9{\cdot}14 \times 21{\cdot}6$

5 $8{\cdot}67 \times 7680$

6 $496 \div 3{\cdot}14$

7 $78{\cdot}2 \div 72{\cdot}8$

8 $1000 \div 8{\cdot}79$

9 $\dfrac{289 \times 76{\cdot}1}{82{\cdot}4}$

10 $\dfrac{100}{3{\cdot}14 \times 17{\cdot}8}$

11 $\dfrac{39{\cdot}7 \times 98{\cdot}3}{10{\cdot}8 \times 7{\cdot}62}$

12 $9{\cdot}84 \times \dfrac{778}{760} \times \dfrac{373}{391}$

13 What percentage is £2·34 of £19·60?

14 Calculate the area of a rectangle 357 m long and 169 m broad.

15 Calculate the breadth of a rectangle of length 58 m and area 2176 m^2.

16 Find the circumference of a circle of radius 6·84 cm.

17 Evaluate the following, expressing the answers in standard form:

a $478\,000 \times 195$ *b* $(7{\cdot}3 \times 10^8) \div 23\,500$

5 *Using logarithms to calculate powers and roots of numbers*

(i) Powers

$3{\cdot}75^4 = 3{\cdot}75 \times 3{\cdot}75 \times 3{\cdot}75 \times 3{\cdot}75$
$\leftrightarrow \log 3{\cdot}75 + \log 3{\cdot}75 + \log 3{\cdot}75 + \log 3{\cdot}75$
i.e. $4 \log 3{\cdot}75$

Raising a number to a power corresponds to multiplication of the logarithm of the number by the index of the power.

Example. Calculate $3{\cdot}75^4$.
$3{\cdot}75^4$
$= 1{\cdot}98 \times 10^2$
$= 198$ to 3 significant figures.

nos	logs
3·75	0·574
	4
198·	2·296

(ii) Roots; the inverse process

$7^2 = 49$, and $\sqrt{49} = 7$. The inverse process to raising a number to a power is taking the corresponding root of the number.

To find the root of a number, divide the logarithm of the number by the order of the root.

Example. Calculate $\sqrt[3]{485}$.
$\sqrt[3]{485} \leftrightarrow (\log 485) \div 3$
$\sqrt[3]{485} = 7{\cdot}85$ to 3 significant figures.

nos	logs
$\sqrt[3]{485}$	3⌊2·686
7·85	0·895

Note that square roots are most easily found from square root tables.

Exercise 5

Use logarithms to calculate the powers and roots in questions *1–8*.

1 $6{\cdot}78^2$ *2* $2{\cdot}97^3$ *3* 34^5 *4* $1{\cdot}03^{10}$

5 $\sqrt{5{\cdot}43}$ *6* $\sqrt[3]{55}$ *7* $\sqrt[4]{8910}$ *8* $\dfrac{16}{\sqrt[5]{10}}$

9 Use logarithmic tables to calculate the square roots of:

a 7·91 *b* 20·8 *c* 347

Compare your answers with those obtained by using square root tables.

10 Calculate the area of:

a a circle of radius 8·95 m *b* a square of side 9·65 cm.

11 Calculate the length of the side of a square of area 1000 m^2.

12 Calculate the radius of a circle of area 30 cm^2.

Evaluate:

13 $\dfrac{4{\cdot}8 \times \sqrt{592}}{9{\cdot}76}$ *14* $\sqrt{\left(\dfrac{4{\cdot}12 \times 71{\cdot}9}{1{\cdot}45 \times 82{\cdot}5}\right)}$ *15* $\sqrt[3]{\left(\dfrac{3 \times 41}{4 \times 3{\cdot}14}\right)}$

6 *Logarithms of numbers less than 1*

Rewriting and extending the sequences of numbers in the multiplication and addition columns, we have:

Multiplication		*Addition*
...		...
1000	⟷	3, or log 1000
100	⟷	2, or log 100
10	⟷	1, or log 10
1	⟷	0, or log 1
0·1	⟷	– 1, or log 0·1
0·01	⟷	– 2, or log 0·01
0·001	⟷	– 3, or log 0·001
...		...

Remembering that $10^{-2} = \dfrac{1}{10^2} = 0{\cdot}01$, we have

$$\log 0{\cdot}0423 = \log 4{\cdot}23 \times 10^{-2} \leftrightarrow \log 4{\cdot}23 + \log 0{\cdot}01$$

i.e. $0{\cdot}626 + (-2)$

i.e. $\bar{2}{\cdot}626$

Logarithms of numbers less than 1

The notation $\bar{2}{\cdot}626$ ('bar 2·626') is used in order to keep the negative and positive parts of the logarithm separate. This allows us to use the same tables for the decimal parts of logarithms less than 1 as well as for those greater than 1.

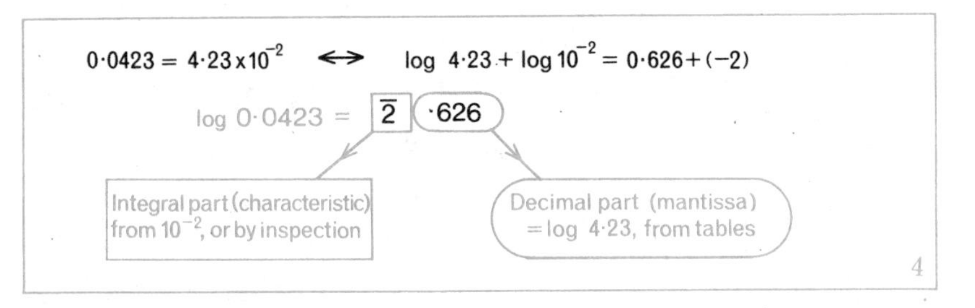

In every logarithm the *integral part* (*characteristic*) is positive, zero, or negative depending on whether the number is greater than 10, between 1 and 10, or less than 1, respectively; the *decimal part* (*mantissa*) is always positive or zero.

Example 1. $\log 0{\cdot}000\,56 = \log (5{\cdot}6 \times 10^{-4}) = \bar{4}{\cdot}748$.

Note that in 0·000 56 the decimal point is 4 places *to the left* of its 'standard position': 0·000 5̓6

Example 2. Find the number whose logarithm is $\bar{3}{\cdot}521$.

From tables, antilog $\bar{3}{\cdot}521 = 3{\cdot}32 \times 10^{-3} = 0{\cdot}003\,32$.

Exercise 6

1 By expressing the following in standard form ($a \times 10^n$), or otherwise, find the characteristics of the logarithms of:

a 0·091 *b* 0·008 *c* 0·673 *d* 0·0009 *e* 0·101

2 Copy and complete the following tables:

a

numbers	logarithms
0·234	
0·017	
0·008	
36·4	
5·86	

b

numbers	logarithms
	$\bar{1}{\cdot}345$
	$\bar{3}{\cdot}707$
	$\bar{2}{\cdot}015$
	2·015
	1·672

3 Given that $\log 3{\cdot}16 = 0{\cdot}500$, write down the logarithms of:

a 0·316 *b* 316 *c* 316 000 *d* 0·003 16

4 Write down the logarithms of:

a 234 *b* 0·234 *c* 0·057 *d* 0·101 *e* 0·0007

(i) Multiplication and division

Example 1. Add:

a
2·7
$\bar{3}$·8
0·5
[7+8 = 15, so carry 1; 1+(−3)+2 = 0]

b
$\bar{2}$·7
$\bar{3}$·8
$\bar{4}$·5
[1+(−3)+(−2) = −4]

Example 2. Subtract:

a
2·9
5·4
$\bar{3}$·5
[2−5 = −3]

b
3·4
3·8
$\bar{1}$·6
[3−(1+3) = −1]

c
$\bar{2}$·3
$\bar{4}$·2
2·1
[−2−(−4) = 2]

d
$\bar{2}$·3
$\bar{4}$·7
1·6
[−2−(1+(−4)) = −2−(−3) = 1]

Note. To subtract one negative characteristic from another it may be found helpful to *add the inverse*. For example, in worked example *2d* above, −2−(−3) = −2+3 = 1.

Exercise 7

Add the logarithms in questions *1–5*:

1	*2*	*3*	*4*	*5*
3·7	1·8	$\bar{2}$·8	1·9	0·6
$\bar{2}$·9	3·4	$\bar{3}$·4	$\bar{3}$·4	$\bar{4}$·7

Subtract the logarithms in questions *6–20*:

6	*7*	*8*	*9*	*10*
1·8	1·8	1·6	$\bar{1}$·6	0·3
3·6	$\bar{3}$·6	3·8	3·8	1·4

11	*12*	*13*	*14*	*15*
$\bar{1}$·6	1·6	4·2	$\bar{4}$·2	0·0
$\bar{3}$·8	$\bar{3}$·8	$\bar{3}$·7	$\bar{3}$·7	$\bar{2}$·4

16	*17*	*18*	*19*	*20*
4·2	$\bar{4}$·2	0·7	0·7	0·0
3·7	3·7	2·9	$\bar{2}$·9	2·4

Use logarithms to calculate the following products and quotients:

21 8·62 × 0·531 *22* 0·765 × 0·921 *23* 58·7 × 0·349

24 $0{\cdot}936 \times 0{\cdot}0682$ *25* $49{\cdot}7 \div 0{\cdot}218$ *26* $289 \div 0{\cdot}684$

27 $0{\cdot}791 \div 4{\cdot}86$ *28* $0{\cdot}345 \div 8{\cdot}72$ *29* $0{\cdot}725 \div 0{\cdot}362$

30 $0{\cdot}184 \div 0{\cdot}728$ *31* $42{\cdot}9 \div 0{\cdot}009\,31$ *32* $0{\cdot}0281 \div 583$

33 $0{\cdot}001\,85 \div 0{\cdot}0124$ *34* $0{\cdot}0275 \div 0{\cdot}000\,836$

35 $37{\cdot}6 \times 0{\cdot}281 \times 0{\cdot}572$ *36* $0{\cdot}819 \times 0{\cdot}0723 \times 647$

37 $\dfrac{4{\cdot}97 \times 0{\cdot}001\,09}{0{\cdot}909}$ *38* $\dfrac{0{\cdot}125 \times 0{\cdot}832}{0{\cdot}723}$

39 $\dfrac{91{\cdot}8}{3{\cdot}12 \times 0{\cdot}0752}$ *40* $\dfrac{0{\cdot}008\,93}{0{\cdot}614 \times 0{\cdot}0106}$

(ii) Powers and roots

Example 1. Multiply $\bar{1}{\cdot}7$ by 5.

$\bar{1}{\cdot}7$ [$5 \times 7 = 35$, so carry 3]
$\underline{5}$
$\underline{\bar{2}{\cdot}5}$

Example 2. Divide $\bar{6}{\cdot}4$ by 3.

$3\,|\,\bar{6}{\cdot}400$
$\underline{\bar{2}{\cdot}133}$

Example 3. Divide $\bar{1}{\cdot}5$ by 2.

Here we cannot divide straight away, as we would obtain a negative decimal part. We replace -1 by $-2+1$ in order that 2 will divide exactly into the negative number.

Then $\bar{1}{\cdot}5 \div 2$
$= (\bar{2}+1{\cdot}5) \div 2$
$= \bar{1}+0{\cdot}75$
$= \bar{1}{\cdot}75$

or $2\,|\,\bar{2}+1{\cdot}5$
$\underline{\bar{1}+0{\cdot}75}$

or $2\,|\,\bar{1}{\cdot}5$
$\underline{\bar{1}{\cdot}75}$

Example 4. Divide $\bar{4}{\cdot}5$ by 3.

$-4 = -6+2$

$3\,|\,\bar{6}+2{\cdot}5$
$\underline{\bar{2}+0{\cdot}833}$

or $3\,|\,\bar{4}{\cdot}5$
$\underline{\bar{2}{\cdot}833}$

Exercise 8

Multiply the logarithms by the numbers in questions ***1–5***:

1 $\bar{1}{\cdot}3$ by 3 *2* $\bar{2}{\cdot}7$ by 3 *3* $\bar{3}{\cdot}9$ by 2 *4* $\bar{1}{\cdot}7$ by 4 *5* $\bar{1}{\cdot}1$ by 10

Divide in questions ***6–17***, to 2 decimal places where appropriate:

6 $\bar{5}{\cdot}7$ by 5 *7* $\bar{2}{\cdot}8$ by 2 *8* $\bar{1}{\cdot}8$ by 2 *9* $\bar{1}{\cdot}3$ by 2

10 $\bar{4}{\cdot}6$ by 2 *11* $\bar{4}{\cdot}6$ by 4 *12* $\bar{3}{\cdot}6$ by 4 *13* $\bar{2}{\cdot}6$ by 4

14 $\bar{1}{\cdot}6$ by 4 *15* $\bar{2}{\cdot}8$ by 3 *16* $\bar{1}{\cdot}7$ by 3 *17* $\bar{3}{\cdot}8$ by 2

Use logarithms to calculate the following:

18 $0{\cdot}87^2$ *19* $0{\cdot}95^5$ *20* $0{\cdot}0736^3$

21 the area of a circle of radius 0·58 m

22 $\sqrt{0{\cdot}836}$ *23* $\sqrt{0{\cdot}0836}$ *24* $\sqrt{0{\cdot}469}$

25 $\sqrt[3]{0{\cdot}18}$ *26* $\sqrt[3]{0{\cdot}0872}$ *27* $\sqrt[3]{0{\cdot}009\,18}$

28 the radius of a circle of area 1·79 m^2

29 $1{\cdot}33 \times 3{\cdot}14 \times 0{\cdot}86^3$ *30* $\sqrt{\left(\frac{84{\cdot}9}{257}\right)}$ *31* $\sqrt[3]{\left(\frac{3 \times 125}{4 \times 3{\cdot}14}\right)}$

Exercise 8B

Use logarithms to calculate the products, quotients, powers and roots in questions ***1–18***.

1 $2{\cdot}97 \times 0{\cdot}563 \times 0{\cdot}83$ *2* $0{\cdot}872 \times 3150 \times 0{\cdot}0345$ *3* $48{\cdot}9 \div 57{\cdot}6$

4 $0{\cdot}56 \div 0{\cdot}93$ *5* $0{\cdot}56 \times 0{\cdot}93$ *6* $7{\cdot}23 \div 0{\cdot}087$

7 $\frac{5{\cdot}4 \times 0{\cdot}0064}{0{\cdot}086}$ *8* $\frac{0{\cdot}48}{0{\cdot}73 \times 0{\cdot}92}$ *9* $0{\cdot}0153^3$

10 $\sqrt{0{\cdot}876}$ *11* $\sqrt[3]{0{\cdot}503}$ *12* $0{\cdot}145^2$

13 $(1{\cdot}85 \times 0{\cdot}027)^2$ *14* $\sqrt{(0{\cdot}632 \times 0{\cdot}819)}$ *15* $\sqrt[3]{(38{\cdot}1 \times 0{\cdot}0072)}$

16 $\left(\frac{52{\cdot}9}{69{\cdot}3}\right)^3$ *17* $\sqrt{\left(\frac{0{\cdot}0106}{2{\cdot}45}\right)}$ *18* $\sqrt[3]{\left(\frac{4 \times 0{\cdot}0278}{3 \times 3{\cdot}14}\right)}$

19 Calculate the area of a square of side 0·85 cm.

20 The radius of a circle is 296 metres. Calculate:

a its perimeter *b* its area.

21 The volume of a cylinder is given by the formula $V = \pi r^2 h$.

a Calculate V when $r = 12{\cdot}5$ and $h = 19{\cdot}6$.

b Calculate h when $V = 89$ and $r = 7{\cdot}4$.

c Calculate r when $V = 23$ and $h = 15{\cdot}6$.

22 The area of a square is 0·92 m^2. Calculate the length of its side to three significant figures.

23 The time of swing T of a pendulum of length l is given by the formula

$T = 2\pi\sqrt{\frac{l}{g}}$. Calculate T when $l = 7$ and $g = 32$.

24 The formula for calculating simple interest is $I = \frac{PRT}{100}$. Calculate I when $P = 76{\cdot}9$, $R = 5{\cdot}75$, and $T = 31$.

25 A formula for finding velocities is $v^2 = u^2 + 2fs$. Calculate v when $u = 19{\cdot}7$, $f = 5{\cdot}5$, and $s = 247$.

26 A formula for finding the length of the longest side in a right-angled triangle is $a^2 = b^2 + c^2$. Calculate a when:

a $b = 7{\cdot}9$ and $c = 9{\cdot}7$ b $b = 0{\cdot}85$ and $c = 1{\cdot}37$.

7 Using a calculating machine

If you have a calculating machine you will be able to add, subtract, multiply, and divide numbers quite quickly, and obtain a greater number of significant figures, where appropriate, than is possible with tables. You can also find the square roots of numbers by the iterative method, already explained in Book 4.

To give you further practice in the use of such a machine, here are some examples from Book 2 (page 175). To help you to check the answers quickly, these are given after question ***20***.

Exercise 9

Add the following:

1 56·8, 37·9, 88·6 2 9·37, 8·97, 4·86, 7·65

3 987·4, 869·3, 578·8 4 9·876, 4·777, 6·895, 0·909, 3·456

Subtract:

5 68 359 from 77 777 6 79·89 from 83·09

7 0·2936 from 0·3091 8 5·765 from 6·014

Evaluate:

9 $84{\cdot}63 + 97{\cdot}28 - 79{\cdot}95$ 10 $125{\cdot}9 + 764{\cdot}8 - 841{\cdot}9$

Multiply:

11 636 by 35 12 6·36 by 3·5 13 94·87 by 9

14 68·57 by 8·5 15 23·56 by 85·2 16 9·999 by 9·999

Divide (giving answers to 1 decimal place):

17 57·60 by 5·90 *18* 234·90 by 88·70

19 0·948 by 0·497 *20* 98·76 by 532

Answers

1	183·3	*2*	30·85	*3*	2435·5	*4*	25·913	*5*	9418
6	3·2	*7*	0·0155	*8*	0·249	*9*	101·96	*10*	48·8
11	22 260	*12*	22·26	*13*	853·83	*14*	582·845	*15*	2007
16	99·98	*17*	9·8	*18*	2·6	*19*	1·9	*20*	0·2

The following examples in this chapter on Logarithms might be worked out by machine:

Exercise 3, questions ***1–14***; Exercise 4, questions ***3–17***;
Exercise 5, questions ***1–4***, and ***10***; Exercise 7, questions ***21–40***;
Exercise 8, questions ***18–21***; Exercise 8B, questions ***1–9***.

Summary

1 *The logarithm of a number* in a multiplication set M is the corresponding number in the addition set A:

Base 2			Base 10		
M	A		M	A	
$1 \longleftrightarrow$	0,	$\log_2 1 = 0$	...	...	...
$2 \longleftrightarrow$	1,	$\log_2 2 = 1$	$100 \longleftrightarrow$	2,	$\log_{10} 100 = 2$
$4 \longleftrightarrow$	2,	$\log_2 4 = 2$	$10 \longleftrightarrow$	1,	$\log_{10} 10 = 1$
$8 \longleftrightarrow$	3,	$\log_2 8 = 3$	$1 \longleftrightarrow$	0,	$\log_{10} 1 = 0$
$16 \longleftrightarrow$	4,	$\log_2 16 = 4$	$0{\cdot}1 \longleftrightarrow$	-1,	$\log_{10} 0{\cdot}1 = -1$
...	...	...	$0{\cdot}01 \longleftrightarrow$	-2,	$\log_{10} 0{\cdot}01 = -2$
			...	...	...

2 *Tables* give logarithms of numbers between 1 and 10, and their antilogarithms, to base 10.

3 Every logarithm consists of two parts, the *integral part* (*characteristic*), and the *decimal part* (mantissa).

Example 1.
$4850 = 4{\cdot}85 \times 10^3 \leftrightarrow \log 4{\cdot}85 + \log 1000 = 0{\cdot}686 + 3 = 3{\cdot}686$

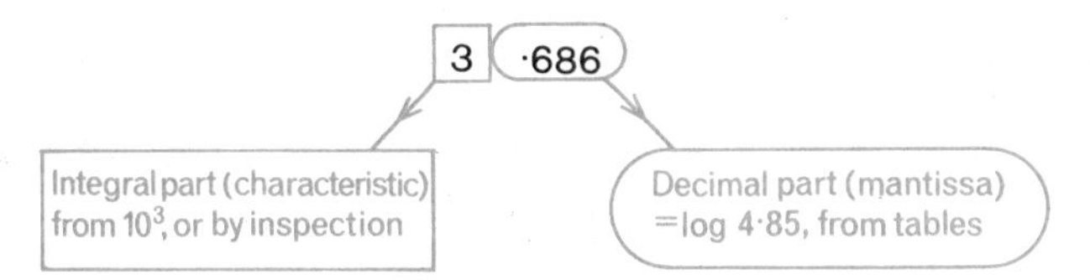

Example 2. $0{\cdot}0423 = 4{\cdot}23 \times 10^{-2}$
so $\log 0{\cdot}0423 = \bar{2}{\cdot}626$.

4 *Multiplication* of numbers corresponds to *addition* of their logarithms; *division* of numbers corresponds to *subtraction* of their logarithms.

$a \times b \leftrightarrow \log a + \log b; \quad a \div b \leftrightarrow \log a - \log b$

5 *Raising a number to a power* corresponds to *multiplication* of the logarithm of the number by the index of the power.

Example. $3{\cdot}75^4 \leftrightarrow 4 \log 3{\cdot}75$

6 *Finding the root of a number* corresponds to *division* of the logarithm of the number by the order of the root.

Example. $\sqrt[3]{485} \leftrightarrow (\log 485) \div 3$

7 *Logarithms* are thus used to reduce the weight of calculation by:

(i) replacing multiplication and division by addition and subtraction

(ii) replacing calculation of powers and roots by multiplication and division.

Areas and Volumes

1 Revision

If we take l, b, h, r, C, A and V to denote length, breadth, height, radius, circumference, area and volume, as in Books 1–4, we have the following formulae:

(i) *Area of rectangle*: $A = lb$

(ii) *Area of square*: $A = l^2$

(iii) *Area of triangle*: $A = \frac{1}{2}bh$

(iv) *Area of circle*: $A = \pi r^2$

(v) *Circumference of circle*: $C = 2\pi r = \pi d$

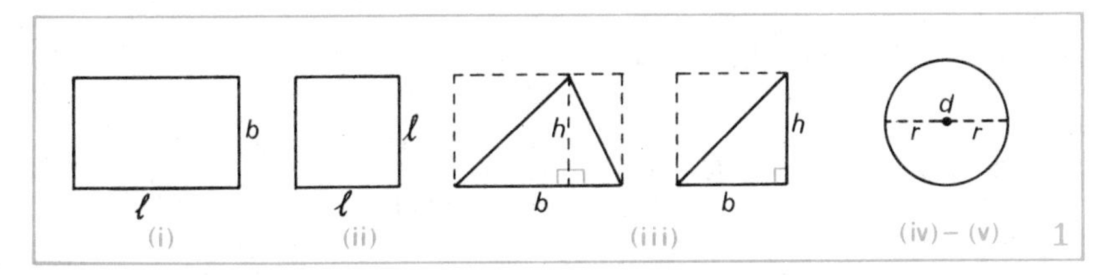

1

(vi) *Volume of cuboid*: $V = Ah = lbh$ (vii) *Volume of cube*: $V = l^3$

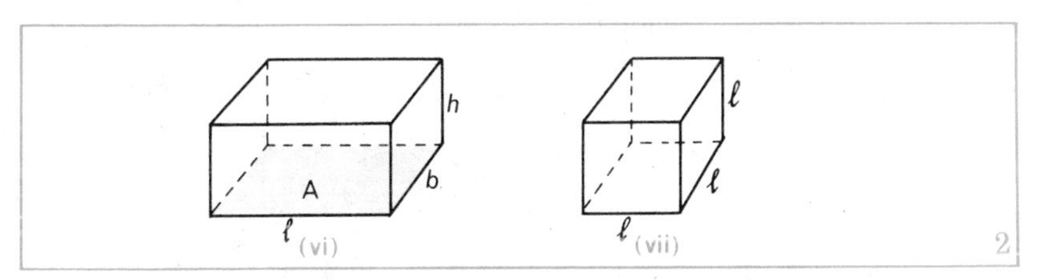

2

Reminders of certain metric measures

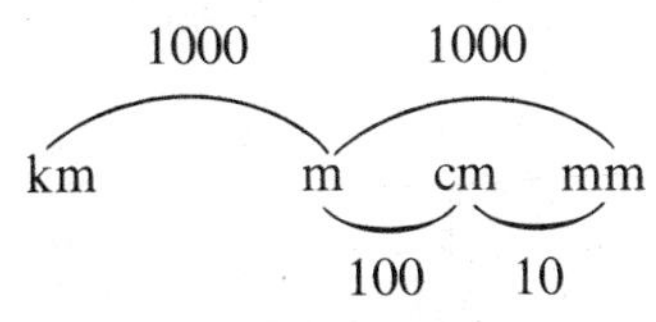

1 litre = 1000 cm^3; 1 tonne = 1000 kg; 1 hectare = 10 000 m^2

Approximations in calculations

In the calculations in this chapter you may use a slide rule, or logarithms, or a calculating machine. Where necessary, take $3\frac{1}{7}$ or 3·14 as an approximation for π, and give your answers to a degree of accuracy consistent with the data.

Exercise 1

1 Complete the following table for rectangles:

	a	*b*	*c*	*d*
l	12 cm	10 m	1·5 km	45 m
b	8 cm	1·25 m	20 m	
A			m^2	1350 m^2

2 Complete the following table for triangles:

	a	*b*	*c*	*d*
b	24 mm	15 m	15 cm	12 m
h	16 mm	28 m	9 cm	
A				18 m^2

3 Complete the following table for circles:

	a	*b*	*c*	*d*
r	14 m	3·5 cm	5 mm	
A				
C				44 m

4 Complete the following table for cuboids:

	a	*b*	*c*	*d*
l	5 m	5 cm	2 m	8 m
b	4 m	2 cm	5 cm	6 m
h	3 m	4 mm	1 mm	
V		cm^3	cm^3	72 m^3

5 Figure 3 shows the shape of a small window consisting of a square of side 28 cm and a semicircle. Find the *perimeter* of the window.

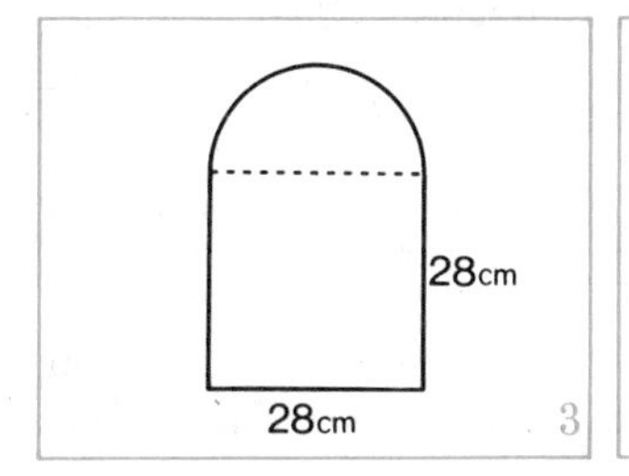

3

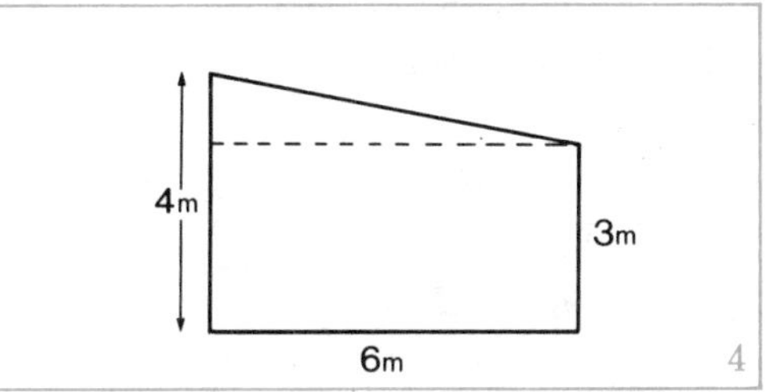

4

6 Figure 4 shows the end view of a lean-to shed. Calculate the area of this end.

7 A rectangular tank is 1·5 m long, 30 cm broad and 20 cm high. How many litres of water can it hold?

8 A lorry weighs 2·5 tonnes when empty. If it carries 45 crates, each of mass 60 kg, calculate the total mass of the lorry and its load.

9 A rectangular field is 340 m long and 250 m broad. Calculate its area in hectares.

10 A cuboid is 4 m long, 1·5 m broad and 0·75 m high. Calculate:

a the total length of its edges *b* its volume *c* its total surface area.

Exercise 1B

1 Calculate the area of each of the following:

a A rectangle with length 14 m and breadth 1·25 m.

b A square of side 3·7 cm.

c A right-angled triangle with sides 4·5 mm, 6 mm and 7·5 mm long.

d A circle of radius 6 cm.

e An isosceles triangle with base 4·8 cm and altitude 1·2 cm long.

2 Calculate the area of the shape in Figure 3.

3 A hollow metal box measures 4 cm by $3\frac{1}{2}$ cm by $2\frac{1}{2}$ cm outside, and 3 cm by $2\frac{1}{2}$ cm by 2 cm inside. How many cubic centimetres of metal are used in making it? (Think of an outer volume minus an inner volume.)

4 The area of a rectangular strip of land is 1 hectare. Find its breadth if its length is: *a* 5 km *b* 1 km *c* 250 m.

5 *a* Calculate the perimeter of a circle with diameter 11·2 cm.

b Find also the area of a square which has the same perimeter as the circle.

6 Calculate the radius of a circle with area:

a 1386 cm^2 *b* 10 m^2

7 A pond of area 2·5 hectares is frozen over to a depth of 2·5 cm. Calculate the mass in tonnes of ice, given that 1 m^3 of ice weighs 1000 kg. (Work in m, m^2 and m^3.)

8 A circular pool of diameter 7 m is constructed in a garden, and it is planned to surround it with a path 1·4 m wide. Find the cost of edging the path on both sides at 10p per metre, and of surfacing the path at 25p per m^2.

2 *The volume of a prism and a cylinder*

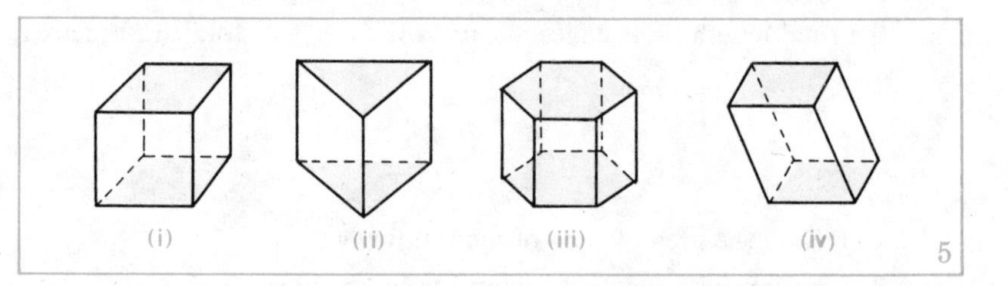

5

Figure 5 shows four *prisms*, in each of which two ends are congruent and in parallel planes. (i), (ii) and (iii) are called *right prisms* as the ends are at right angles to the edges.

If the cuboid in Figure 6 (i) is cut vertically down the diagonal plane as shown, two equal *triangular prisms* are obtained; these can be placed together to form the triangular prism in 6 (ii). Clearly the volume, area of base and height of this prism are the same as the cuboid in 6 (i), so that its volume is given by the formula $V = Ah$.

Figure 7 (i) shows a hexagonal prism, consisting of six equal triangular prisms. The volume of each of these triangular prisms is

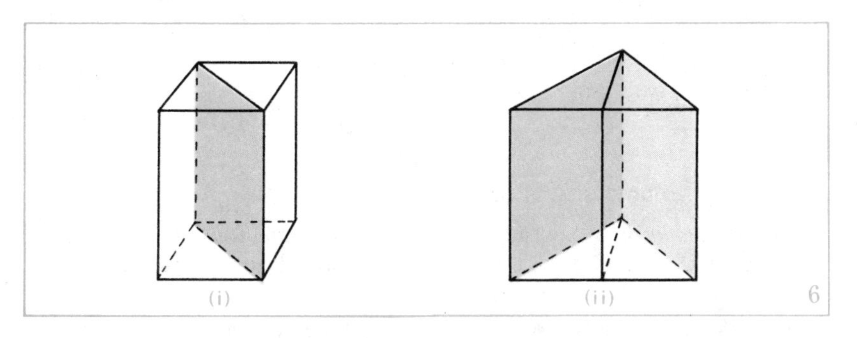

given by $V = Ah$, so the volume of the hexagonal prism is given by $6Ah$, or $(6A)h$. Since $6A$ is the area of the base of the hexagonal prism, its volume is also given by *area of base* $\times$ *height*.

In fact for every prism, $V = Ah$.

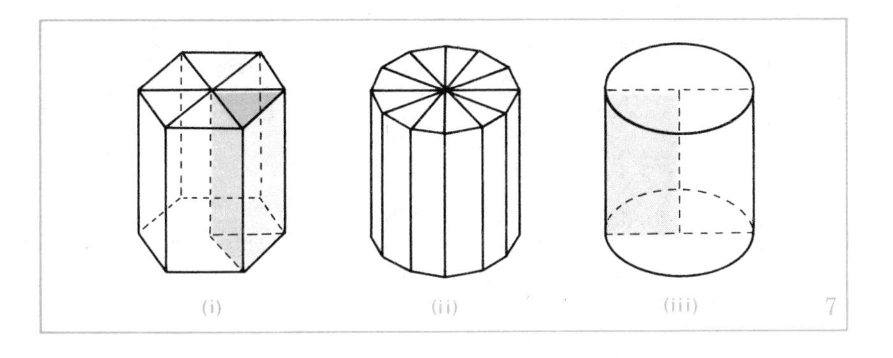

Imagine that we continue to increase the number of equal sides on the ends of such a prism, as shown in Figure 7 (ii). Ultimately we will obtain the prism shown in 7 (iii), where the ends have become indistinguishable from circles. In this case the prism is indistinguishable from a *cylinder*.

The formula for the volume, $V = Ah$, is still true. Since the area of the base is now the area of a circle, πr^2, we have:

The volume of a cylinder is given by the formula $V = \pi r^2 h$.

Note. In calculations care must be taken with all the units used.

Example 1. Calculate the volume of a prism 1·5 m high, with area of base 40 cm^2.

In cm, $h = 150$.

$$V = Ah = 40 \times 150 = 6000$$

The volume is 6000 cm^3.

Example 2. Find the mass of a wire 1 km long with a circular cross-section of diameter 3 mm, given that 1 cm^3 of the wire weighs 7·5 g.

We can think of the wire as a long thin cylinder.

The length of 1 km = 1000 m = 100 000 cm; in cm, $h = 100\,000$

The radius is 1·5 mm = 0·15 cm ; in cm, $r = 0{\cdot}15$

$$V = \pi r^2 h = 3{\cdot}14 \times (0{\cdot}15)^2 \times 100\,000$$

$$\text{Mass} = 3{\cdot}14 \times (0{\cdot}15)^2 \times 100\,000 \times 7{\cdot}5 = 53\,000 \text{ g}$$

The mass is about 53 kg.

nos	logs
3·14	0·497
0·15	$\bar{1}{\cdot}176$
0·15	$\bar{1}{\cdot}176$
100 000	5·000
7·5	0·875
53 000	4·724

Exercise 2

1 Complete the following table for prisms:

	a	*b*	*c*	*d*
Area of base	1·2 m^2	2·5 cm^2	9 cm^2	5 m^2
Height	3·5 m	11 cm	3 m	
Volume				13·5 m^3

2 Calculate the volume of a prism which is 3·6 cm high, and has a base in the form of a right-angled triangle with sides 5 cm, 12 cm and 13 cm long.

3 Sketch each of the following prisms, and then calculate its volume:

	a	*b*	*c*	*d*
Base	A square of side 6 cm	A rectangle 5 cm by 2·5 cm	An isosceles triangle, 'base' 8 cm, height 6 cm	A hexagon of area 75 m^2
Height	4 cm	8 cm	12 cm	2·4 m

4 Complete the following table for cylinders (in ***c*** and ***d*** take 3·14 for π):

	a	*b*	*c*	*d*	*e*
Radius	7 cm	14 cm	2 m	3 cm	14 m
Height	5 cm	2 cm	1 m	10 cm	
Volume					308 m^3

5 A cylindrical tin of instant coffee is 10 cm tall and has a diameter of 7 cm. What volume of coffee can it hold?

6 A cylindrical tin of drinking chocolate is 6 cm in diameter and is 7 cm tall. How many such tins can be filled from a cubical box of side 30 cm?

7 How many litres of petrol can be stored in a cylindrical tank of diameter 2 m and length 3·5 m?

8 A cylindrical tower is 200 m high, and its diameter is 20 m. Find its volume. (Use 3·14 for π and give your answer in the form $a \times 10^n$.)

9 It is required to make a cylindrical can which will hold 1 litre of liquid.

a If the diameter is to be 10 cm, what must the height of the can be?

b If the height is to be 10 cm, what must the diameter of the can be?

10 Find the height of a cylinder of volume 231 cm^3 and radius 1·75 cm.

11 Calculate the mass of a wire 200 m long, with a circular cross-section having a diameter of 5 mm. 1 cm^3 of the metal in the wire has a mass of 8·9 g.

12 Calculate the radius of a cylindrical rod of volume 484 cm^3 and length 1·4 m.

3 *The area of a cylinder*

Introductory Exercise

Examine a cylindrical can which has a paper wrapping, or a cylindrical packet of sweets. You will see that it has three surfaces, two of which are flat (or plane), and one curved. Use a knife to slit the wrapping down the cylinder; remove the wrapping, and lay it out flat. What shape is it now?

Take another cylinder, and cut a piece of paper to fit exactly round the curved surface. What shape is the paper when flattened out? What are the connections between the length and breadth of this sheet of paper and the dimensions of the cylinder?

What shape of paper would you need to cover each end of the cylinder? Can you now draw a net for making a cylinder from thin card?

Figure 8 (ii) shows a net for making the cylinder in 8 (i).

It can be seen that the area of the curved surface of the cylinder is given by the formula $A = 2\pi rh$.

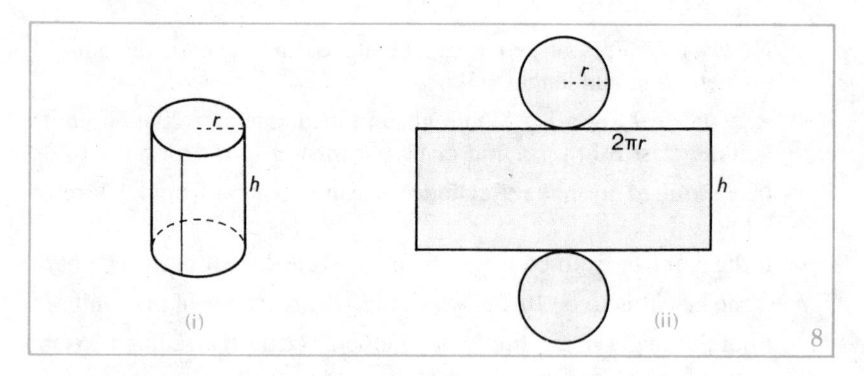

Exercise 3

1 Suppose you have to make a cylinder from thin card.

a If the radius of one end is 3·5 cm, what is the circumference of that end?

b If the height of the cylinder is 10 cm, what are the dimensions of the rectangle in the net?

c Sketch the net and mark in its dimensions, or draw it accurately, cut it out, and make the cylinder. (Tabs on each part will help you to join the three parts together).

2 a Calculate the *total* area of the cylinder in question ***1***.

b What would the area be if the cylinder had no 'lid'?

3 Repeat questions ***1*** and ***2*** for a cylinder of height 5 cm and radius of base 5 cm.

4 A cylinder has height h, and radius of base r. Write down formulae for:

a the area of one end

b the curved surface area

c the (outside) area of the cylinder if it is open at one end

d the area if both ends are closed.

5 Calculate the *curved surface area* of the following cylinders (in ***b*** and ***d*** take 3·14 for π):

	a	*b*	*c*	*d*
Radius of base	7 m	4 cm	1 cm	20 cm
Height	10 m	5 cm	14 mm	1 m

6 What area of cardboard is needed to make a closed cylinder of height 20 cm and radius of base 14 cm?

7 A cylindrical water tank is open at the top. If its height is 1·2 m and its diameter is 40 cm, calculate its total (external) area in square metres.

8 A closed cylindrical tank can hold 7700 litres of petrol. The radius of its base is 70 cm. Calculate the total area of the tank.

4 Pyramids and cones

(i) The volume of a pyramid

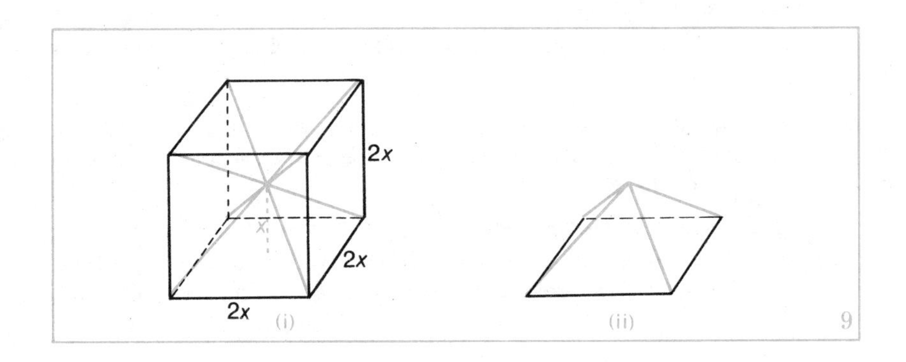

Figure 9 (i) shows a cube of side $2x$, with its space diagonals drawn.

You should be able to see six equal pyramids inside the cube, each with a square base on a side of the cube; one is illustrated in 9 (ii).

If the volume of each pyramid is V, the total volume of the six pyramids is the same as the volume of the cube.

$$\begin{aligned} \text{So } 6V &= (2x)^3 \\ \text{i.e. } V &= \tfrac{1}{6}(2x)^2.2x \\ &= \tfrac{1}{3}(2x)^2.x \\ &= \tfrac{1}{3} \text{ area of base} \times \text{height} \end{aligned}$$

It can be shown that for every pyramid,

$$\text{volume} = \tfrac{1}{3} \text{ area of base} \times \text{height}$$

Exercise 4B

1 Notice that in a diagram of a pyramid, a square or rectangular base appears as a parallelogram; the altitude of the pyramid can be drawn through the point of intersection of the diagonals of the parallelogram. Sketch pyramids whose bases are:

a square *b* rectangular *c* triangular. (A triangular pyramid is called a *tetrahedron*.)

2 Calculate the volumes of the following pyramids:

	a	*b*	*c*
Base	A square of side 5 m	A square of side 6 cm	A rectangle 5·5 cm by 3·3 cm
Height	6 m	8 cm	10 cm

3 The roof of a bungalow 25 m long and 15 m broad is in the form of a pyramid 7 m high. How many cubic metres of air are there in this roof space?

4 A pyramid has a right-angled triangular base, and a volume of 135 cm^3. If the shorter sides of the base are 4 cm and 9 cm long, find the height of the pyramid.

5 Figure 10 (i) shows the net of a pyramid, consisting of a square of side 10 cm and four congruent isosceles triangles, each of altitude 13 cm. When the pyramid is constructed, as shown in 10 (ii), calculate:

a its total area *b* its height *c* its volume.

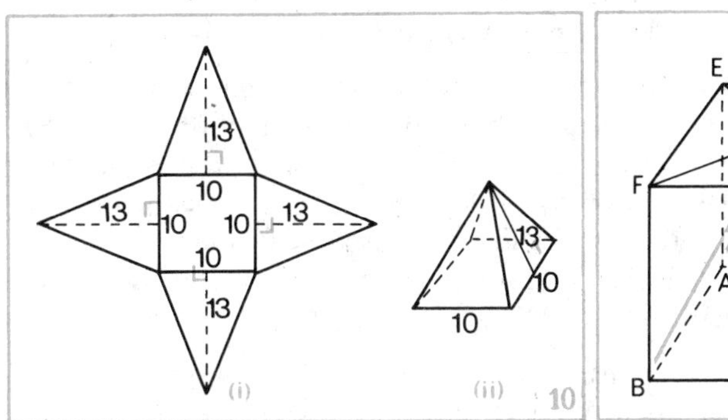

10

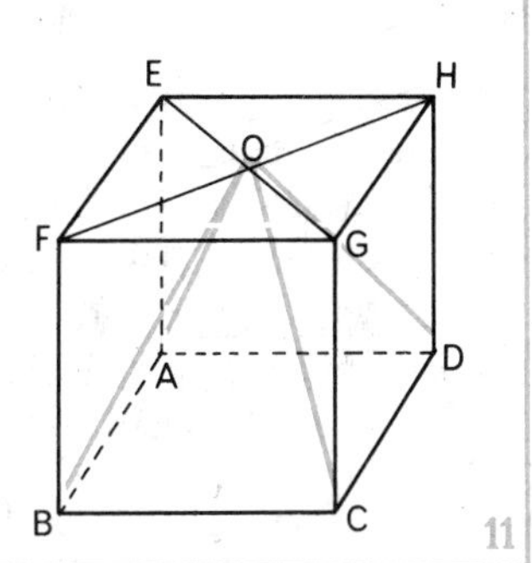

11

6 In Figure 11, ABCDEFGH is a cube of edge 4 m. O is the centre of the face EFGH.

a What is the volume of the pyramid OABCD?

b What is the volume of the pyramid OGCDH? Name three more pyramids with the same volume.

7 Find out about the dimensions and construction of the famous pyramids in Egypt.

(ii) The volume of a cone

A cone, as illustrated in Figure 12 (i), can be thought of as a pyramid with a circular base.

So the volume of a cone $= \frac{1}{3}$ area of base × height

i.e. $V = \frac{1}{3}\pi r^2 h$.

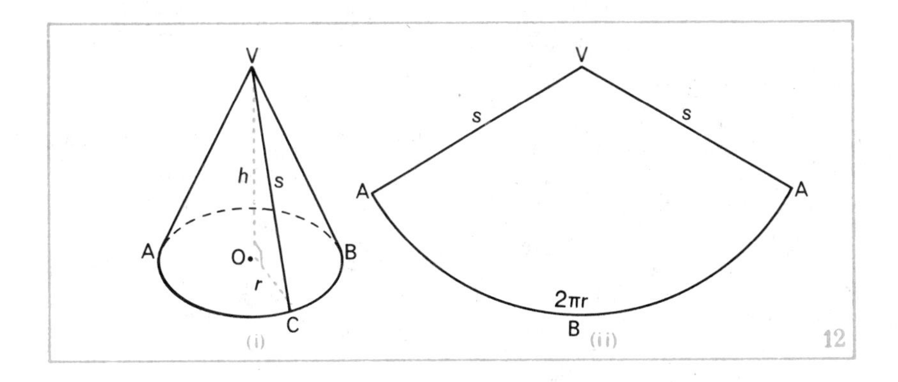

12

(iii) The curved surface area of a cone

A right circular cone can be formed by rotating a right-angled triangle like VOC in Figure 12 (i) about one of its shorter sides, VO. Thus every point on the circumference of the base is the same distance s from the apex V; it follows that if a cut is made along VA, the slant surface when opened out would form a sector of a circle as shown in Figure 12 (ii). It can be shown as below that the area of this sector is πrs.

$$\frac{\text{arc ABA}}{\text{circumference of circle, centre V}} = \frac{2\pi r}{2\pi s} = \frac{r}{s}$$

so
$$\frac{\text{area of sector VABA}}{\text{area of circle, centre V}} = \frac{r}{s}$$

and
$$\text{area of sector VABA} = \frac{r}{s} \times \pi s^2 = \pi rs$$

It follows that the area of the curved surface of the cone $= \pi rs$, where r is the radius of the base and s is the *slant height* of the cone.

Note. From Figure 12 (i), $s^2 = h^2 + r^2$.

Example. A solid cone has a circular base of diameter 12 cm and a slant height of 10 cm. Calculate: ***a*** its total area ***b*** its volume.

a Area of curved surface $= \pi rs$
$= 3{\cdot}14 \times 6 \times 10$
$= 188 \text{ cm}^2$

Area of base $= \pi r^2$
$= 3{\cdot}14 \times 6^2$
$= 113 \text{ cm}^2$

Total area $= 301 \text{ cm}^2$

nos	logs
3·14	0·497
36	1·556
113	2·053

b $h^2 = 10^2 - 6^2 = 64$, so $h = 8$

Volume of cone $= \frac{1}{3}\pi r^2 h$
$= \frac{1}{3} \times 3{\cdot}14 \times 6^2 \times 8$
$= 3{\cdot}14 \times 96$
$= 301 \text{ cm}^3$

nos	logs
3·14	0·497
96	1·982
301	2·479

Exercise 5B

1 Calculate the volumes of the following cones (in ***c*** and ***d*** take 3·14 for π):

	a	*b*	*c*	*d*
Radius of base	6 m	21 cm	10 m	2·87 m
Height	7 m	10 cm	12 m	9·34 m

2 The interior diameter and depth of an 'ice-cream cone' are 6 cm and 10 cm respectively. Calculate the volume of ice-cream it would hold if full to the top.

3 Calculate the volume of a cone with height 12 m and diameter of base 12 m.

4 A cone has a circular base of radius 7 cm, and a slant height of 25 cm. Calculate: *a* the area of the curved surface *b* the area of the base *c* the height *d* the volume of the cone.

5 A cone has a circular base of diameter 10 cm, and a slant height of 13 cm. Find: *a* the total area *b* the volume of the cone.

6 Find the depth of a conical flask which holds 200 ml of liquid, the diameter of its base being 12 cm.

7 A fire extinguisher, which is approximately conical in shape, has a capacity of 7 litres and a base diameter of 22 cm. Calculate its height.

8 A cone is formed from a semicircular sheet of tin of diameter 7 m. Find the radius of the base, and the volume of the cone.

5 *Spheres*

It can be shown that the area and volume of a sphere of radius r are given by the formulae:

$$A = 4\pi r^2 \text{ and } V = \tfrac{4}{3}\pi r^3$$

Every plane section of a sphere is a circle. If the plane passes through the centre of the sphere, as in Figure 13, it divides the sphere into two *hemispheres*.

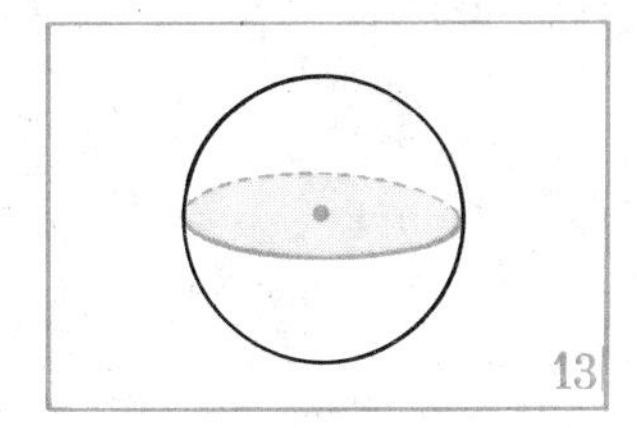
13

Example 1. A solid hemisphere has a diameter of 6 cm. Calculate:
a its volume ***b*** its total area.

a Volume of hemisphere
$= \frac{1}{2} \times \frac{4}{3}\pi r^3$
$= \frac{2}{3} \times 3{\cdot}14 \times 3^3$
$= 3{\cdot}14 \times 18$
$= 56{\cdot}5 \text{ cm}^3$

b Total area
= curved surface area + area of base
$= (\frac{1}{2} \times 4\pi r^2) + \pi r^2$
$= 3\pi r^2$
$= 3 \times 3{\cdot}14 \times 3^2$
$= 84{\cdot}8 \text{ cm}^2$

Example 2. Calculate the radius of a sphere of volume 25 m³.

$$V = \tfrac{4}{3}\pi r^3$$
$$\Leftrightarrow 3V = 4\pi r^3$$
$$\Leftrightarrow r^3 = \frac{3V}{4\pi}$$
$$\Leftrightarrow r = \sqrt[3]{\frac{3V}{4\pi}}$$

$$r = \sqrt[3]{\frac{3 \times 25}{4 \times 3{\cdot}14}} = \sqrt[3]{\frac{75}{12{\cdot}6}} = 1{\cdot}81 \text{ m}$$

nos	logs
75	1·875
12·6	1·100
	3|0·775
1·81	0·258

Exercise 6B

1. Calculate the areas of spheres with radii: ***a*** 3·5 cm ***b*** 10 m.
2. Calculate the volumes of spheres with radii: ***a*** 1 m ***b*** 7 mm.
3. Calculate the area and volume of a spherical football of *diameter* 21 cm.
4. What is the volume of the largest sphere that can be put in a cubical box of edge 6 m?
5. Calculate the radius of a sphere of volume 10 cm³.
6. At an exhibition, one of the pavilions had a hemispherical dome of diameter 35 m. Find the cost of gilding it at £1·20 per square metre.
7. A boiler is in the shape of a cylinder with hemispherical ends. Its total length is 16 m, and its diameter is 6 m. Find its cubic content.
8. A lead 'sinker' has the shape of a hemisphere topped by a cone. The diameter of the hemisphere and the height of the cone are each 1·4 cm. Find the mass of the sinker, given that 1 cm³ of lead weighs 11·4 g.
9. Calculate the area of a sphere which has a volume of 850 cm³.
10. Calculate the mass of 5000 ball bearings, each of diameter 0·7 cm, made of steel, 1 cm³ of which weighs 7·8 g.

11 Three spheres with surface areas A_1, A_2, A_3 and volumes V_1, V_2, V_3 respectively have radii 1 m, 2 m and 3 m. Without calculating their areas and volumes, give values of:

a $A_1 : A_2$ *b* $A_2 : A_3$ *c* $A_1 : A_2 : A_3$ *d* $V_1 : V_2$
e $V_2 : V_3$ *f* $V_1 : V_2 : V_3$

12 A cylindrical measuring jar has diameter 5 cm, and contains water to a depth of 6 cm. A sphere of diameter 3 cm is dropped in and sinks to the bottom. What is the water depth now?

6 Miscellaneous questions

Exercise 7

1 Calculate the perimeter and area of a circle of radius 35 mm.

2 Calculate the perimeter and area of a circle of diameter 8·4 cm.

3 A circular running track is 440 m long. Find the radius of the track.

4 The area of a circle is 38·5 cm^2. Calculate its radius and circumference.

5 A brick measures 22·5 cm by 11 cm by 7·5 cm. How many bricks, all lying the same way, can be packed into a cubical box of side 1 metre?

6 A rectangular lawn 9 metres long and 7 metres broad has a circular rose bed of diameter 2 metres in the centre of it. Find the area of the grass.

7 A window is in the shape of a rectangle 4 m by 2 m surmounted by a semicircle of diameter 4 m. Find the area of the glass in the window.

8 A cylindrical tin of powder is 14 cm high and has a diameter of 7 cm. How many such tins can be filled from a rectangular box, 2·2 m by 0·7 m by 0·1 m, which is full of the powder?

9 A cylindrical tin holds $\frac{1}{2}$ litre of liquid and has a diameter of 7 cm. Calculate its height.

10 A cylindrical water tank holds 88 litres of water. If the water is 70 cm deep, calculate the radius of the base of the tank.

11 Find the area of a right-angled triangle with sides 8 cm, 4·8 cm and 6·4 cm long.

12 A swimming pool is 40 m long and 15 m wide. The depth of water at the shallow end is 1 m, increasing uniformly to 3 m at the deep end.

a Make a sketch as in Figure 14, and calculate the area of one side of the pool.

b Taking this side as the 'base' of a prism, calculate the volume of water in the pool, in cubic metres.

Exercise 7B

1 Calculate the perimeter and area of a circle of diameter 8 cm.

2 Find the altitude and area of an equilateral triangle of side 6 cm.

3 A rectangular lawn 8 metres long and 6 metres broad has a circular rose bed of diameter 2·4 metres in the centre of it. Find the area of the grass.

4 A cylindrical tin of powder is 10 cm high and has a diameter of 14 cm. How many such tins can be filled from a rectangular box, 1·5 m by 0·3 m by 0·1 m, which is full of the powder?

5 Water from a rectangular roof 18 m by 15 m drains into a cylindrical water barrel of radius 0·75 m. By how much will the water level in the barrel rise as the result of a rainfall of 1·6 mm?

6 A cylindrical hot-water tank has to hold 150 litres of water and has to have a height of 60 cm. Calculate the radius of its base.

7 Find the area of the triangle with vertices A (3, 1), B (9, 1), and C (7, 6).

8 Calculate the volume of *a* a prism *b* a pyramid of height 16 units and with triangle ABC (in question 7) as base.

9 Calculate the mass of 2000 ball bearings each of diameter 0·5 cm, made of steel, 1 cm^3 of which weighs 7·8 g.

10 A solid cone has a circular base of diameter 14·4 cm and a slant height of 12 cm. Calculate: *a* its total area *b* its volume.

11 A swimming pool is 50 m long and 16 m wide. The depth of water at the shallow end is 1 m, increasing uniformly to a depth of 3 m at the deep end, as shown in Figure 14. Calculate the number of cubic metres of water in the pool.

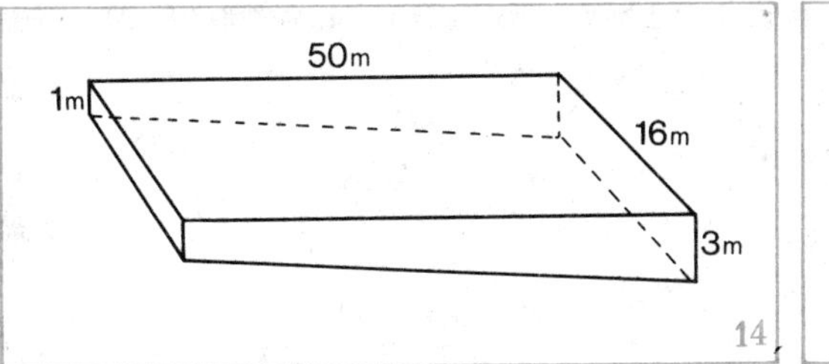

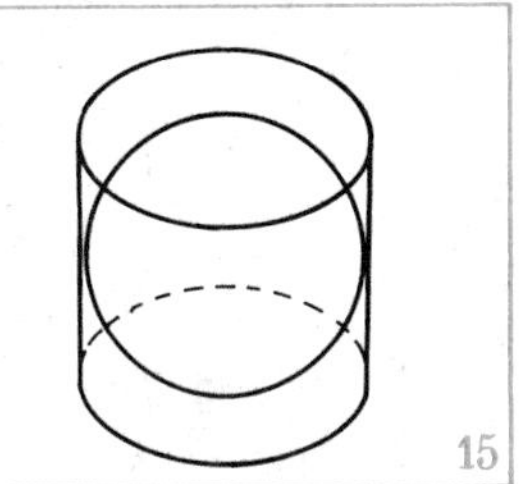

12 Figure 15 shows a sphere of radius r fitting exactly into a cylinder, i.e. the sphere touches the cylinder at the top, bottom and sides.

a Show that the area of the sphere is equal to the curved surface area of the cylinder.

b Find the ratio of the volume of the sphere to the volume of the cylinder.

c Find the volume of a cone with the same circular base and height as the cylinder, and show that the volumes of the cone, sphere, and cylinder are in the ratio $1:2:3$.

Summary

1 Area of rectangle: $A = lb$

2 Area of square: $A = l^2$

3 Area of triangle: $A = \frac{1}{2}bh$

4 (i) Area of circle: $A = \pi r^2$

(ii) Circumference of circle: $C = 2\pi r = \pi d$

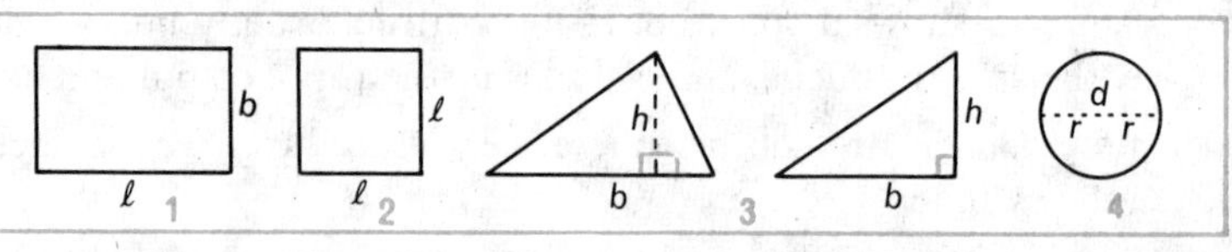

5 Volume of cuboid: $V = Ah = lbh$

6 Volume of cube: $V = l^3$

7 Volume of prism: $V = Ah$

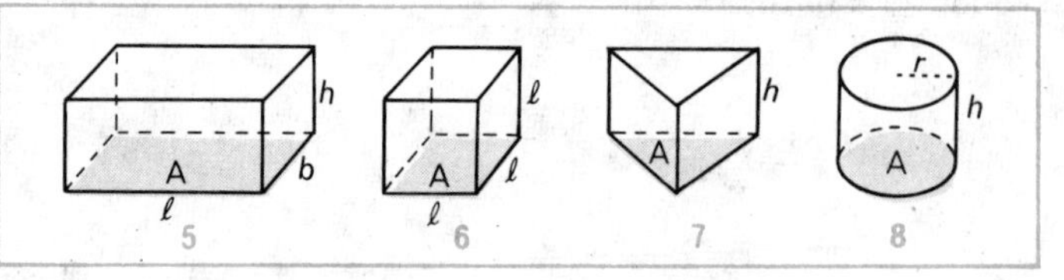

8 (i) Volume of cylinder: $V = Ah = \pi r^2 h$

(ii) Curved surface area: $2\pi rh$

9 Volume of pyramid: $V = \frac{1}{3}Ah$

10 (i) Volume of cone: $V = \frac{1}{3}Ah = \frac{1}{3}\pi r^2 h$

(ii) Curved surface area: πrs

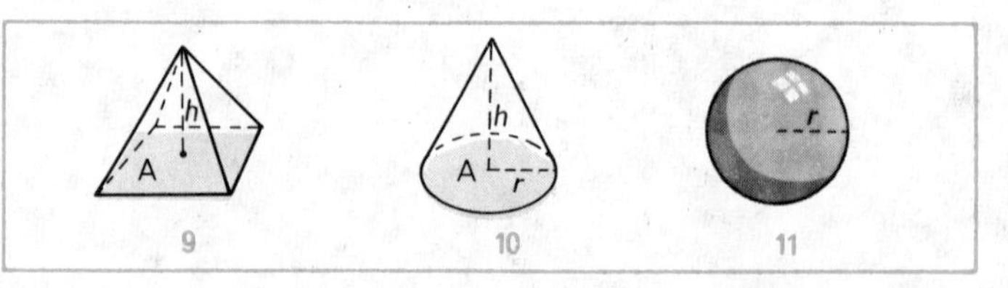

11 (i) Volume of sphere: $V = \frac{4}{3}\pi r^3$

(ii) Area of sphere: $A = 4\pi r^2$

12 *Metric measures*

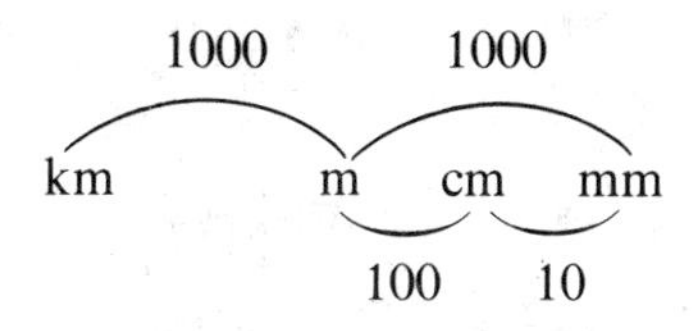

1 litre = 1000 cm^3; 1 tonne = 1000 kg;
1 hectare = 10 000 m^2

Number Patterns and Sequences

1 *Patterns in an array of numbers*

Exercise 1

Copy this *array* of whole numbers on to a sheet of squared paper:

0	1	2	3	4	5	6	7	8	9
10	11	12	13	14	15	16	17	18	19
20	21	22	23	24	25	26	27	28	29
30	31	32	33	34	35	36	37	38	39
40	41	42	43	44	45	46	47	48	49
50	51	52	53	54	55	56	57	58	59
60	61	62	63	64	65	66	67	68	69
70	71	72	73	74	75	76	77	78	79
80	81	82	83	84	85	86	87	88	89
90	91	92	93	94	95	96	97	98	99

1 What do you notice about all the numbers in:

a the third row *b* the eighth row *c* the sixth column?

2 In which row and column is the number 55? 99?

3 What do you notice about the numbers in:

a the main diagonal *b* the other diagonal?

4 Where do all the even numbers occur?

5 a Colour the squares containing numbers which are multiples of 3.

b Colour differently all the squares containing multiples of 7.

c What can you say about numbers in the squares with both these colours?

6 Make other patterns by colouring squares containing:

a prime numbers *b* numbers whose digits add up to 5

c any other interesting set of numbers you can find.

2 *Patterns and sequences*

In Chapter 1 of Book 1 we studied many sets of numbers, including:

the set of whole numbers {0, 1, 2, 3, . . .}
the set of natural numbers {1, 2, 3, . . .}
the set of even numbers {0, 2, 4, 6, . . .}
the set of odd numbers {1, 3, 5, 7, . . .}
the set of prime numbers {2, 3, 5, 7, 11, . . .}
the set of multiples of 6 {0, 6, 12, 18, . . .}

When we arrange numbers in order according to some rule the numbers form a *sequence*. Each number is called a *term* of the sequence.

Examples of sequences

a 0, 2, 4, 6, . . . *b* 1, 10, 100, 1000 . . . *c* 50, 45, 40, 35, . . .

Many sequences can be shown quite simply by means of patterns. Figure 1 (i) represents the sequence of *natural numbers* 1, 2, 3, . . . and (ii) shows the sequence of *triangular numbers* 1, 3, 6, 10, . . .

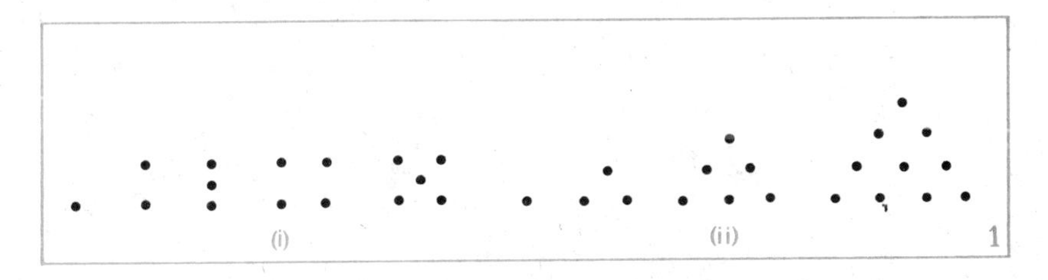

Exercise 2

1 *a* Copy Figure 1, and continue each sequence for two more patterns of dots.

b List the first six terms of each sequence.

2 Draw patterns of dots to illustrate the sequence of squares of natural numbers $1^2, 2^2, 3^2, 4^2, \ldots$

3 Write down numbers which can fill the two spaces in each of these sequences:

a 1, 3, 5, 7, ..., ... *b* 2, 4, 8, 16, ..., ...

c 4, 9, 16, 25, ..., ... *d* 1, 2, 1, 3, 1, 4, ..., ...

e 0, 5, 10, 15, ..., ... *f* $0 \times 3, 1 \times 4, 2 \times 5, 3 \times 6, \ldots, \ldots$

4 Which numbers should be omitted, or replaced, in the following to form simple sequences?

a 1, 5, 9, 11, 17, 21 *b* 1, 4, 9, 16, 20, 25

c 91, 84, 77, 71, 63 *d* 1, 3, 6, 10, 15, 20

5 Find a suitable missing number in each of these sequences:

a 4, ..., 12, 16, 20 *b* 2, 1, 3, 2, 4, ..., 5, 4.

c 16, 8, 4, 2, ... *d* 99, 87, 75, ..., 51

6 Make up some sequences, and see if your neighbour can provide a suitable next term for each.

7 Figure 2 shows a 'tree diagram' for a knockout football, or tennis, competition.

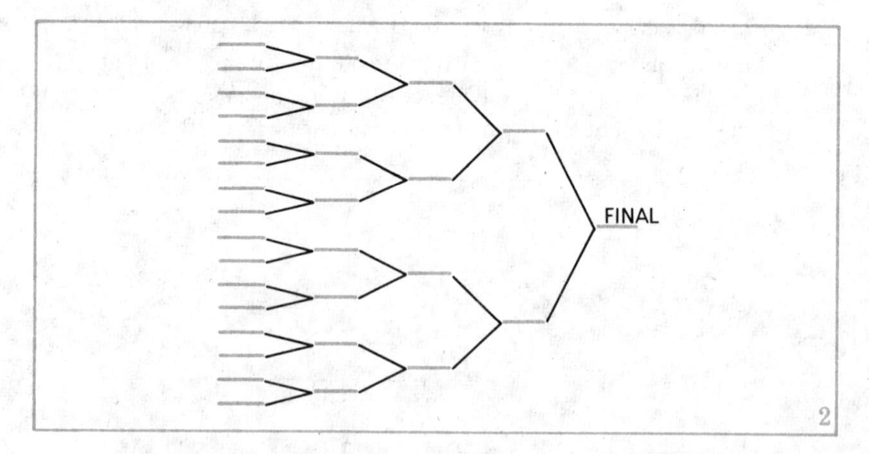

2

a How many teams are there in the final? in the semi-finals? in the quarter-finals?

b Write down the sequence of the number of teams in each round.

c How many rounds would be necessary if 128 teams entered the competition? What would happen if 130 entered?

8

	January					
S	–	3	10	17	24	31
M	–	4	11	18	25	–
T	–	5	12	19	26	–
W	–	6	13	20	27	–
T	–	7	14	21	28	–
F	1	8	15	22	29	–
S	2	9	16	23	30	–

February				
–	7	14	21	28
1	8	15	22	–
2	9	16	23	–
3	10	17	24	–
4	11	18	25	–
5	12	19	26	–
6	13	20	27	–

a Add up the numbers in the *columns* headed January 3, 10, and 17. What is the difference between each pair of successive totals? Why should these differences be the same? Without adding the numbers, write down the total for the column headed January 24.

b Will the difference between the totals for the first two columns in February be 49? Give a reason for your answer. Add up the numbers in the column headed February 14, and then write down the total for the column headed February 21.

c Can you find similar patterns in the *rows* of the calendar?

9 *a* Assuming that the large square in Figure 3 has an area of 1 square unit, write down the sequence of numbers which gives the areas (in fractions of 1 square unit) of the coloured portions of the other large squares.

b What will the sixth term of this sequence be? Which term will be $\frac{1}{128}$?

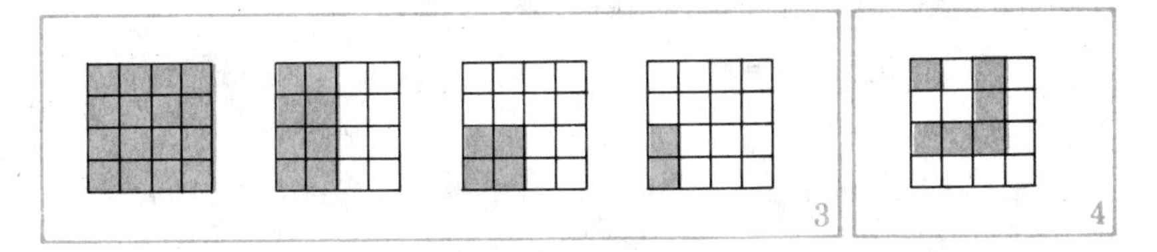

10 Write down possible fifth and eighth terms for each of the following:

a 3, 5, 7, 9, . . .

b 2, 3, 5, 8, . . .

c 3, 6, 12, 24, . . .

d 1, 0·1, 0·01, 0·001, . . .

e 1, 4, 9, 16, . . .

f prime numbers (starting at 2)

g $1 \times 2, 2 \times 3, 3 \times 4, 4 \times 5, \ldots$

h $1, \frac{1}{2}, \frac{1}{4}, \frac{1}{8}, \ldots$

11 Figure 4 contains a pattern of white and coloured squares illustrating the sequence 1, 3, 5, 7.

a What do you notice about the terms of this sequence?

b How can you tell by looking at the diagram that when we add these four numbers, the sum must be 4×4?

c Draw a similar diagram on squared paper and extend it to show the sum of the first five, six, seven odd numbers; state the sum in each case.

d Write down the sum of the first hundred odd numbers. How could you find the sum of the first 497 283 odd numbers? (Do not find the sum.)

e If n represents a natural number, write down an expression for the sum of the first n odd numbers.

---

3 *The order of terms in a sequence*

In question ***10*** of Exercise 2 you were able to supply the fifth and eighth terms of sequences in which only four terms were given, because of the fact that there is *order* in the terms of a sequence. This is the essential difference between a *sequence* and a *set* in which the order of the elements does not matter.

For example, we may consider a mapping of the natural numbers to the sequence of odd numbers as shown:

Natural numbers	1	2	3	4	5	...
	↓	↓	↓	↓	↓	
Odd numbers	1	3	5	7	9	...

It can be seen that the third odd number is 5, and the fifth is 9.

Again, we have:

Natural numbers	1	2	3	4	5	...
	↓	↓	↓	↓	↓	
Square numbers	1	4	9	16	25	...

These mappings are illustrated graphically in Figures 5 and 6.

It is interesting to note that the wavelength of a musical note is double the wavelength of the note one octave higher, i.e. the wavelengths of notes an octave apart are proportional to the numbers in the sequence 1, 2, 4, 8, 16, ... Intermediate notes on a guitar, for

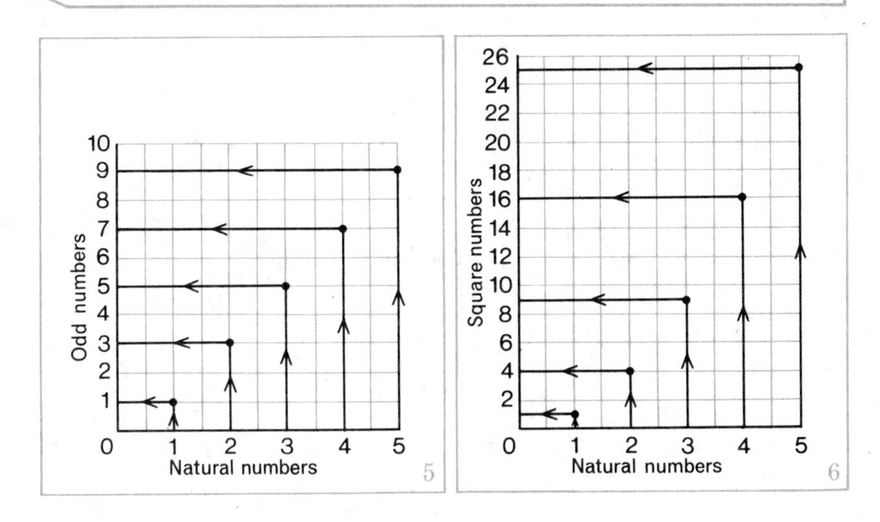

5

6

example, can be obtained by mapping the appropriate 'number positions' into the ridges of the guitar, and placing the ridges at the correct places (see Figure 7).

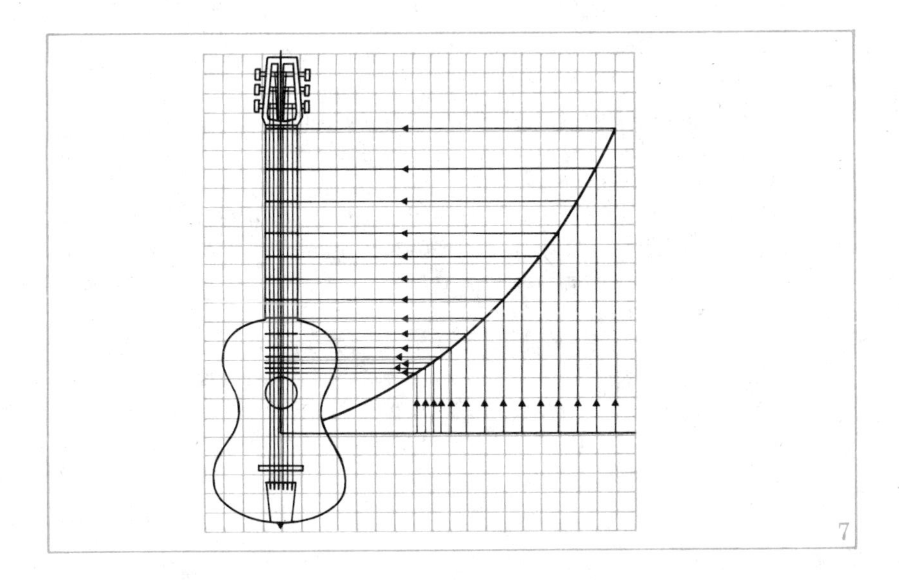

7

Exercise 3

1 Write down:

a the sixth odd number
b the fifth even number
c the fourth square number
d the third triangular number
e the seventh prime number
f the twelfth multiple of 12.

2 Show a mapping of the first five natural numbers into the first five terms of the following sequences, and illustrate each case by drawing a graph like Figure 5:

a even numbers
b cubes of the natural numbers (i.e. $1^3, 2^3, 3^3, \ldots$)
c 2, 3, 5, 8, ...
d $1, \frac{1}{2}, \frac{1}{4}, \frac{1}{8}, \ldots$

3 Here is a sequence in which each term after the second is the sum of the preceding two terms:

0, 1, 1, 2, 3, 5, 8, 13, ...

Write down the next four terms.

It is called a Fibonacci sequence, after the man who discovered it; you can make up different Fibonacci sequences by choosing different numbers as the first two terms.

Continue to 5 terms as Fibonacci sequences:

a 1, 3, ...
b 1, 4, ...

Fibonacci sequences have many interesting properties; here is one of them. Calculate, in decimal form, the ratio of each term to the following term in the sequence 0, 1, 1, 2, 3, ... i.e. $\frac{0}{1}$ or 0; $\frac{1}{1}$ or 1; $\frac{1}{2}$ or 0·5; $\frac{2}{3}$ or 0·667; and so on. Draw a mapping of natural numbers to the first 9 terms of this sequence of decimals, and you will see the effect of the 'settling down' of the ratio at about 0·618; this is true for every Fibonacci sequence.

If you are interested in architecture you will find that this ratio is considered to be the most pleasing ratio of the breadth to the height of a rectangular object (window, etc.). Is the breadth–height ratio for any of your textbooks close to this number?

4 *The next term in a sequence*

Example 1. The next term in the sequence 5, 8, 11, 14, ... is obtained by adding 3 to the preceding term.

Example 2. The next term in the sequence 3, 1, $\frac{1}{3}$, ... is obtained by multiplying the preceding term by $\frac{1}{3}$.

Exercise 4

1 Write down a rule for obtaining the next term in each of the following:

a 100, 96, 92, 88, ...

b 3, 6, 12, 24, ...

c 1, 3, 5, 7, ...

d 64, 32, 16, 8, ...

e 9·9, 8·8, 7·7, 6·6, ...

f 1, 10, 100, 1000, ...

g 0, $\frac{1}{2}$, 1, $1\frac{1}{2}$, ...

h 2, 2^2, 2^3, 2^4, ...

2 Figure 8 shows a sequence of patterns composed of matches. Copy the figure and draw the next two diagrams in the sequence.

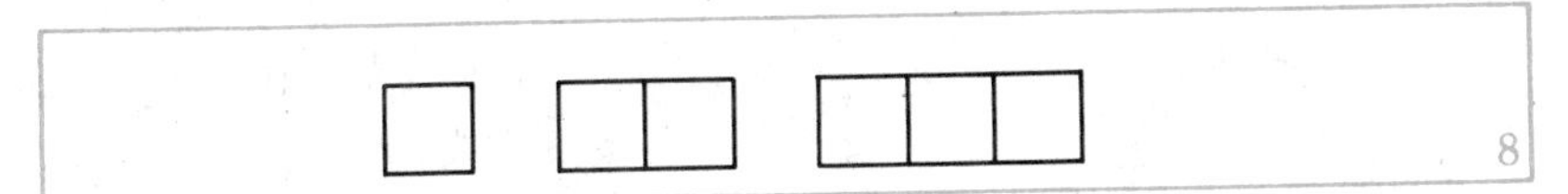

8

Write down the sequence of numbers given by each of the following, and add two terms to each sequence:

a the number of squares in the patterns

b the number of matches in the patterns

c the number of squares and rectangles in the patterns. Do you recognize this sequence?

3 Figure 9 shows another sequence of patterns composed of matches. Copy it and draw the next pattern.

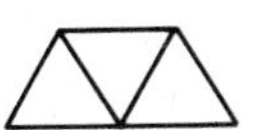
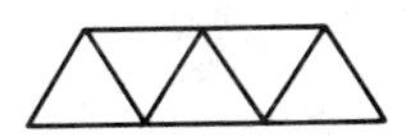

9

Write down the sequence of numbers given by each of the following, and add two terms to each sequence:

a the number of triangles in the patterns

b the number of matches in the patterns

c the number of triangles and parallelograms in the patterns.

4 Figure 10 shows a 'pin-ball' machine, in which a steel ball moves down the board, striking one pin and then striking either of the two pins immediately below. Thus after striking A the ball may strike B or C, after striking B it may strike D or E, and so on.

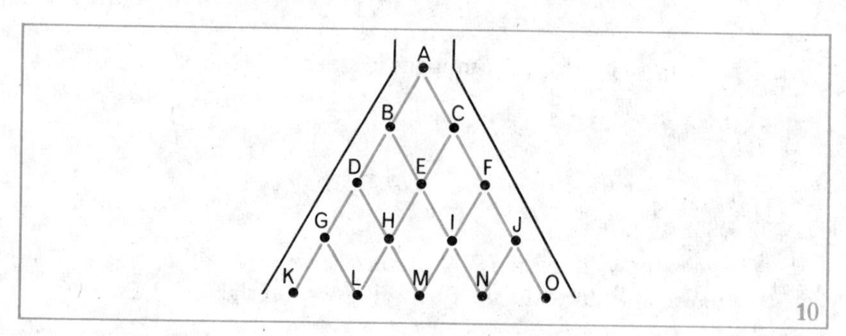

10

There are three routes from A to H, viz. ABDH, ABEH, ACEH. Give the routes from A to I, from A to J, from A to N.

Now replace each letter by the number of routes to it from A, replacing A by 1.

The pattern of numbers you now have is called *Pascal's Triangle.*

a What are the first and last numbers in each row?

b How is each number related to the two nearest numbers in the row above it?

c Copy down the pattern and add two more rows to it.

d Calculate the sum of the numbers in each of the first four rows starting with A. Now write down the sum of the numbers in each of the next four rows. Write all these as powers of 2, and hence give the sum of the numbers in (*1*) the twentieth row (*2*) the thousandth row.

e Reading down the diagonals we obtain the sequences:

1, 1, 1, . . . ; 1, 2, 3, . . . ; 1, 3, 6, . . .

Copy these and give the first six terms in each. Do the same for the next three diagonals.

5 The pattern of numbers in Pascal's Triangle occurs frequently in mathematics. For example, in Book 3 you investigated various probabilities connected with the tossing of coins.

a In tossing a penny what is the probability of obtaining (*1*) a head (*2*) a tail?

b In tossing two pennies, what is the probability of obtaining (*1*) two heads (*2*) one head and one tail (*3*) two tails?

c In tossing three pennies, what is the probability of obtaining (*1*) three heads (*2*) two heads and one tail (*3*) one head and two tails (*4*) three tails?

You should see that the numerators of the fractions that you have written down (assuming that you did not simplify the fractions) are the rows of numbers in Pascal's Triangle.

d Can you now *write down* the probability, when tossing four pennies, of obtaining (*1*) three heads and one tail (*2*) a head?

6 Tennis balls are laid close together in the pattern of dots giving the sequence of triangular numbers.

a How many tennis balls could be placed on top of the second pattern? the third pattern?

b Write down the number of balls in each of the first five pyramids so formed (include the first ball on its own).

c Can you find this sequence in Pascal's Triangle?

d Can you give a rule which enables you to obtain the sequence of 'pyramid' numbers from the sequence of triangular numbers?

7 Repeat question **6** for tennis ball pyramids with square bases.

5 *The nth term of a sequence*

It is often possible to express the rule for forming the terms of a sequence by means of an algebraic formula.

Example 1

Natural numbers	1	2	3	4	......	n	......
	↓	↓	↓	↓		↓	
Even numbers	0	2	4	6	......	?	......
i.e.	2×0	2×1	2×2	2×3	......	$2\times(n-1)$	......

Clearly the term of the sequence to which n maps is $2\times(n-1)$, i.e. $2(n-1)$.

So the nth term of the sequence of even numbers is $2(n-1)$.

Notice that if we know the nth term of a sequence we can obtain every term in the sequence. Thus the hundredth even number is 2×99, i.e. 198.

Example 2. Give the first four terms of the sequence whose nth term is $\frac{1}{2}n(n+1)$.

To find the *first* term we replace n by 1, obtaining $\frac{1}{2}\times 1\times 2 = 1$.
To find the second term we replace n by 2, obtaining $\frac{1}{2}\times 2\times 3 = 3$.
The third term is $\frac{1}{2}\times 3\times 4 = 6$.
The fourth term is $\frac{1}{2}\times 4\times 5 = 10$.
So the sequence is 1, 3, 6, 10, ...

Exercise 5B

1 a Write down the nth term of the sequence 1, 2, 3, 4, ...
Now write the nth term of the sequence 2, 3, 4, 5, ...
Can you now see how to write down the nth term of the sequences 5, 6, 7, 8, ... and 11, 12, 13, 14, ... ?

b Write down the nth term of the sequence 3, 6, 9, 12, ...
Now write down the nth term of the sequence 4, 7, 10, 13, ...
Can you now write down the nth term of the sequences 8, 11, 14, 17, ... and 12, 15, 18, 21, ... ?

c Using the ideas of parts ***a*** and ***b***, write down a formula for the nth term of (*1*) 6, 11, 16, 21, ... (*2*) 3, 7, 11, 15, ... (*3*) 0, 6, 12, 18, ...

2 Write down a formula for the nth term of each of the following sequences:

a 1, 2, 3, 4, ... *b* 1, 4, 9, 16, ...
c $1\times 2, 2\times 3, 3\times 4, \ldots$ *d* 1, 8, 27, 64, ...
e 3, 9, 27, 81, ... *f* 5, 9, 13, 17, ...
g $1\times 2\times 3, 2\times 3\times 4, 3\times 4\times 5, \ldots$ *h* $\frac{1}{2}, \frac{2}{3}, \frac{3}{4}, \frac{4}{5}, \ldots$

3 Give the first four terms of sequences whose nth terms are given by:

a $3n+2$ *b* 5×2^n *c* $195-6n$
d $n(n+1)$ *e* 2^n+1 *f* $\dfrac{2n-1}{n}$
g $(n-1)(2n+1)$ *h* $\frac{1}{2}n(n-1)$

4 Find the terms indicated in this table, in which T_n denotes the nth term:

T_n	$n+3$	n^4	3^n	$n(n+1)$	$4n-1$	$n(n+1)(n+2)$
Find	T_7	T_5	T_4	T_{100}	T_6	T_{12}

5 A Siamese cat has been taught to retrieve ping-pong balls.

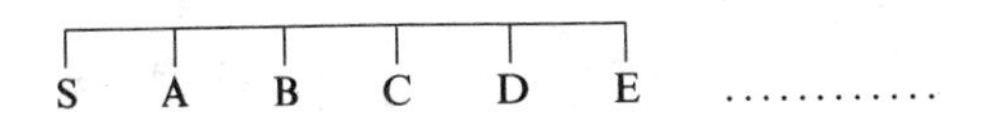

His master places a ball at each of A, B, C, . . . , where SA = AB = BC = . . . = 1 metre. The cat starts at S and brings each ball in turn back to S before fetching the next one; write down the distance he has to run in order to retrieve 1 ball; 2 balls; 3 balls; 4 balls; 5 balls.

Which formula quoted in question ***3*** appears to give the distance he has to run to retrieve n balls? Assuming this to be the correct formula, how far would he run in retrieving 25 balls?

6 *The 'snowflake' pattern*

Figure 11 shows a famous sequence of patterns. The first shape is an equilateral triangle. In the second, the centre third of each side is replaced by the two sides of an equilateral triangle, drawn outwards. Similarly for the third diagram.

Draw the three patterns, and also the fourth if time permits. (Isometric paper is very useful, and the first triangle should have sides of length 9 or 27 units.)

Write down the sequence of numbers of sides. Hence give: *a* the next two numbers in the sequence *b* the nth term of the sequence.

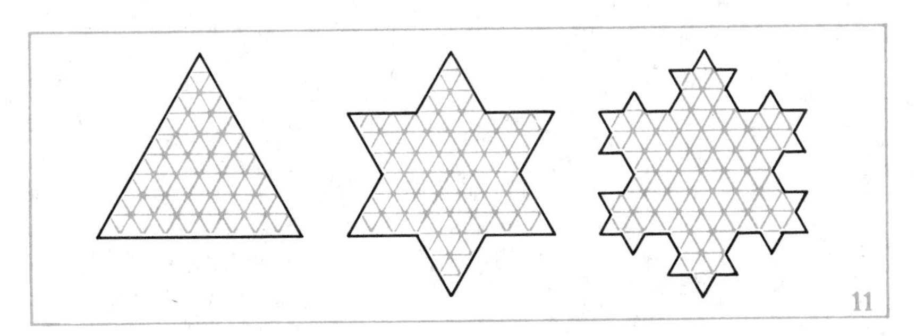
11

7 The 'anti-snowflake' pattern may be obtained by drawing the equilateral triangles in the second, third, ... patterns *inwards* instead of outwards. You could answer the same questions as in ***6***.

Exercise 6B Miscellaneous questions

1 Write down a possible formula for the nth term of each of these sequences:

a 5, 10, 15, 20, ... *b* 2, 4, 8, 16, ...
c 3, 4, 5, 6, ... *d* 0, 1, 2, 3, ...
e 2, 5, 8, 11, ... *f* $1, \frac{1}{2}, \frac{1}{3}, \frac{1}{4}, \ldots$

2 Find the first term and the tenth term of the sequences whose nth terms are given by the formulae:

a $n+5$ *b* $2n-1$ *c* n^3
d $(n+1)^2$ *e* $n(n-1)$ *f* $100-10n$

3 In a school hall there are twenty seats in the front row, and each row contains two seats more than the one in front of it. If there are ten rows of seats, how many people can be seated in the hall?

4 If you look carefully at Figure 12 you will see that the patterns contain one, five, and fourteen squares (of various sizes). It is said that if you continued the sequence of patterns the number of squares contained in the tenth one would be $1^2+2^2+3^2+\ldots+10^2$. Do you think that this is true?

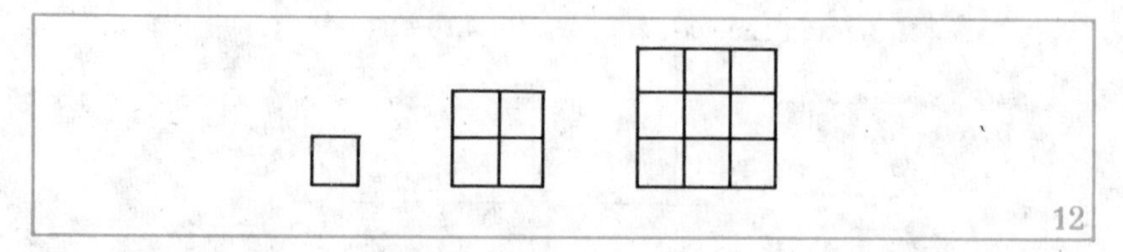
12

5 In Figure 12, if you count rectangles as well as squares the numbers are 1, 9, and 36. (Check that these are correct.) Notice that these are square numbers. It is said that the number of squares and rectangles in the tenth pattern would be $(1+2+3+\ldots+10)^2$, which is too many to try to count. Instead, guess the answer for the fourth pattern, and write down the answer for the nth pattern.

6 As each guest arrives at a party he shakes hands with the host and with each of the other guests who has arrived before him. Write down in a table, as below, the number of handshakes when the first guest arrives (two people present), the second guest arrives (three people present), and so on.

Number present	*Number of handshakes*
2	1
3	1+...
4	1+...+...
5	1+...+...+...

How many handshakes are there when fifty people are present? How many when n people are present?

Now find the answers in another way, by considering how many times each person shakes hands, and the total number of handshakes.

Then find a formula for the sum $1+2+3+ \ldots\ldots\ldots\ldots +n$.

7 A triangular framework is made of 10-cm strips as shown in Figure 13. How many strips are required for each framework? How many for frameworks with sides made of 4, 5, 6 strips?

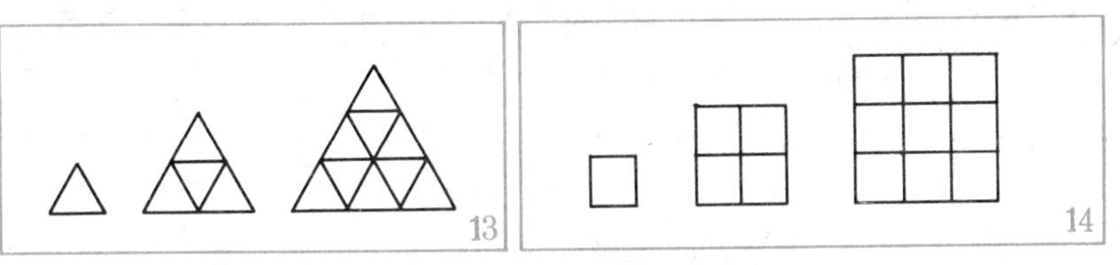
13 14

If square frameworks were made how many strips would be needed for each of the first six patterns (see Figure 14)?

Is there a number of strips that could be used for either a triangular or square framework?

8 Consider the sum of the numbers in the sequence $1, \frac{1}{2}, \frac{1}{4}, \frac{1}{8}, \ldots$ Draw a fairly long line 2 units in length.

For the first term, colour one of these units.

For the addition of the second term, colour half of the other unit. What fraction of the length of the whole line is still uncoloured?

For the addition of the third term, colour a further quarter of a unit. What fraction of the total length is still uncoloured?

Continue to colour in fractions as far as you can.

What is the sum of the first *a* 4 terms *b* 6 terms *c* 8 terms?

How many terms added together would have 2 as their sum?

Summary

1 There is *order* in the terms of a sequence (unlike a set).

2 A *sequence* can be defined by:

(i) *patterns* of dots or shapes. For example,

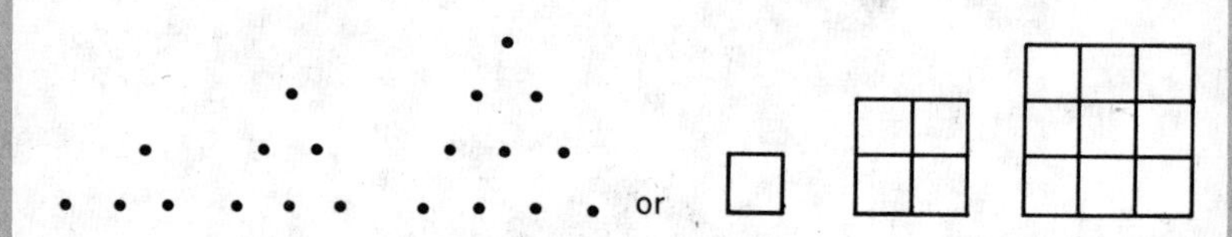

(ii) a *rule.* For example,
for 2, 4, 8, 16, . . . the rule could be 'Multiply by 2'.

(iii) a *formula.* For example,
for 3, 5, 7, 9, . . . the formula for the nth term could be $2n+1$.

Topics to explore

Constructing a table of logarithms to base 3; the original inventors

1 *a* Write down three more terms in the multiplication set $M = \{1, 3, \ldots, \ldots, \ldots\}$ corresponding to the elements in the addition set $A = \{0, 1, 2, 3, 4\}$. Hence write down values of $\log_3 1$, $\log_3 3$, etc.

b On 2-mm squared paper draw a graph showing the correspondence between sets M and A, taking a scale of 2 cm to 10 units on the horizontal axis for M and 2 cm to 1 unit on the vertical axis for A.

c Reading values of the logarithms of whole numbers to base 3 from the graph, make your own table of logarithms to base 3.

d Use your table to carry out calculations of products and quotients of numbers.

2 John Napier ('Napier of Merchiston') (1550–1617), a Scot, and Henry Briggs (1561–1631), an Englishman, invented and developed the first system of logarithms. Find out about this in reference books.

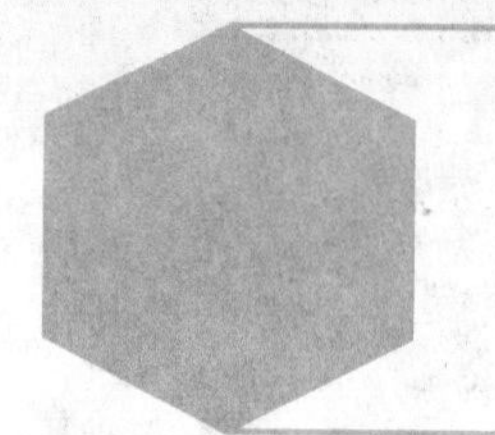

Revision Exercises

Revision Exercises on Chapter 1
Logarithms and Calculating Machines

Revision Exercise 1A

Use logarithms to calculate the products, quotients, powers and roots in questions ***1–18***:

1 $48{\cdot}9 \times 7{\cdot}63$ *2* $560 \times 10{\cdot}6$ *3* $18{\cdot}9 \times 28{\cdot}3 \times 3{\cdot}14$

4 $79{\cdot}8 \div 3{\cdot}67$ *5* $48\,900 \div 863$ *6* $27{\cdot}6 \div 21{\cdot}9$

7 $1{\cdot}15^{10}$ *8* $\sqrt{67\,200}$ *9* $\sqrt[3]{586}$

10 $57{\cdot}3 \times 0{\cdot}125$ *11* $0{\cdot}912 \times 0{\cdot}764$ *12* $21{\cdot}7 \times 0{\cdot}009\,32$

13 $28{\cdot}6 \div 0{\cdot}584$ *14* $0{\cdot}321 \div 575$ *15* $0{\cdot}006\,39 \div 0{\cdot}0925$

16 $0{\cdot}843^3$ *17* $\sqrt[3]{0{\cdot}009\,21}$ *18* $\sqrt[3]{0{\cdot}0921}$

19 Calculate 93·5% of 5·17 million.

20 What percentage is 1·56 of 9270, to 2 significant figures?

Revision Exercise 1B

Use logarithms to calculate the products, quotients, powers and roots in questions ***1–12***.

1 $\dfrac{13{\cdot}6 \times 523}{2830}$ *2* $\dfrac{489}{2{\cdot}83 \times 9{\cdot}76}$ *3* $\dfrac{0{\cdot}383 \times 0{\cdot}257}{0{\cdot}581 \times 0{\cdot}932}$

4 $\sqrt{(59{\cdot}8 \times 21{\cdot}7 \times 1{\cdot}3)}$ *5* $\sqrt[3]{\left(\dfrac{0{\cdot}482}{12{\cdot}6}\right)}$ *6* $\sqrt[3]{\left(\dfrac{1{\cdot}46 \times 0{\cdot}73}{0{\cdot}64 \times 2{\cdot}89}\right)}$

7 $(1{\cdot}03)^{10}$ *8* $(0{\cdot}65 \times 0{\cdot}248)^3$ *9* $1{\cdot}76^2 \times 0{\cdot}265^3$

10 $3{\cdot}14 \times 2{\cdot}73^3$ *11* $6{\cdot}28 \times \sqrt{0{\cdot}867}$ *12* $\sqrt{(6{\cdot}5 \div 32{\cdot}2)}$

13 Calculate the value of $\frac{1}{2}mv^2$ when $m = 2{\cdot}63$ and $v = 1270$, giving the answer in standard form.

14 Repeat question ***13*** for the case when $m = 0{\cdot}135$ and $v = 0{\cdot}046$.

15 From the formula $A = PR^n$, calculate A when $P = 375$ for the cases when: *a* $n = 5$ *b* $n = 25$ *c* $n = 50$. Assume the logarithm of R is 0·012 65, and give the answers to 3 significant figures.

16 The volume of a cylinder is given by $V = \pi r^2 h$. Take 3·14 for π.

a Calculate V when $r = 0{\cdot}865$ and $h = 0{\cdot}942$.

b Calculate h when $V = 4{\cdot}76$ and $r = 1{\cdot}25$.

c Calculate r when $V = 100$ and $h = 4{\cdot}75$.

d Calculate r when $V = 0{\cdot}0056$ and $h = 0{\cdot}037$.

17 The volume of a sphere is given by $V = 4{\cdot}19r^3$.

a Calculate V when (1) $r = 4{\cdot}64$ (2) $r = 0{\cdot}73$.

b Calculate r when (1) $V = 10$ (2) $V = 2{\cdot}97$.

18 From the formula $F = \dfrac{m_1 m_2}{D^2}$, calculate:

a F when $m_1 = 6{\cdot}25$, $m_2 = 2{\cdot}75$, $D = 6{\cdot}7$

b F when $m_1 = 2{\cdot}3 \times 10^8$, $m_2 = 6{\cdot}42 \times 10^4$, $D = 9{\cdot}3 \times 10^7$

c D when $m_1 = 3{\cdot}7 \times 10^{12}$, $m_2 = 8{\cdot}6 \times 10^1$, $F = 0{\cdot}436$.

Revision Exercises on Chapter 2
Areas and Volumes

Revision Exercise 2A

1 Calculate the circumference and area of a circle of radius 8·4 cm.

2 A classroom is 10 m long, 6 m wide and 5 m high. If there are 29 pupils in the room with one teacher, how many cubic metres of air space is available per person?

3 A clock face is in the form of a circle. If the hour hand is 3·5 cm long, find:

a the distance the tip of the hour hand travels

b the area swept out by the hour hand, between 1500 hours and 1800 hours.

4 A rectangular floor is 6·3 m long and 3·9 m broad. How many tiles each 30 cm by 30 cm would be required to cover the floor?

5 *a* Sketch a prism 12 cm high on a rectangular base 5 cm long and 2·5 cm broad. Calculate the volume of the prism.

b Sketch one of the triangular prisms obtained when the prism in ***a*** is cut down a diagonal plane. What is the volume of this triangular prism?

6 Use logarithms or slide rule to calculate:

a the area of a circle of radius 3·69 cm

b the volume of a cuboid of length 15·7 cm, breadth 12·4 cm and height 8·8 cm

c the volume of a cylinder of radius 2·5 cm and height 7·6 cm.

7 A rectangular metal tank (with no lid) is 4 m long, 3 m wide and 75 cm deep. Calculate: *a* its total external surface area in square metres *b* its capacity in litres.

8 *a* 360 tiles, each 25 cm by 25 cm, completely cover a rectangular floor of breadth 7·5 m. Find the length of the floor.

b How many tiles would be required for a floor twice as broad and one and one half times as long?

9 A cylindrical copper tank, closed at top and bottom, is 14 m high, and the diameter of its base is 4 m. Calculate:

a the volume of the tank

b the area of copper sheet required to make it.

10 A cylindrical candle 20 cm high, and with a diameter of 5 cm, burns at the rate of 2·5 cm^3 per hour. How long will it burn?

11 Calculate the cost of laying plain carpet at £2·65 per square metre on a floor shaped as in Figure 1.

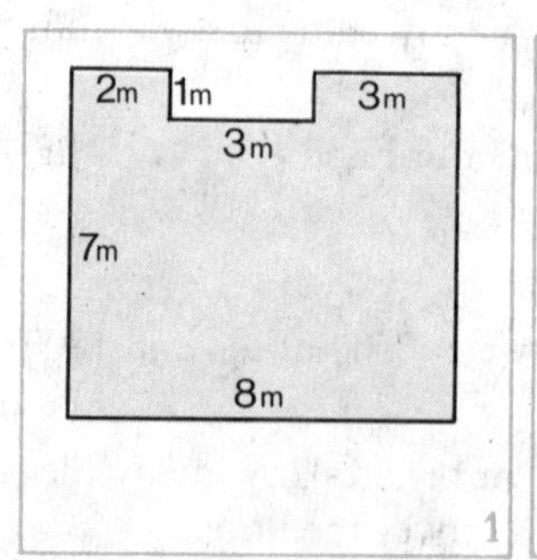

1

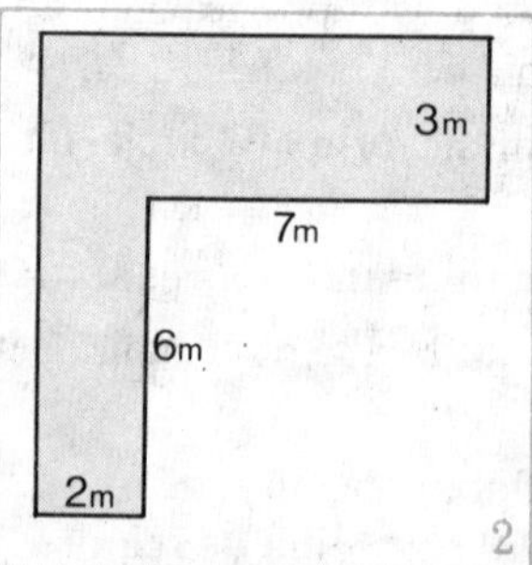

2

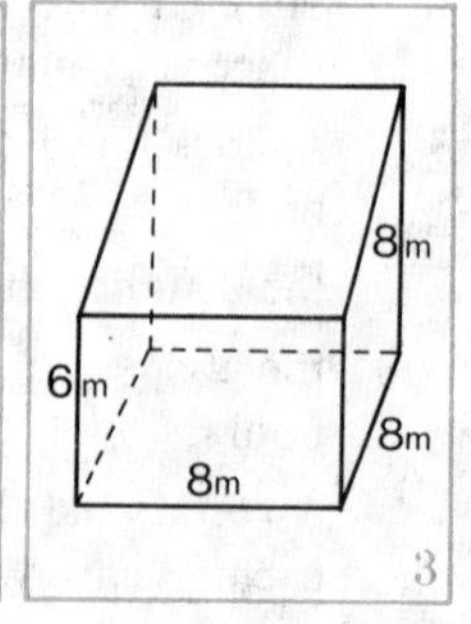

3

12 The **L**-shaped hall shown in Figure 2 is to be fitted with plain self-coloured carpet costing £3·60 per square metre. What is the cost of the carpet?

13 Figure 3 illustrates a shed standing on a square base of side 8 m. The front is 6 m and the back is 8 m high. Calculate:

a the area of the four walls of the shed

b the volume of the shed.

14 A swimming pool is 25 m long and 15 m broad. The bottom slopes uniformly from a depth of 1 m at the shallow end to 3 m at the deep end. Make a sketch, and calculate:

a the area of one 'side' of the pool

b the volume of water in the pool, in cubic metres.

Revision Exercise 2B

1 A classroom is 8 m long, 7 m wide and 4·5 m high. If there are 35 pupils and one teacher in the room, how many cubic metres of air are available per person?

2 A clockface is circular. If the hour hand is 14 cm long, find:

a the distance travelled by the tip of the hand between 1500 and 1700 hours

b the area swept out by the hand in this time.

3 A rectangular tank, open at the top, is 3·5 m long, 2 m wide and 1·5 m deep. Calculate:

a the total area of its sides and base

b its volume, in litres.

4 A floor 4·8 m by 3·6 m is to be covered with carpet 60 cm wide.

a How many strips would be required each way?

b What length of carpet would be needed?

5 A cylindrical candle of diameter 4 cm and height 21 cm burns at the rate of 11 cm^3 per hour. How long will it burn?

6 The net of a pyramid consists of a square of side 14 cm, with four congruent isosceles triangles of base 14 cm and height 25 cm. Calculate:

a the total area of the pyramid

b the height and volume of the pyramid.

7 Taking $\frac{22}{7}$ as an approximation for π, calculate the curved surface area of:

a a cylinder with radius of base 5 m and height 21 m

b a cone with radius of base 7 cm and slant height 12 cm

c a sphere with radius 21 mm.

8 Taking 3·14 as an approximation for π, use slide rule or logarithms to calculate the volume of:

a a cylinder with *diameter* of base 9 mm and height 18·7 mm

b a cone with radius of base 5·87 m and height 4·73 m

c a sphere with radius 3·69 cm.

9 Sand falls from a chute and forms a cone of height 3 m and diameter 5 m. Find the weight of the sand in tonnes, given that 1 m^3 of sand weighs 5 tonnes.

10 36 tins of fruit, each a cylinder of diameter 6 cm and height 8 cm, are packed in two layers in a rectangular box which is just big enough to hold them. Calculate the volume of the space in the box which is not occupied by the tins.

11 Assuming that the earth is a sphere of radius 6400 km rotating about its axis once a day (24 hours), find the speed in km/h at which a point on the equator is moving due to the earth's rotation.

12 How many spherical lead shot of diameter 3 mm can be made from 1 kg of lead which weighs 11·6 g per cm^3?

Revision Exercises on Chapter 3 Number Patterns and Sequences

Revision Exercise 3A

1 Write down three more terms of each of the following sequences, and give a rule for finding the next term in each case:

a 0, 3, 6, 9, 12, . . .

b 2, 3, 5, 7, 11, 13, . . .

c 53, 48, 43, 38, ... *d* 729, 243, 81, 27, ...

e 1, 1, 2, 3, 5, 8, ... *f* 2, 5, 10, 17, 26, ...

2 Write down a possible tenth term for each of these sequences:

a $\frac{1}{2}, \frac{1}{3}, \frac{1}{4}, \frac{1}{5}, \ldots$ *b* $0 \times 1, 1 \times 2, 2 \times 3, 3 \times 4, \ldots$

c 0, 11, 22, 33, ... *d* 1, 4, 9, 16, ...

3 Which number should be omitted, or replaced, in each of the following to form simple sequences?

a 95, 89, 82, 77, 71 *b* 0, 3, 6, 9, 12, 18

c $\frac{1}{2}, \frac{2}{3}, \frac{3}{4}, \frac{3}{5}, \frac{4}{5}, \frac{5}{6}$ *d* 3, 9, 15, 24, 27

4 Write down the first four terms of sequences defined as follows.

a The first term is 3, and successive terms are formed by adding 2 to the previous term.

b The first two terms are 1 and 2, and successive terms are formed by adding the previous two terms.

c The first term is 5, and successive terms are formed by multiplying the previous term by 2.

5 Continue the pattern of dots in Figure 4 for two more terms. Count the number of dots in each, and find the number of dots in the sixth member of the pattern.

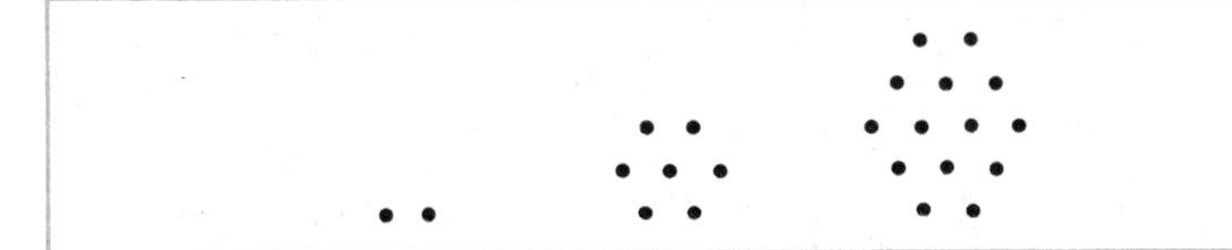

4

6 27 players enter a knockout golf competition in which the competitors play in groups of 3. Draw a 'tree-diagram' for the competition and write down the sequence of the number of players taking part in each round.

7 The owner of a factory reckons that the value of one of his machines at the end of a year's work is only half what it was at the beginning of that year. If the machine cost £512 when it was new, what was its value after 1, 2, 3, 4 years? If its scrap value is £5, after how many years should he sell it?

8 Consider the sequence of natural numbers 1, 2, 3, 4, ...

a Find the average of the first 9 terms of the sequence.

b Is the average a term of the sequence?

c Find the average of the first 10 terms of the sequence.

d How is this average related to the first and last terms?

e Can you use the ideas in parts ***c*** and ***d*** to find the sum of the first 99 terms of the sequence, and of the first 144 terms?

9 Hang a tape measure vertically on the wall. Drop a golf ball, or better still, a round lump of silica putty, from a height of, say, 2 metres and measure the height to which it rebounds after each bounce (try to use a concrete floor). Do the successive heights form a sequence for which you can find a rule for obtaining the next term?

Revision Exercise 3B

1 Write down the first five terms of the sequences whose nth terms are:

a $n(n-1)$ *b* $\dfrac{n}{n+1}$ *c* $\dfrac{120}{n}$

d $\dfrac{n(n+1)(2n+1)}{6}$ *e* $\dfrac{n^2(n+1)^2}{4}$

2 Continue each of the letter sequences shown in Figure 5 to five terms, and find a formula for the number of dots in the nth term of each.

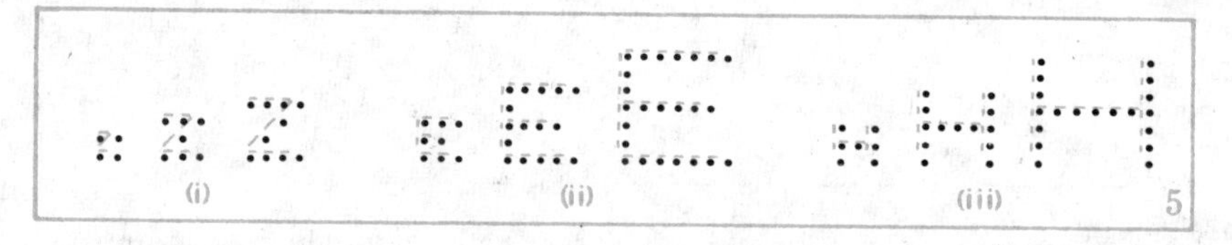
(i) (ii) (iii) 5

3 Find a formula for the nth term of each of the following sequences:

a 7, 9, 11, 13, ... *b* $\frac{1}{2}, \frac{1}{3}, \frac{1}{4}, \frac{1}{5}, \ldots$

c 1, 2, 4, 8, ... *d* $\frac{1}{3}, \frac{1}{9}, \frac{1}{27}, \frac{1}{81}, \ldots$

4 In Figure 6 (i), ABCD is a square. PQRS is a second square made by joining the mid-points of the sides of ABCD. What can you say about the area of square PQRS compared with the area of ABCD? (If necessary draw the diagram on 5-mm squared paper.) Express the areas of the squares formed by joining the mid-points of sides as a sequence of numbers, taking the area of ABCD as 1 unit. Can you write a formula for the area of the nth square?

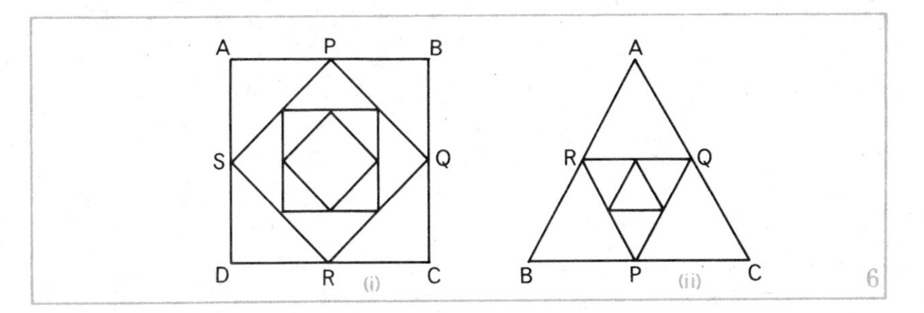

5 Repeat the investigation for the sequence of equilateral triangles in Figure 6 (ii).

6 *a* Write down the next term of the sequence
$1, (3+5), (7+9+11), (13+15+17+19), \ldots$
Simplify each term by adding the numbers in brackets. Do you recognize the resulting sequence? Write down the simplified form of the sixth and tenth terms. How could you find the ninety-ninth term?

b Repeat the above for each of the following, in which the terms are written underneath each other.

(1)	(2)
1	1^2
$(1+2)+1$	$(1+2)^2-1^2$
$(1+2+3)+(1+2)$	$(1+2+3)^2-(1+2)^2$
$(1+2+3+4)+(1+2+3)$	$(1+2+3+4)^2-(1+2+3)^2$
... ...	... ...

7 Although it is now illegal to do so, it was at one time permissible for a person to originate a 'chain letter' by sending out, say, one letter to each of four people; these four people were each asked to send out a second round of four letters, and so on. How many rounds would be required before every man, woman and child *a* in Scotland *b* in the United Kingdom had received a letter, assuming that no one received more than one letter? You should estimate the answer, then build up the sequence and find the answer. It may surprise you. (Population of Scotland 5 200 000, of the United Kingdom 55 500 000).

8 A sheet of paper is 0·001 cm thick. It is torn in half and the parts are placed one on top of the other. This doubled sheet is torn in half and the parts again placed one on top of the other; how thick will the pile now be? The process is repeated again and again. How often will you require to tear the paper to make a pile one metre thick?

Trigonometry

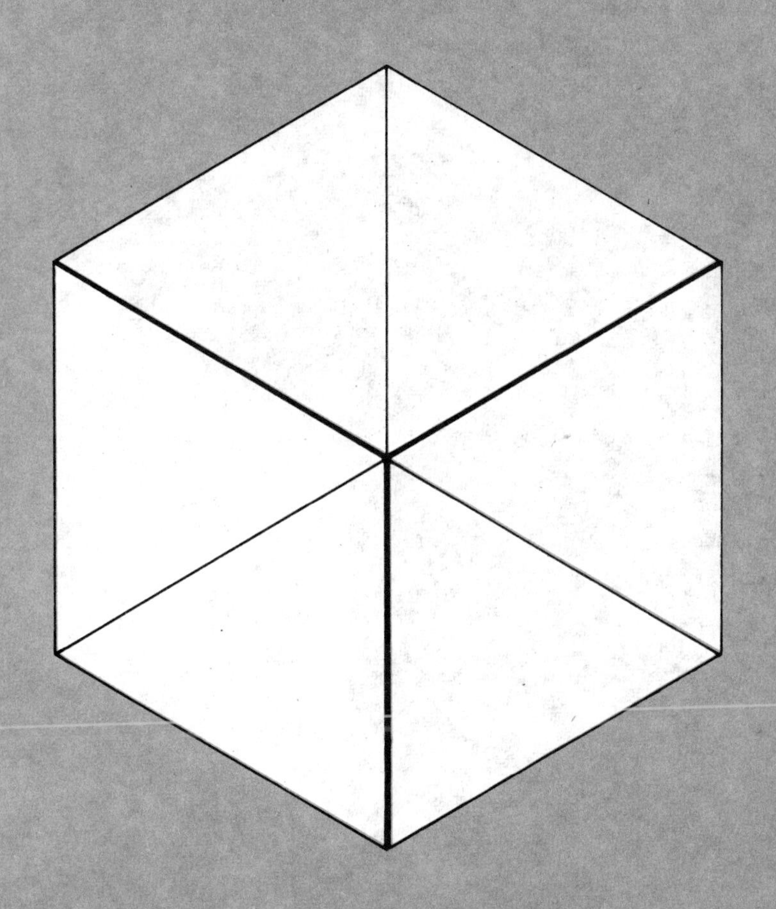

Trigonometry

The Cosine, Sine and Tangent Functions

1 *The definitions of the cosine, sine and tangent functions*

Figure 1 represents the Big Wheel at a funfair. Suppose you start in the chair at A, and the wheel turns in the direction indicated by the arrow.

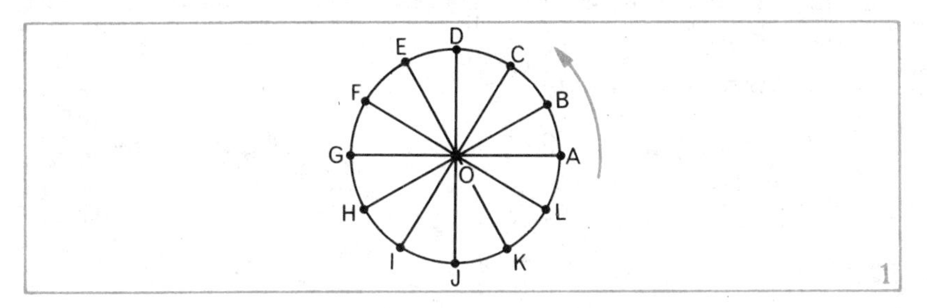

1

Exercise 1

1 At which positions would you be:

a highest *b* lowest

c farthest to the left of O *d* farthest to the right of O?

2 Describe how the size of the angle between OA and the horizontal changes as the wheel turns, taking:

a OA to OB *b* OA to OD *c* OA to OF

d OA to OG *e* OA to OJ *f* OA to OA.

3 Copy Figure 1 and draw in coloured lines to show:

a the distances of all the chairs to the right and left of the vertical through O

b the distances of all the chairs above and below the horizontal through O.

4 Complete the following:

As OA rotates from the horizontal to the position OD:—

the size of the angle between OA and the horizontal increases from 0° to ... ;
the distance of A to the right of O decreases from OA to ... ;
the distance of A above O increases from zero to ....
Do you see that the distances of A from the vertical and horizontal through O depend on the size of the angle between OA and the horizontal?

In this chapter we investigate various relationships between the lengths of lines and the sizes of angles associated with them.

5 In Figure 2, points P_1, P_2, and P_3 are taken on OA, and perpendiculars are drawn to OX and OY as shown.

a In Figure 2(i) name two triangles which are similar to triangle OM_1P_1.

b Give two ratios equal to $\frac{OM_1}{OP_1}$.

c What construction could you make in the diagram to give more ratios equal to $\frac{OM_1}{OP_1}$?

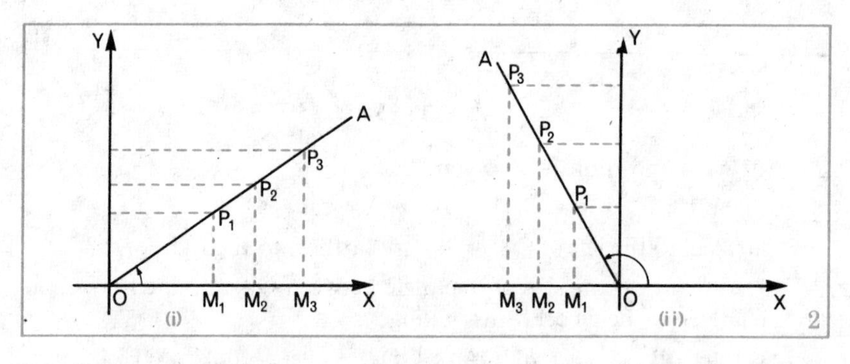

2

6 a In Figure 2 (ii), name two triangles which are similar to triangle OM_1P_1.

b Give two ratios equal to $\frac{OM_1}{OP_1}$.

c What construction could you make in the diagram to give more ratios equal to $\frac{OM_1}{OP_1}$?

Note. For a given size of $\angle XOA$ there is only one value of $\frac{OM}{OP}$

for all positions of P on OA, where M is the foot of the perpendicular from P to the x-axis.

7 From Figure 2 (i), give two ratios equal to $\frac{M_1P_1}{OP_1}$.

Note. For a given size of $\angle$XOA there is only one value of $\frac{MP}{OP}$ for all positions of P on OA.

8 From Figure 2 (ii), give two ratios equal to $\frac{M_1P_1}{OM_1}$.

Note. For a given size of $\angle$XOA there is only one value of $\frac{MP}{OM}$ for all positions of P on OA.

9 In Figure 2 (i), OA is in the *first quadrant*; in Figure 2 (ii), OA is in the *second quadrant*. Draw corresponding diagrams in which OA is in the third and fourth quadrants. Notice that for all sizes of angle XOA there is only one value of the ratio $\frac{OM}{OP}$, of $\frac{MP}{OP}$, and of $\frac{MP}{OM}$ for all positions of P on OA.

In this Exercise we have been thinking of OA as a line which can rotate about the origin in order to give any size of angle XOA. Let P be a point on OA such that the length of OP is r, where $r > 0$. For a given angle size $a°$ there is only one position of OA, and only one pair of coordinates for P; let P be the point (x, y), as shown in Figure 3.

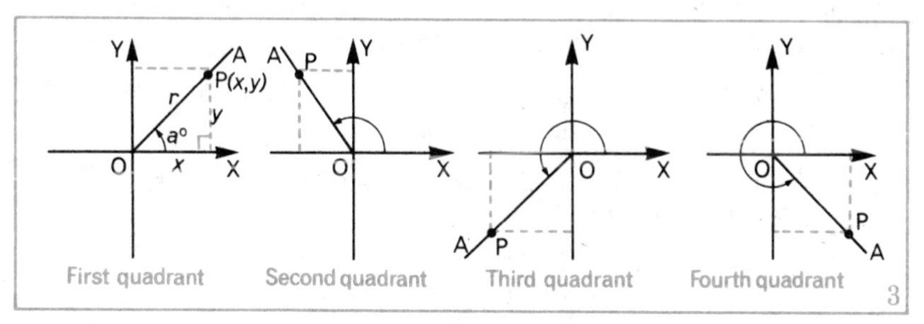

3

We define:

the cosine of $a°$ to be $\frac{x}{r}$ (e.g. $\frac{OM_1}{OP_1}$ in questions 5 and 6) the sine of $a°$ to be $\frac{y}{r}$ (e.g. $\frac{M_1P_1}{OP_1}$ in question *7*)

the tangent of $a°$ to be $\frac{y}{x}$ (e.g. $\frac{M_1P_1}{OM_1}$ in question *8*).

We write:

$$\cos a° = \frac{x}{r}, \qquad \sin a° = \frac{y}{r}, \qquad \tan a° = \frac{y}{x}.$$

From questions ***6***, ***7*** and ***8*** we know that to each angle size $a°$ there is one, and only one, value of $\cos a°$, of $\sin a°$, and of $\tan a°$. So we have *mappings* from the set of angle sizes $A = \{a° : a \in R\}$ to the set of real numbers R, as indicated in Figure 4.

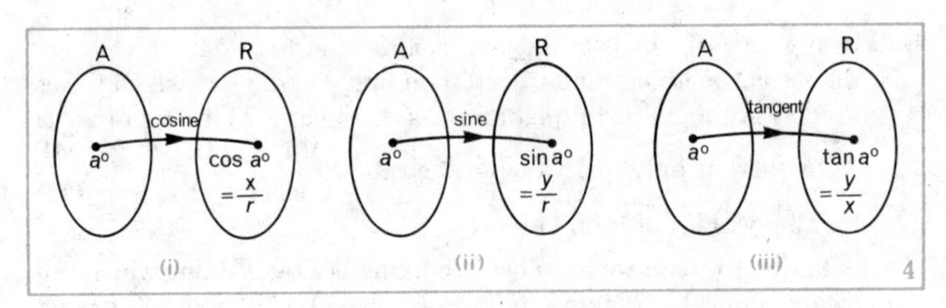

4

We say that the cosine, sine and tangent *functions* map the set of angle sizes to the set of real numbers.

As P may lie in any quadrant its x- and y-coordinates may be positive, negative or zero, and hence the values of cosine, sine and tangent may be positive, negative or zero.

Note. We usually abbreviate 'the cosine of angle XOP' to 'cos XOP' as in the following example.

Example 1. In Figure 5, P is the point (8, 6). Find the values of cos XOP, sin XOP and tan XOP.

By Pythagoras' theorem, $OP^2 = 8^2 + 6^2 = 64 + 36 = 100$.

$$\text{So } OP = 10.$$

$$\cos XOP = \frac{x}{r} = \frac{8}{10} = 0{\cdot}8; \ \sin XOP = \frac{y}{r} = \frac{6}{10} = 0{\cdot}6;$$

$$\tan XOP = \frac{y}{x} = \frac{6}{8} = 0{\cdot}75.$$

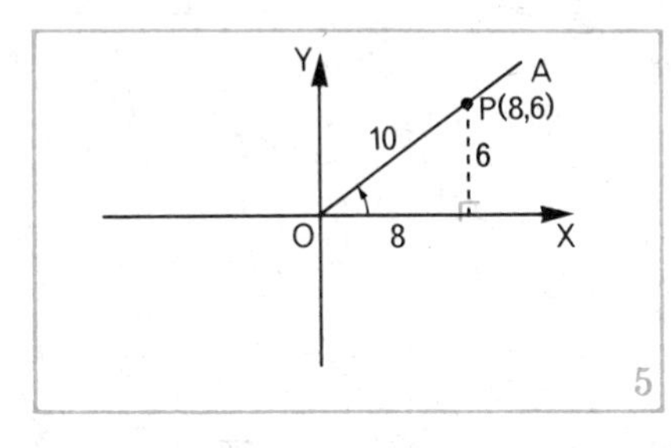

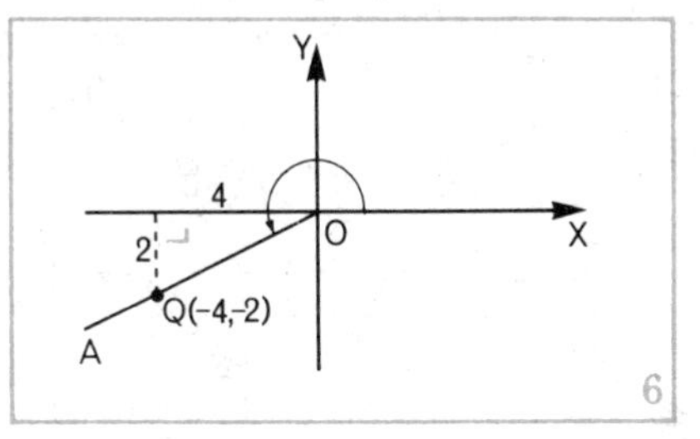

Example 2. In Figure 6, Q is $(-4, -2)$. Find cos XOQ, sin XOQ, tan XOQ.

By Pythagoras' theorem, $OQ^2 = 4^2+2^2 = 16+4 = 20$.

So $OQ = \sqrt{20}$.

$\cos XOQ = \frac{x}{r} = -\frac{4}{\sqrt{20}}$; $\sin XOQ = \frac{y}{r} = -\frac{2}{\sqrt{20}}$;

$\tan XOQ = \frac{y}{x} = \frac{-2}{-4} = \frac{1}{2}$.

Exercise 2

1 Make a sketch for each of the following points, and calculate the values of cos $a°$, sin $a°$ and tan $a°$, where $a°$ is the size of ∠XOA, then ∠XOB, etc.

A (4, 3), B $(-6, 8)$, C $(-12, -5)$, D $(8, -6)$

2 Without using a sketch calculate the values of the cosine, sine and tangent of ∠XOP, given that P is the point (7, 24).

3 From Figure 7 give the coordinates of A, and the length of OA. Write down the values of cos XOA, sin XOA and tan XOA.

4 Repeat question ***3*** for B and ∠XOB, leaving square roots in your answers.

5 Repeat question ***3*** for C and ∠XOC.

6 Repeat question ***3*** for D and ∠XOD.

7 Repeat question ***3*** for E and ∠XOE.

8 a List the values of cos XOP, sin XOP and tan XOP for each of the points P_1 (2, 2), $P_2(-2, 2)$, P_3 $(-2, -2)$, P_4 $(2, -2)$.

b In which quadrants are the various functions positive?

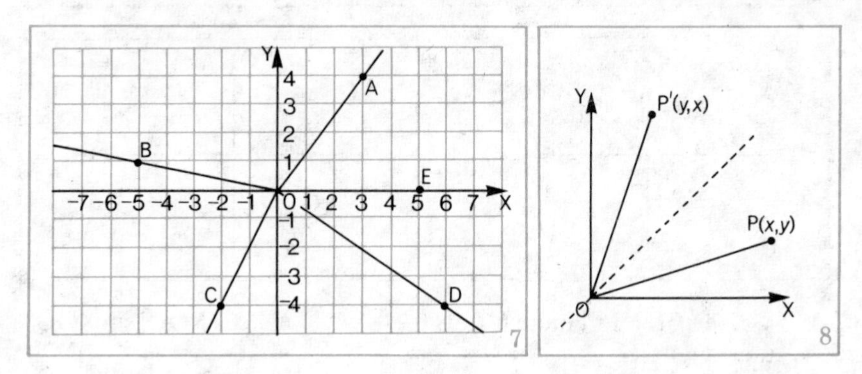

Note. In Figure 8, the image of P (x, y) in the bisector of $\angle$XOY is P′ (y, x), so cos XOP $= \frac{x}{r} =$ sin XOP′. Thus there are pairs of angles such that the cosine of the one angle is the sine of the other; when $\angle$XOP is acute, $\angle$XOP′ is its complement. This explains why one of the functions is called the *co-function* of the other function.

2 *Calculating values of cos a°, sin a° and tan a°*

Copy Figure 9 on to 2-mm squared paper; note that the scale is 1 cm per unit on each axis, and that the radius r is 10 cm long. $\angle XOP_1 = 10°$, $\angle XOP_2 = 20°$, $\angle XOP_3 = 30°$, etc.

Your diagram will enable you to calculate approximate values of cos $a°$, sin $a°$ and tan $a°$ for $a = 0, 10, 20, 30, \ldots, 90$.

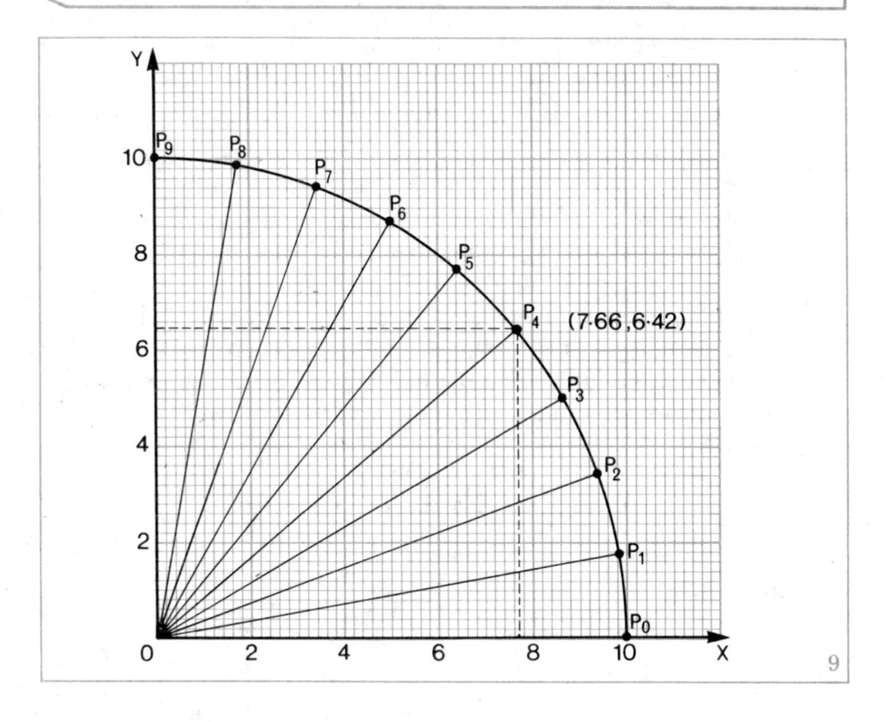

In the table below, the coordinates of P_4, where $\angle XOP_4 = 40°$, have been filled in, and from these the values of cos 40°, sin 40° and tan 40° have been calculated from the definitions:

$$\cos a° = \frac{x}{r} \qquad \sin a° = \frac{y}{r} \qquad \tan a° = \frac{y}{x}$$

Angle $a°$	P		cos $a°$	sin $a°$	tan $a°$
a	x	y			
0					
10					
20					
30					
40	7·66	6·42	0·77	0·64	0·84
50					
60					
70					
80					
90					

Copy and complete the table, using a slide rule to ease the calculation where necessary.

Note. (i) At one place in the table, when finding tan 90°, we are faced with $\frac{y}{x}$ and $x = 0$. Division by zero is meaningless and undefined, so tan 90° is also undefined.

(ii) Using more advanced mathematical methods it is possible to *calculate* cos a°, sin a°, and tan a° to any required degree of accuracy for all angle sizes, and some of these values are given with three-figure accuracy in mathematical tables. Look up the appropriate entries in the natural cosine, natural sine, and natural tangent tables, and compare the values given there with those in your table.

Exercise 3

In questions *1–3* use mathematical tables.

1 Read off, or write down, the values of:

a	sin 27°	*b*	sin 43·5°	*c*	sin 30°	*d*	cos 8°
e	cos 56·8°	*f*	cos 88·1°	*g*	tan 45°	*h*	tan 67·7°
i	tan 87·0°	*j*	cos 60°	*k*	sin 0·7°	*l*	tan 23·4°

2 Find the size of the acute angle in each of the following:

a	$\sin a° = 0{\cdot}437$	*b*	$\cos b° = 0{\cdot}009$	*c*	$\tan c° = 0{\cdot}488$
d	$\cos d° = 0{\cdot}866$	*e*	$\sin e° = 0{\cdot}866$	*f*	$\tan f° = 1{\cdot}732$
g	$\sin g° = 0{\cdot}162$	*h*	$\tan h° = 300$	*i*	$\cos i° = 0{\cdot}076$

3 Which of the following are true and which are false (considering acute angles only, and accepting the values given in your tables)?

a	If $\tan p° = 0{\cdot}213$, $p = 12$	*b*	If $\sin q° = 0{\cdot}648$, $q = 40{\cdot}4$
c	If $\sin r° = 0{\cdot}973$, $r = 69{\cdot}5$	*d*	If $\tan s° = 4{\cdot}586$, $s = 77{\cdot}7$
e	If $\cos t° = 0{\cdot}334$, $t = 70{\cdot}5$	*f*	If $\cos u° = 0{\cdot}999$, $u = 89{\cdot}5$
g	If $\sin v° = 1$, $v = 90$	*h*	If $\tan w° = 1$, $w = 45$

4 Figure 10 shows a square OMPN with unit side.

a Calculate the length of OP, leaving the square root in your answer.

b Using the definitions $\cos a° = \frac{x}{r}$, $\sin a° = \frac{y}{r}$, $\tan a° = \frac{y}{x}$, write down ratios for cos 45°, sin 45° and tan 45°.

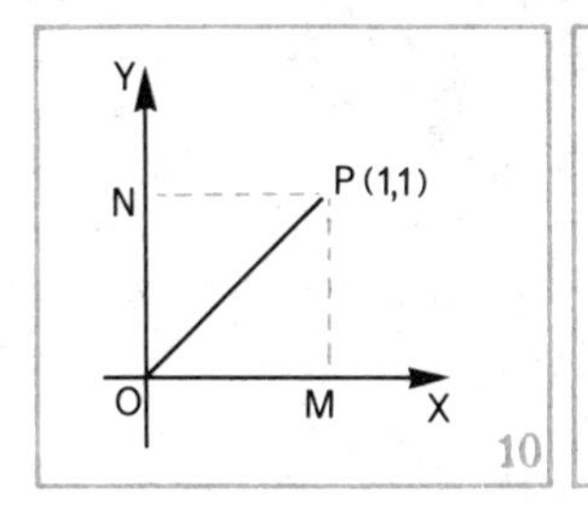

10

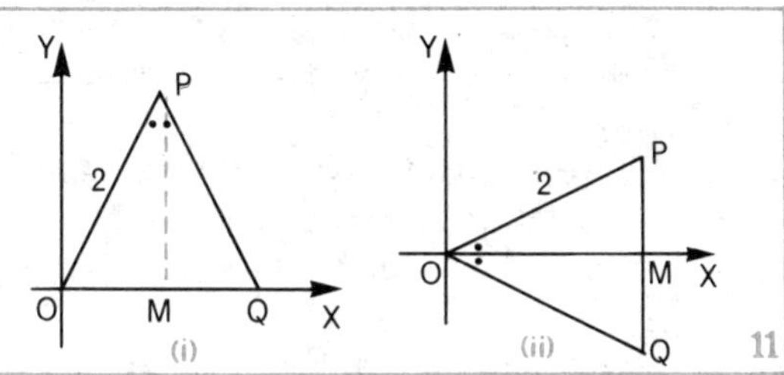

11

5 Each part of Figure 11 contains an equilateral triangle OPQ of side 2 units, with one of its angles bisected.

a From Figure (i) find the coordinates of P (leave the square root where necessary). Write down ratios for cos 60°, sin 60° and tan 60°.

b From Figure 11 (ii) find the coordinates of P.

Write down the ratios for cos 30°, sin 30° and tan 30°.

6 Using your answers to questions ***4*** and ***5*** complete this table:

a°	30°	45°	60°
cos a°	.	.	.
sin a°	.	.	.
tan a°	.	.	.

3 Calculations concerning heights and distances

In Figure 12 (i), ∠XOA is an angle in the first quadrant and P is the point (x, y).

We know that $\cos XOA = \frac{x}{r}$; $\sin XOA = \frac{y}{r}$; $\tan XOA = \frac{y}{x}$.

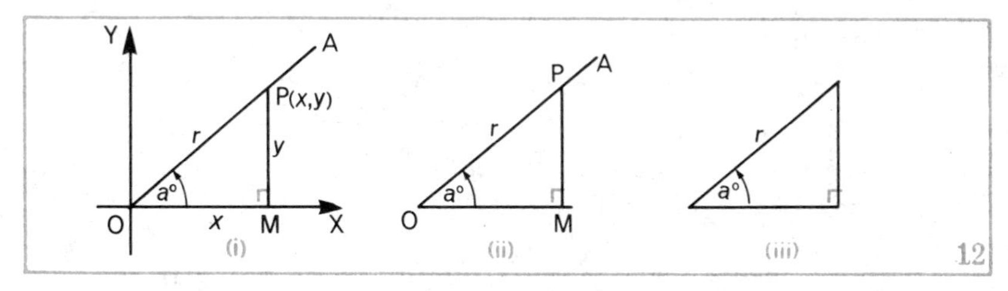

12

Imagine the axes to fade out of the diagram, leaving Figure 12 (ii), so that x and y now have no meaning. What shall we call the angle now?

If we wanted to use the diagram to work out approximate values of cos MOP, sin MOP, and tan MOP, what measurements would we have to make in each case, and what calculations would we have to perform?

Note. For a given angle in a right-angled triangle we can always place r, x, and y by observing that r is *opposite* the right angle, x is the other arm of the angle, and y is the remaining side. It might be helpful at first to mark these in the diagrams as shown in Figure 13, in the order r, x, y.

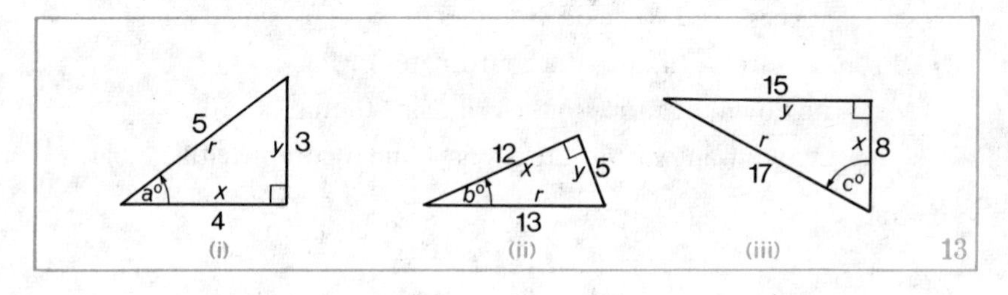

13

Exercise 4

1. Use Figure 13 to write down ratios for cos $a°$, sin $a°$, tan $a°$; cos $b°$, sin $b°$, tan $b°$; and cos $c°$, sin $c°$, tan $c°$.
2. Write down ratios for cos $z°$, sin $z°$ and tan $z°$ in each triangle in Figure 14.

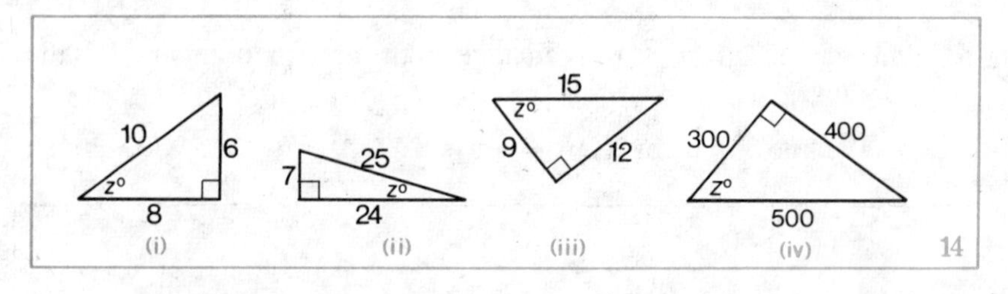

14

3. Write down ratios for cos $u°$, sin $u°$ and tan $u°$ in each triangle in Figure 15.

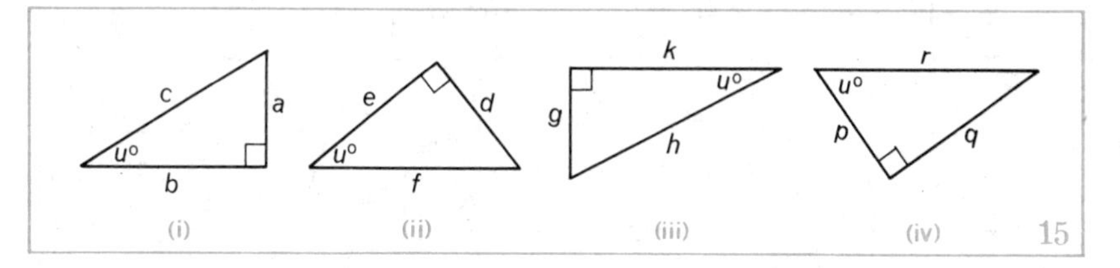

(i) (ii) (iii) (iv) 15

Example. Calculate h in Figure 16 (i).

In the triangle we mark the positions of r (opposite the right angle), then x (the other arm of the given angle), then y (the remaining side). We see that the two sides we 'know' are y and r, so we choose the ratio $\frac{y}{r}$ which is the sine ratio.

Then $\sin 40° = \frac{h}{12}$

$$\Leftrightarrow h = 12 \sin 40°$$
$$= 12 \times 0{\cdot}643$$
$$= 7{\cdot}72$$

We can use logarithmic *sines* for the calculation.

nos	logs
12	1·079
sin 40°	$\bar{1}$·808
7·71	0·887

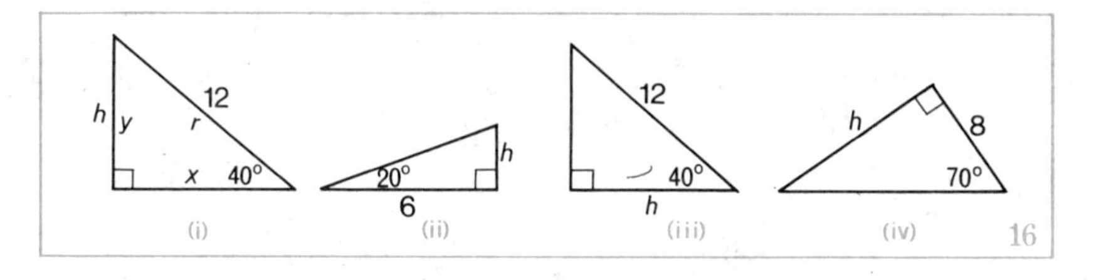

(i) (ii) (iii) (iv) 16

4 Calculate h in Figure 16 (ii), (iii) and (iv).

5 Calculate d in each triangle in Figure 17.

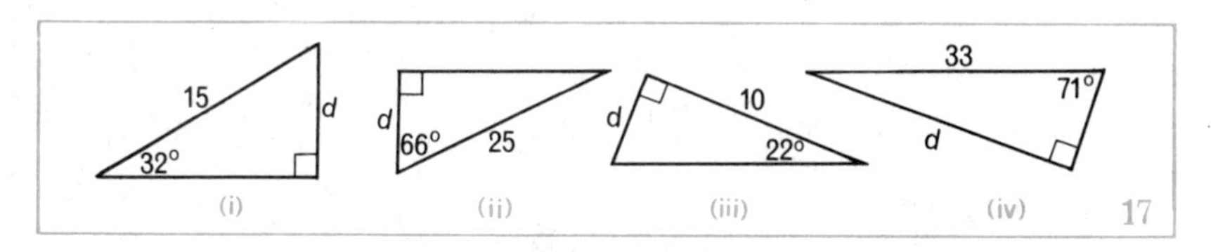

(i) (ii) (iii) (iv) 17

6 Calculate a in each triangle in Figure 18.

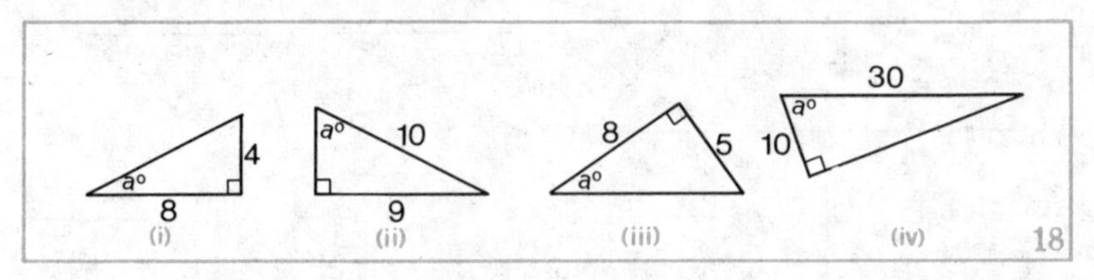

18

Example. The pilot of a helicopter hovering 1200 m above a level stretch of land 'spots' a landmark at an *angle of depression* of 35°. How far is the landmark from the point on the ground directly below the helicopter?

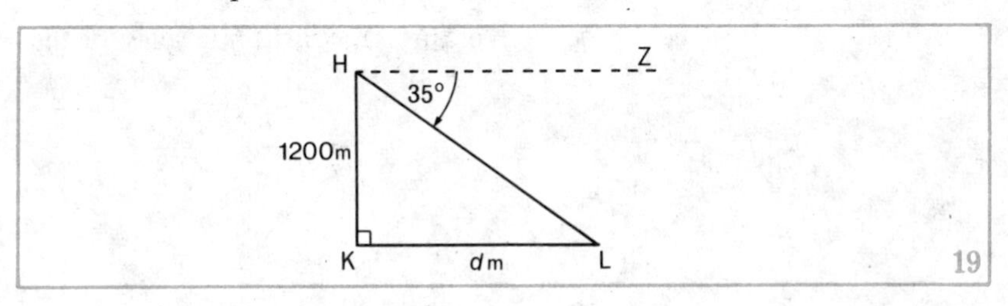

19

In Figure 19, H represents the helicopter and L the landmark, so that ∠ZHL is the angle of depression of the landmark. We have to calculate KL.

Method 1

∠HLK = 35° (alternate to ∠ZHL).

In △HKL, we now have a $\frac{y}{x}$ situation with respect to ∠HLK.

$$\tan 35^\circ = \frac{1200}{d}$$

$$\Leftrightarrow d \times \tan 35^\circ = 1200$$

$$\Leftrightarrow \quad d = \frac{1200}{\tan 35^\circ}$$

$$= 1710$$

nos	logs
1200	3·079
tan 35°	$\bar{1}$·845
1710	3·234

Method 2

∠LHK = 55° (∠ZHK = 90°).

$$\tan 55^\circ = \frac{d}{1200}$$

$$\Leftrightarrow \quad d = 1200 \times \tan 55^\circ$$

$$= 1710$$

nos	logs
1200	3·079
tan 55°	0·155
1710	3·234

The distance is about 1710 metres.

Exercise 5

1 A boy flying a kite lets out 200 m of string. If the string is taut and makes an angle of 72° with the horizontal, what is the height of the kite?

2 A ladder is 15 m long. The top rests against the wall of a house, and the foot rests on level ground 2 m from the wall. Calculate the angle between the ladder and the ground.

3 A ladder 12 m long is set against the wall of a house, and makes an angle of 75° with the ground.

a How far up the wall will the ladder reach?

b How far is the foot of the ladder from the wall?

4 From a point 400 m from the base of a tower, and on the same level as the base as shown in Figure 20, the angle of elevation of the top of the tower is 39°. Calculate the height of the tower.

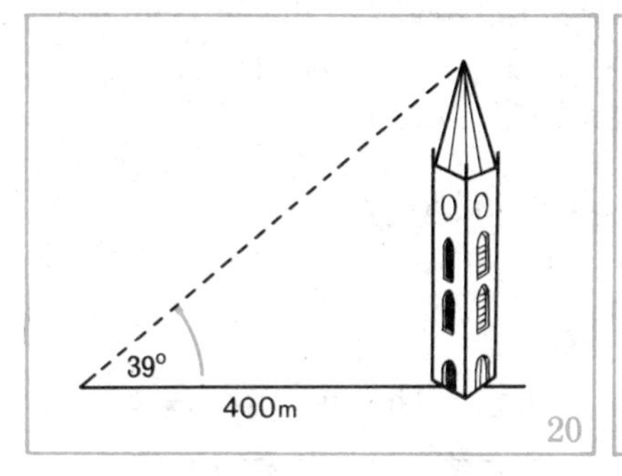

20

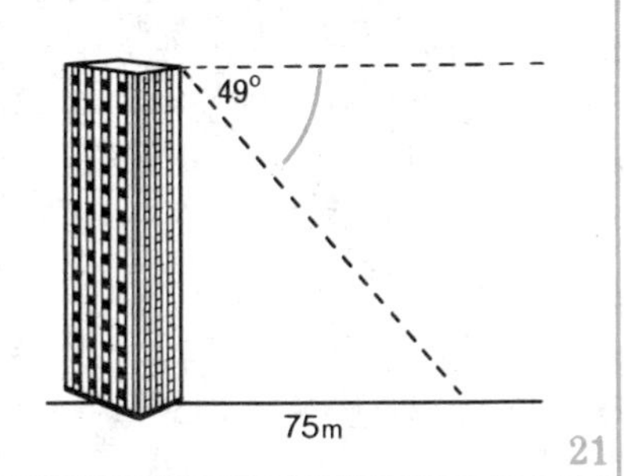

21

5 From the top of a block of high flats, as shown in Figure 21, the angle of depression of a landmark 75 m from the foot of the flats is 49°. Calculate the height of the flats.

6 An aircraft flies 200 km on a north-easterly course. How far:

a north *b* east, will it be from its starting point?

7 A telegraph pole standing on horizontal ground is 9 m high, and is supported by a wire 10 m long fixed to the top of the pole and to the ground. Calculate:

a the angle between the wire and the ground

b the distance of the point on the ground from the foot of the pole (*1*) using Pythagoras' theorem (*2*) using trigonometry.

8 Find the height above the centre of a clockface of the tip of the hour hand, which is 10 cm long, at the following times:

a 1 pm *b* 2 pm *c* 3 pm.

9 In a rectangle ABCD, AB = 10 cm and BC = 7 cm. Calculate the size of ∠BAC. Hence write down the sizes of the other angles at A and C.

10 PQRS is a rhombus. Its diagonals PR and QS are 16 cm and 10 cm long respectively. Calculate the sizes of the angles of the rhombus.

11 EFGH is a kite. Diagonal EG bisects diagonal HF at M. EM = 9 cm, MG = 5 cm and HF = 12 cm. Calculate the sizes of the angles of the kite, and the lengths of its sides.

12 From ship A, ship B bears 000° at a distance of 25 km. From B, ship C bears 270° at a distance of 20 km. Calculate the bearing and distance of C from A.

13 A straight road 350 m long rises 10 m vertically from one end to the other. Calculate the angle between the road and the horizontal.

14 A boy at a point P on one edge of a straight stretch of river observes two posts Q and R on the opposite bank. Q is directly across from P, and the distance between Q and R is 60 m. If the boy estimates angle QPR to be 40°, calculate the width of the river.

Exercise 5B

1 Calculate d in each diagram in Figure 22. The units are metres.

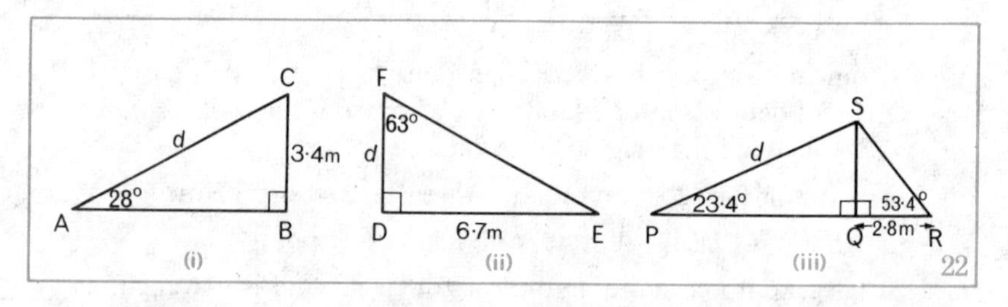

22

2 ABCD is a rectangle with its diagonals intersecting at O. If AB = 5 cm and BC = 3 cm, calculate the sizes of angles BAC and AOB.

3 In △ABC, AB = 3·25 cm, ∠ABC = 43·6° and ∠ACB = 58·3°. Draw the altitude AD of the triangle, and calculate its length. Hence calculate the length of AC.

4 A regular hexagon is inscribed in a circle of radius 8·5 cm. Calculate the distance from the centre of the circle to a side of the hexagon.

5 A horizontal concrete floor is 30 cm thick. A small hole is bored through it at an angle of 35° to the vertical. Calculate the length of the hole to the nearest cm.

6 A mine-working $\frac{1}{2}$ km long rises steadily at 5° to the horizontal, and then for $\frac{3}{4}$ km it rises steadily at 17° to the horizontal. What are the total horizontal and vertical distances covered by the mine-working?

7 A ship sails 48 km on a bearing 065°, then 112 km on a bearing 012°. How far east and how far north is the ship from its starting point?

8 An observer whose eye is 2·5 m above the ground observes the angle of depression of the foot of the building to be 5°, and the angle of elevation of the top to be 20°. Calculate the horizontal distance of the observer from the building, and the height of the building, each to the nearest metre.

9 From the top of a cliff 125 m high the angles of depression of two boats, both due east of the observer, are 16° and 23°. Calculate the distance between the boats.

10 A mast 75 m high stands at corner A of a square horizontal field ABCD of side 160 m. Find to the nearest degree the angles of elevation of the top of the mast from B and from C.

4 *Extending the use of the trigonometrical tables*

In geometrical diagrams it is never necessary to consider sizes of angles outside the range 0° to 360°, but in practical situations, e.g. concerning a turning wheel as in Section 1, it may be useful to consider angles of all sizes. To do this we think of an angle formed by the rotation of a line from the position of the x-axis.

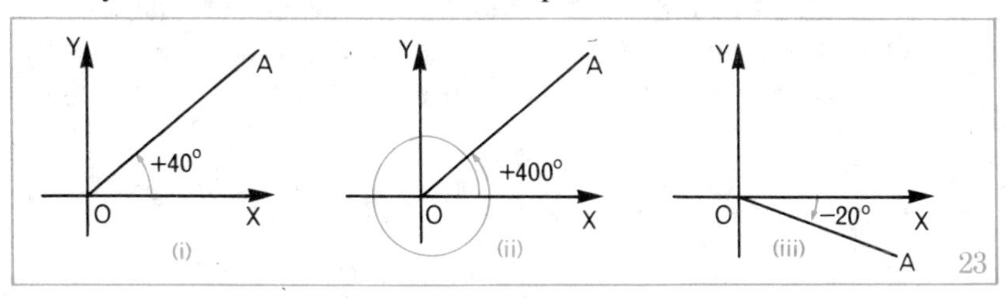

In Figure 23 a straight line is rotated about O from OX to OA. If the rotation is *anticlockwise* (as in (i) and (ii)) the angle formed by the rotation, $\angle$XOA, is *positive*. In (i) the angle is $+40°$; in (ii) the angle is $+400°$, and it can be seen that the same position of OA is given by $400°$, i.e. $40° + 360°$, as by $40°$.

If the rotation is *clockwise* (as in (iii)) then $\angle$XOA is *negative*. In (iii) the angle is $-20°$.

Exercise 6

1 Suppose that in Figure 23 (i) $\angle$XOA $= 80°$. Give two more measures for $\angle$XOA.

2 Draw a sketch to show a rotation of $120°$ from OX. Write down two more angle sizes that would give the same final position of the rotating line.

3 Repeat question *2* for an initial angle size of $-15°$.

4 Repeat question *2* for an initial angle size of $200°$.

5 Draw a sketch to illustrate a rotation of $60°$ from OX. If a rotation of $60° + 360k°$ gives the same final position of the rotating line, to what set of numbers must k belong?

6 a If the magnitude of a rotation is given (e.g. $50°$), how many final positions are there of the rotating line?

b If the final position of the rotating line is given, how many magnitudes can the rotation have?

* * *

In Section 1 we had the definition of the cosine, sine, and tangent functions for every size of angle. Figure 24 shows one way of relating an angle in the first quadrant to angles in the other three quadrants; this enables us to use the cosine, sine, and tangent tables for angles of any size.

Figure 24 (i) illustrates $\angle$XOA in the first quadrant. We now reflect this figure in the axes so that there is line symmetry about each axis, as shown in Figure 24 (ii).

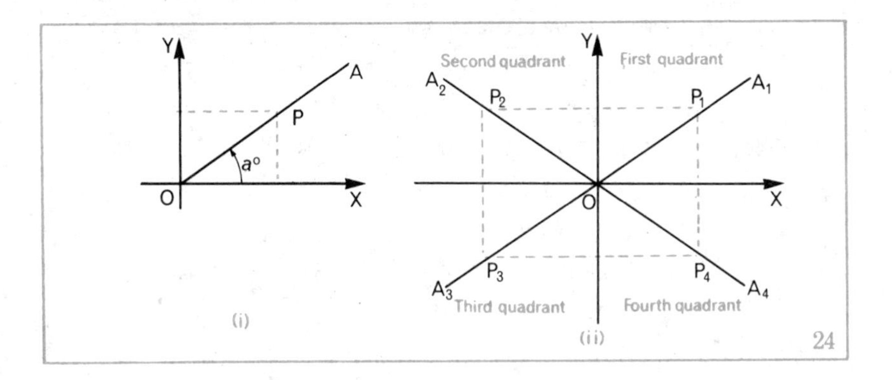

24

Exercise 7

1 a P_1 is the point (4, 3). Calculate r, the length of OP_1.

b Copy and complete this table, referring to Figure 24.

P	x	y	r	cos XOP	sin XOP	tan XOP
P_1	4	3				
P_2						
P_3						
P_4						

c From your table, in which quadrants is each function positive?

2 Repeat question *1* for the point P_1 (5, 12).

Remembering that $\cos a^\circ = \frac{x}{r}$, $\sin a^\circ = \frac{y}{r}$, $\tan a^\circ = \frac{y}{x}$, we see that:

$\cos a^\circ$ is positive $\Leftrightarrow x > 0 \Leftrightarrow (x, y)$ is in first or fourth quadrants
$\sin a^\circ$ is positive $\Leftrightarrow y > 0 \Leftrightarrow (x, y)$ is in first or second quadrants
$\tan a^\circ$ is positive $\Leftrightarrow \frac{y}{x} > 0 \Leftrightarrow (x, y)$ is in first or third quadrants.

These results are summarized in Figure 25.

Second quadrant	SINE POSITIVE	ALL POSITIVE	First quadrant
Third quadrant	TANGENT POSITIVE	COSINE POSITIVE	Fourth quadrant

25

3 State whether the value of each of the following is positive or negative:

a sin 73° *b* cos 158° *c* sin 285° *d* tan 141°
e sin 200° *f* cos 300° *g* tan 350° *h* sin 400°

4 Referring to Figure 24 (ii), calculate the size of $\angle XOA_1$, given that:

a $\angle XOA_2 = 100°$ *b* $\angle XOA_3 = 230°$ *c* $\angle XOA_4 = 333°$

Example. Use mathematical tables to find the values of:

a cos 150° ***b*** sin 200° ***c*** tan 315° ***d*** cos (−45°)

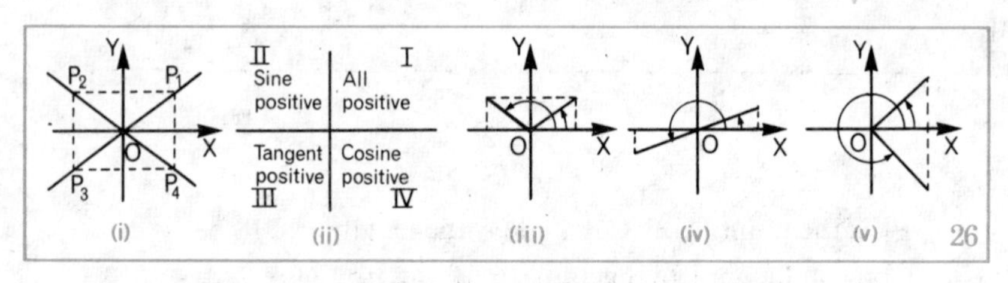

26

In each case we use the ideas in Figure 26.

a From Figure 26 (i) or (iii), the related acute angle is 30°.
From Figure 26 (ii), the sign of cos 150° is −.
So cos 150° = −cos 30° = −0·866.

b In the same way, sin 200° = −sin 20° = −0·342.

c tan 315° = −tan 45° = −1.

d cos (−45°) = cos 45° = 0·707.

Exercise 8

1 Use mathematical tables to give the values of:

a sin 40° *b* sin 140° *c* sin 220° *d* sin 320°
e cos 150° *f* cos 200° *g* cos 300° *h* cos 400°
i tan 100° *j* tan 247° *k* tan 0° *l* tan 380°
m sin (−10°) *n* cos (−20°) *o* tan (−30°) *p* sin 234°

2 For each of the following, find two replacements for x between 0° and 360°:

a $\sin x° = 0{\cdot}766$ *b* $\cos x° = 0{\cdot}565$ *c* $\tan x° = 4{\cdot}915$
d $\cos x° = -0{\cdot}906$ *e* $\sin x° = -0{\cdot}707$ *f* $\tan x° = -2{\cdot}050$
g $\sin x° = 0{\cdot}415$ *h* $\tan x° = 0{\cdot}193$ *i* $\cos x° = -0{\cdot}174$

3 In Exercise 3 you found values of cos 45°, sin 45° and tan 45°.

a Using a triangle like that in Figure 27 (i), write down values of cos 45°, sin 45°, tan 45°.

b Hence write down values of cos 135°, cos 225° and cos 315°.

c Also write down values of sin 135°, sin 225° and sin 315°.

d Also write down values of tan 135°, tan 225° and tan 315°.

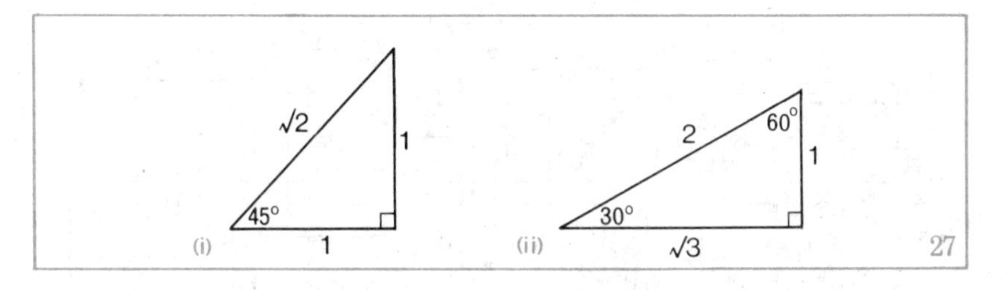

27

4 a Using Figure 27 (ii), write down values of cos 30°, sin 30°, tan 30°.

b Hence write down values of cos 150°, cos 210°, cos 330°.

c Also write down values of sin 150°, sin 210°, sin 330°.

d Also write down values of tan 150°, tan 210°, tan 330°.

5 a Using Figure 27 (ii), write down values of cos 60°, sin 60°, tan 60°.

b Hence write down values of cos 120°, cos 240°, cos 300°.

c Also write down values of sin 120°, sin 240°, sin 300°.

d Also write down values of tan 120°, tan 240°, tan 300°.

6 Find the height above the centre of a clockface of the tip of an hour-hand 2·5 m long at *a* 1 am *b* 4 pm *c* 10 am *d* midday.

7 Suppose that the radius of the Big Wheel in Figure 1 (page 209) is 4 m, and that O is 6 m above the ground. Calculate the height above the ground of each place at which you might stop.

5 Graphs of the trigonometrical functions

Exercise 9

1 *The sine graph*

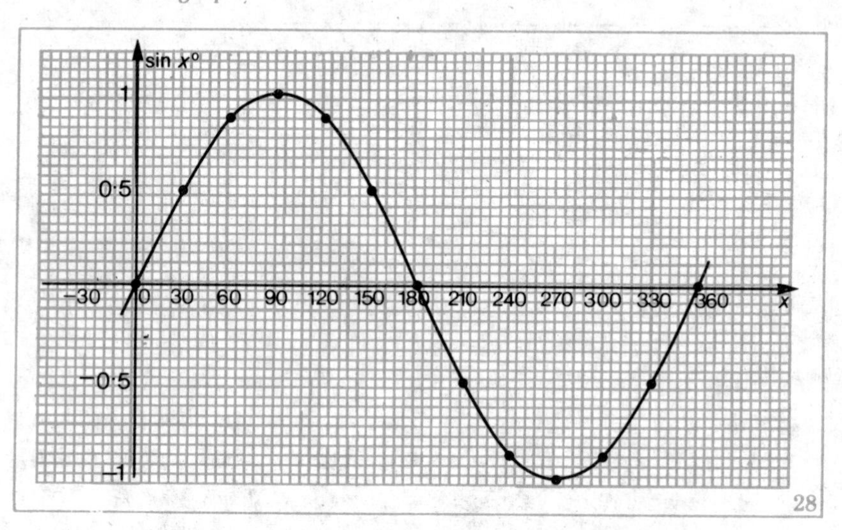

28

With the aid of mathematical tables use ***a***, ***b***, ***c***, ***d*** below to draw part of the sine graph as shown in Figure 28. Use 2-mm squared paper, and take scales of 30 units to 1 cm horizontally and 1 unit to 4 cm vertically.

a $\sin 0° = 0$. Plot the points (0, 0), (180, 0), (360, 0).

b $\sin 30° = 0{\cdot}5$. Plot (30, 0·5), (150, 0·5), (210, −0·5), (330, −0·5).

c Proceed as in ***b*** by 30° intervals, plotting four points at a time.

d $\sin 90° = 1$ gives (90, 1) and (270, −1).

e Draw the smooth curve of the sine graph from 0° to 360°.

f What are the maximum and minimum values of the sine function?

2 *The cosine graph*

Using the same axes and scales as in Figure 28, draw the cosine graph.

a $\cos 0° = 1$. Plot (0, 1) and (360, 1).

b $\cos 30^\circ = 0{\cdot}87$. Plot $(30, 0{\cdot}87)$, $(150, -0{\cdot}87)$, $(210, -0{\cdot}87)$, $(330, 0{\cdot}87)$.

c Proceed as in ***b*** by 30° intervals.

d $\cos 90^\circ = 0$ gives $(90, 0)$ and $(270, 0)$; $\cos 180^\circ = -1$ gives $(180, -1)$.

e Draw the smooth curve of the of the cosine graph.

f What are the maximum and minimum values of the cosine function?

3 *The tangent graph*

Use a scale of 30 units to 1 cm horizontally and 1 unit to 2 cm vertically.

a $\tan 0^\circ = 0$. Plot $(0, 0)$, $(180, 0)$, $(360, 0)$.

b $\tan 30^\circ = 0{\cdot}58$. Plot $(30, 0{\cdot}58)$, $(150, -0{\cdot}58)$, $(210, 0{\cdot}58)$, $(330, -0{\cdot}58)$.

c Proceed as in ***b*** by 30° intervals.

d Draw the 'branches' of the tangent graph. Both $\tan 90^\circ$ and $\tan 270^\circ$ have no meaning; there is a break in the graph at each odd multiple of 90°. At these points the curve is said to be discontinuous.

4 Draw a set of axes on plain paper and sketch in roughly the graph whose equation is $y = \sin x^\circ$, from $x = 0$ to $x = 360$. Continue it beyond $x = 360$ until you have made another complete 'sine wave'. What is the angle difference corresponding to a *wavelength*?

5 Repeat question ***4*** for the graph of the cosine function.

Note that the original curves are preserved by translations through 360 units on the x-axis; only an infinite pattern can be preserved in this way.

A graph which fits on to itself after translation is called a *periodic graph*; if it is the graph of a function, the function is called a *periodic function*. The sine and cosine functions are periodic, with a period of 360°; we already know that for all a, $\sin a^\circ = \sin(360+a)^\circ$ and $\cos a^\circ = \cos(360+a)^\circ$.

6 a Is the x-axis an axis of symmetry of either graph?

b Is the y-axis an axis of symmetry of either graph?

c Is the origin a centre of symmetry of either graph?

7 Figure 29 shows part of the sine and cosine graphs. Use these graphs to give the values of:

a (*1*) $\sin 0^\circ$ (*2*) $\sin 90^\circ$ (*3*) $\sin 180^\circ$

b (*1*) $\cos 180^\circ$ (*2*) $\cos 270^\circ$ (*3*) $\cos 360^\circ$.

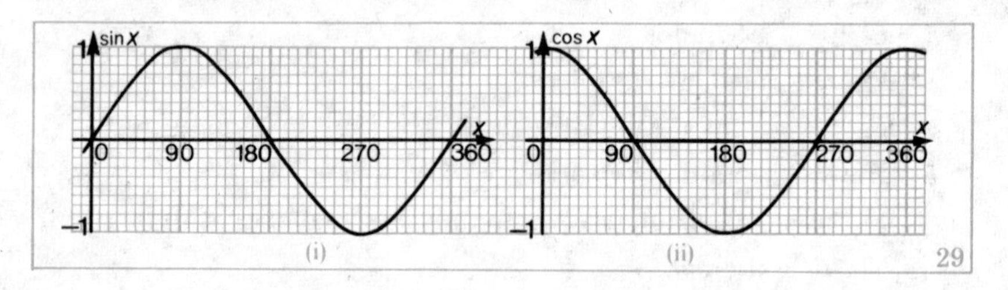

(i) (ii)
29

8 Use the graphs in Figure 29 to find a, given that:

a (1) $\sin a^\circ = 1$ (2) $\sin a^\circ = 0$ (3) $\sin a^\circ = -1$

b (1) $\cos a^\circ = 0$ (2) $\cos a^\circ = 1$ (3) $\cos a^\circ = -1$.

9 Place a tracing of the sine graph on top of the cosine graph. Slide the sine graph to the right, keeping the x-axis in one line. What do you observe?

The sine and cosine graphs are identical in form; it is only their positions relative to the y-axis that allow them to be distinguished by name.

The graph of mains voltage against time in seconds is a cosine curve exhibiting 50 periods between $t = 0$ and $t = 1$ second.

Perhaps you can study some sine and cosine curves on an oscilloscope.

6 Periodicity

Many things in the world around us show periodicity, i.e. they have a pattern which is regularly repeated. Here are some examples:

(i) The school timetable usually has a period of 5 (school) days.

(ii) The swing of a pendulum may have a period of 1 second.

(iii) The number of figures in the period of 123071230712307 . . . is 5.

(iv) The wavelength of B.B.C. Radio 2 Programme is 1500 metres.

(v) The phases of the moon are repeated every 29 days 12 hours 44 minutes 2·78 seconds.

Exercise 10

Explain how each of the following shows periodicity, and where possible say what the period is:

1 $\frac{1}{7} = 0{\cdot}142\,857\,142\,857\,142\,857\,142\ldots$

2 *a* A see-saw *b* a swing in the park.

3 The rotation of the earth on its axis.

4 A weight on a spring.

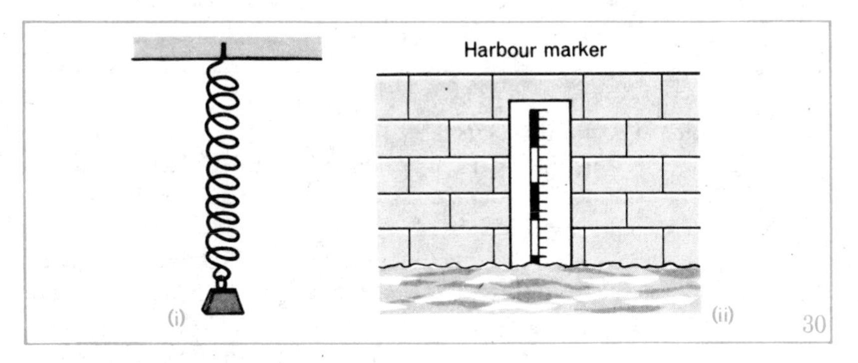

(i) (ii) 30

5 The depth of water in a harbour.

6 A piston driven by a rotating wheel.

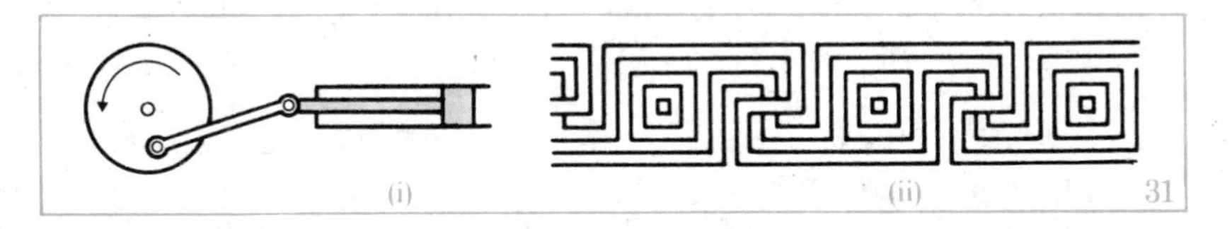

(i) (ii) 31

7 A wallpaper pattern

We shall now study the sine, cosine, and tangent functions as illustrations of periodicity in mathematics.

8 *The graph of the sine function—another method*

Figure 32 (i) shows a circle of unit radius with $\angle XOP_1 = 30°$, $\angle XOP_2 = 60°$ and so on. (Here we have a mathematical model of the Big Wheel of Section 1.)

Now $\sin XOP = \frac{y}{r} = y$ here, since $r = 1$.

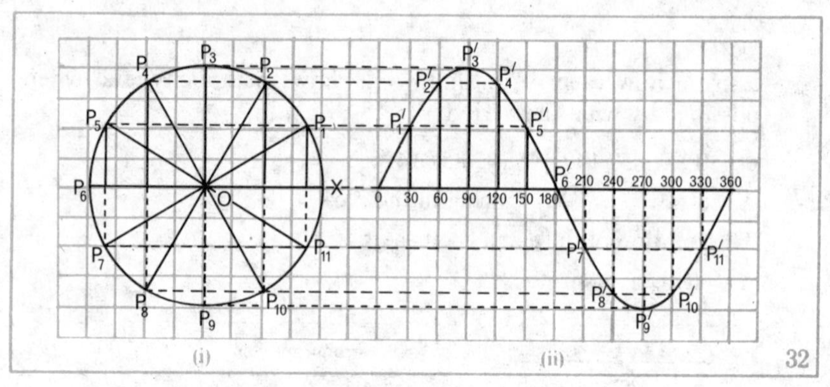

32

Thus the y-coordinates of P_1, P_2, ... give the values of the sine function, sin $a°$, for $a = 30, 60, \ldots, 360$.

The y-coordinates have been drawn in Figure 32 (ii) and the graph of the sine function for all angle sizes from 0° to 360° is shown.

What is the range of values of the sine function?

Draw the graph of the sine function as described above. Start by drawing a circle of radius 2 cm on 5-mm squared paper and make $\angle XOP_1 = 30°$, $\angle XOP_2 = 60°$, etc.; then draw horizontal construction lines to give P_1', P_2', etc., and finally draw the smooth curve of the sine graph. Extend the curve both to the right and to the left.

9 *The graph of the cosine function—another method*

In Figure 32 (i) the x-coordinates of P_1, P_2, ..., give the values of the cosine function, cos $a°$, for $a = 30, 60, \ldots 360$. Read off these values.

Mark off the horizontal axis as in Figure 32 (ii), and from each of the points measure the second coordinate equal to the corresponding value of the cosine function. Draw the smooth curve of the cosine graph. Extend it both to the right and to the left. (Instead of reading off the values to obtain the ordinates for the cosine graph, these may be set off by using a pair of compasses or dividers.)

10 *The graph of the tangent function—another method*

Use your slide rule to calculate $\frac{y}{x}$ from Figure 32 (i) for P_1, P_2, ..., thus obtaining the values of the tangent function, tan $a°$, for $a = 30, 60, \ldots, 360$.

Using the same horizontal scale as in Figure 32 (ii), draw this graph. Try to extend the graph both to the right and to the left. Remember that tan 90° and tan 270° have no meaning; there is a break in the curve at each odd multiple of 90.

Summary

1 *The definitions of the cosine, sine and tangent functions*

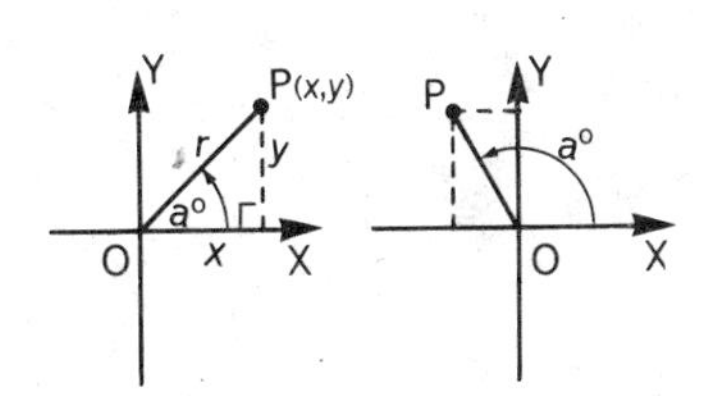

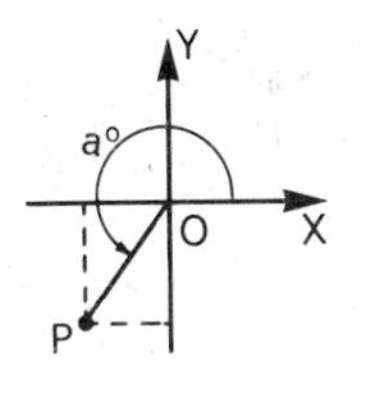

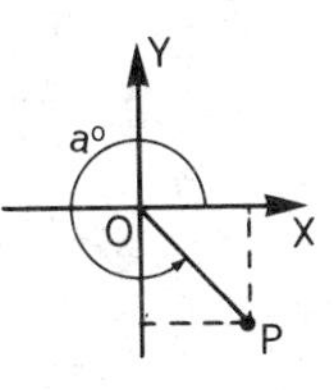

$$\cos a^\circ = \frac{x}{r} \qquad \sin a^\circ = \frac{y}{r} \qquad \tan a^\circ = \frac{y}{x}$$

2 *Values of the trigonometrical functions at 45°, 30°, 60°*

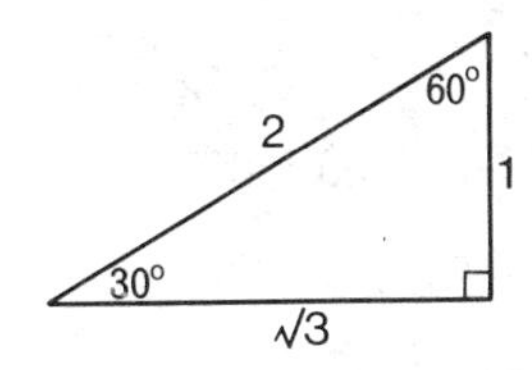

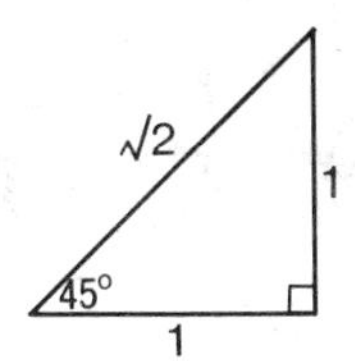

a°	30°	45°	60°
$\cos a^\circ$	$\frac{\sqrt{3}}{2}$	$\frac{1}{\sqrt{2}}$	$\frac{1}{2}$
$\sin a^\circ$	$\frac{1}{2}$	$\frac{1}{\sqrt{2}}$	$\frac{\sqrt{3}}{2}$
$\tan a^\circ$	$\frac{1}{\sqrt{3}}$	1	$\sqrt{3}$

3 Extending the use of the trigonometrical tables

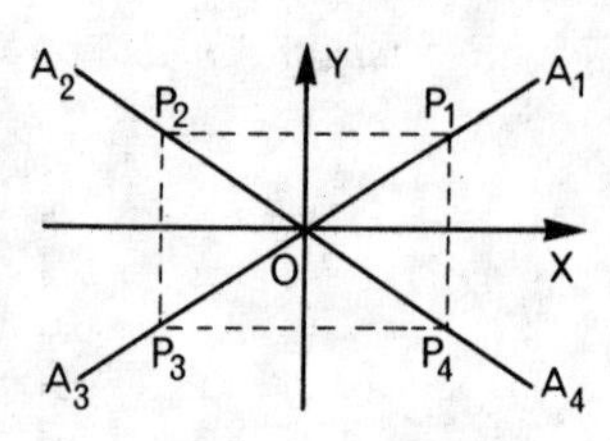

II Sin	I All
Tan III	Cos IV

Example 1. $\sin 405° = \sin 45° = \dfrac{1}{\sqrt{2}}$

Example 2. $\tan(-20°) = -\tan 20° = -0{\cdot}364$

4 Graphs of the trigonometrical functions

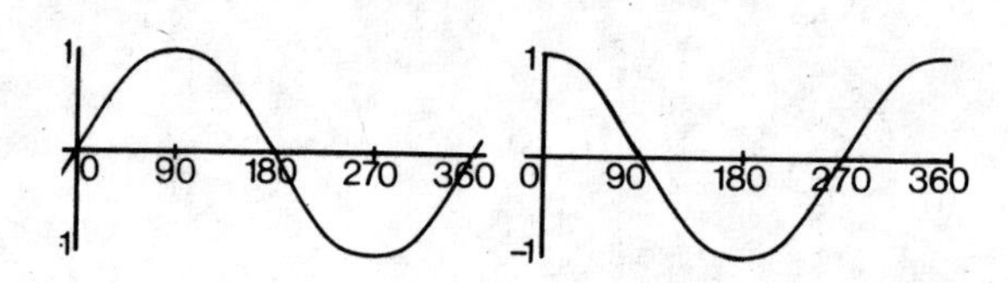

The sine graph
(period 360°)

The cosine graph
(period 360°)

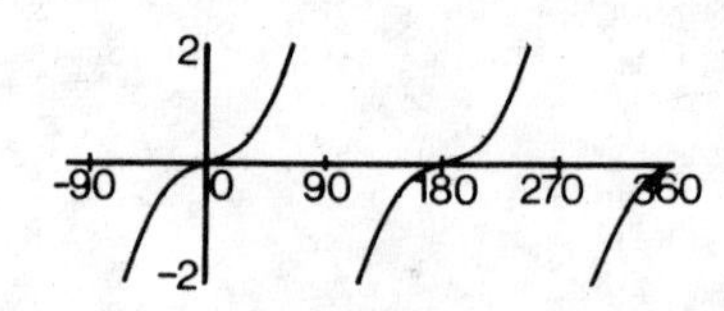

The tangent graph
(period 180°)

Revision Exercises on Chapter 1
The Cosine, Sine and Tangent Functions

Revision Exercise 1A

1 Calculate the values of cos XOA, sin XOA and tan XOA, where A is the point (8, 15).

2 Repeat question *1* for the point A $(-5, -5)$.

3 Use your tables to determine which of the following are true and which are false:

a $\sin 30° = \cos 60°$ *b* $\tan 18{\cdot}1° = \tan 53°$ *c* $\sin 0° = \sin 90°$

d $\cos 45° = \sin 45°$ *e* $\cos 25{\cdot}3° = \tan 42{\cdot}1°$ *f* $\tan 88{\cdot}9° > 50$

4 Calculate *d* in each diagram in Figure 1.

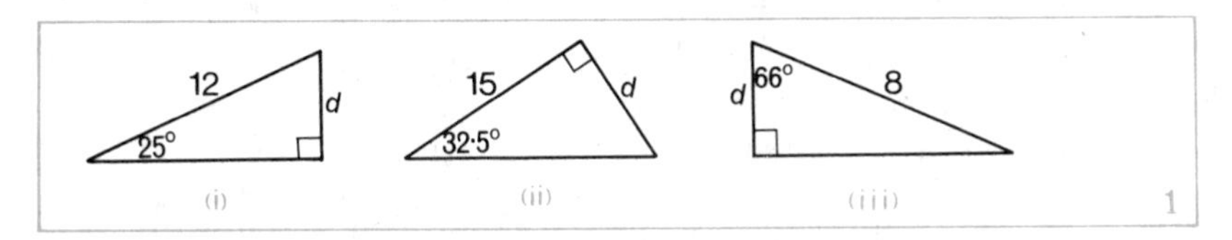

5 The diagonals of a kite ABCD intersect at O. AD = DC = 2 m, and AO = BO = CO = 1 m. Calculate:

a ∠DAO *b* ∠ADO *c* ∠ADC *d* OD *e* the area of ABCD.

6 In Figure 2, calculate: *a* AC *b* ∠CED.

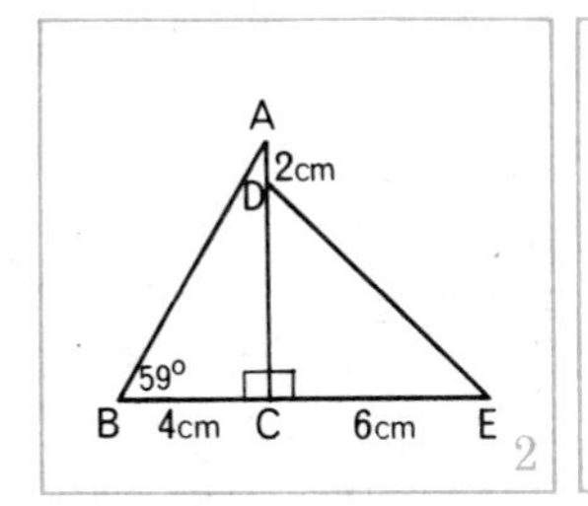

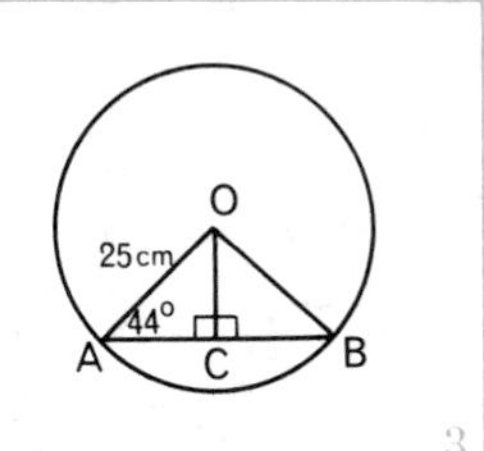

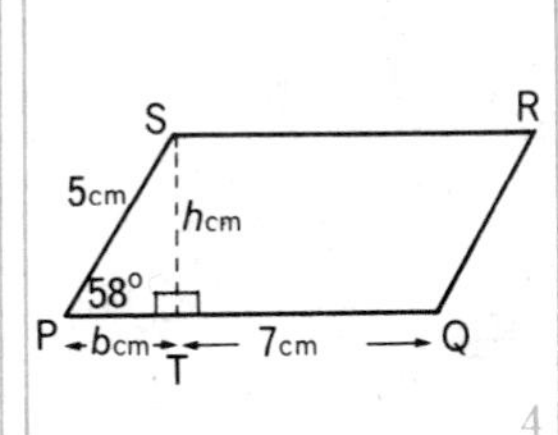

7 O is the centre of the circle in Figure 3. Calculate:

a AC *b* AB *c* ∠AOB

8 In Figure 4, PQRS is a parallelogram. Calculate:

a h *b* b *c* the area of PQRS.

9 The maximum gradient on a certain motorway is 1 in 50, i.e. the road rises 1 m for 50 m along its surface. Calculate the angle between the road and the horizontal, at its steepest part.

10 State whether the value of each of the following is positive or negative:

a cos 81° *b* tan 98° *c* sin 115° *d* cos 300°

e tan 143° *f* sin 317° *g* tan 563° *h* sin (−32°)

11 Use tables to find the values of the following:

a cos 81·3° *b* tan 304° *c* sin 147° *d* cos 115·5°

e cos 249·2° *f* sin (−27°)

12 Which of the following are true, for $0 \leqslant a \leqslant 360$?

a If $\cos a° = 0{\cdot}5$, then $a = 60$ or 300.

b If $\sin a° = 0{\cdot}707$, then $a = 45$ or 315.

c If $\tan a° = 1$, then $a = 45$ or 225.

13 Sketch the sine graph from 0° to 360°, and then write down the values of sin 0°, sin 90° and sin 270°.

14 Sketch the cosine graph from 0° to 720°, and then write down replacements for a $(0 \leqslant a \leqslant 720)$ for which $\cos a° = 1$.

15 The curve with equation $y = \sin x°$ is given a translation $\begin{pmatrix} a \\ b \end{pmatrix}$ in the positive direction of the x-axis. State the values of a and b when the curve first coincides with its original position.

What further translation makes the curve coincide with the curve $y = \cos x°$?

Give the equation of the curve which is the image of $y = \sin x°$ under a half turn about the origin.

Revision Exercise 1B

1 Given $\tan x° = 0{\cdot}882$, and $0 < x < 90$, use tables to find the values of $\cos x°$ and $\sin x°$.

2 Given $\cos a^\circ = \frac{3}{5}$, and $0 < a < 90$, make a sketch with coordinate axes from which you can calculate $\sin a^\circ$ and $\tan a^\circ$.

Is $\dfrac{\sin a^\circ}{\cos a^\circ} = \tan a^\circ$? Check with values taken from tables.

3 Calculate d for each diagram in Figure 5.

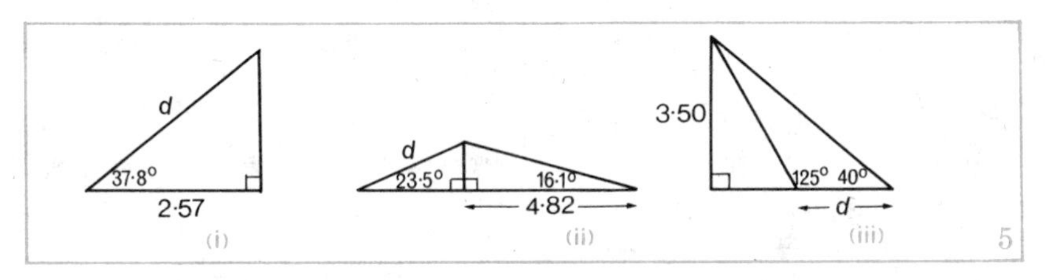

4 A regular 12-sided polygon is inscribed in a circle of radius 20 cm.

a What angle does each chord subtend at the centre?

b Calculate the circumference of the circle and the perimeter of the polygon.

5 From the summit of a mountain at an elevation 1000 metres above a rocket-launching site, the angle of depression of the site is 16° (i.e. the angle between the horizontal and the line from the summit to the site is 16°). Find the direct distance from the summit to the launching site.

6 A path up a hillside makes an angle of 8·4° with the horizontal for 100 metres along the path and an angle of 4·8° for a further 300 metres. In these 400 metres, how far does the path ascend vertically? (Give the answer to the nearest metre.)

7 A man stands on one bank of a small lake and sees in the surface of the water the reflection of a tree which is 10 metres high, growing on the opposite bank. His eye level is 2 metres above water level, and the distance from where he stands to the tree is 23 metres. Use similar triangles to help you to calculate the angle through which his line of vision is turned by reflection in the water.

8 An aeroplane flies at constant height from a point O to A, a distance of 95 km on a bearing of 320°; then from A to B, a distance of 230 km on a bearing of 125°. Find how far B is east and south from O.

9 Which of the following are true and which are false? (Only angles greater than 0° and less than 360° need be considered.)

a If $\sin x^\circ = 0{\cdot}213$, then $x^\circ = 12{\cdot}3^\circ$ or $167{\cdot}7^\circ$.
b If $\tan y^\circ = -4{\cdot}872$, then $y^\circ = 258{\cdot}4^\circ$.
c If $\cos z^\circ = 0{\cdot}841$, then $z^\circ = 57{\cdot}2^\circ$ or $302{\cdot}8^\circ$.
d If $\tan a^\circ = 2{\cdot}646$, then $a^\circ = 69{\cdot}3$ or $249{\cdot}3^\circ$.
e If $\cos b^\circ = -0{\cdot}367$, then $b^\circ = 111{\cdot}5^\circ$ or $291{\cdot}5^\circ$.
f If $\sin c^\circ = -0{\cdot}368$, then $c^\circ = 201{\cdot}6^\circ$ or $338{\cdot}4^\circ$.

10 Find the error in $\angle XOP$ if P is written $(-4, 5)$ instead of $(4, -5)$.

11 Draw the graph of $y = \cos x^\circ$ from $x = 0$ to $x = 400$ on 2-mm squared paper. On the same diagram, draw the graph of $y = \sin x^\circ$ from $x = 0$ to $x = 400$.

Using a pair of compasses, add y-coordinates to obtain points of the locus $\{(x, y): y = \cos x^\circ + \sin x^\circ\}$, being careful in the addition of positive and negative values. Complete the graph of the function $y = \cos x^\circ + \sin x^\circ$.

Compare the curves in the range $0 \leqslant x \leqslant 40$ with the corresponding curves in the range $360 \leqslant x \leqslant 400$. State your conclusions in the form of relations such as $\cos x^\circ = \cos (360 + x)^\circ$. How do you describe functions for which such relations are true?

How closely do the graphs of $\cos x^\circ$, $\sin x^\circ$, and $\cos x^\circ + \sin x^\circ$ resemble one another?

12 Using another page of graph paper similarly laid out, draw the graph of the function $y = \cos x^\circ - \sin x^\circ$.

Compare this graph with the previous one of $\cos x^\circ + \sin x^\circ$ with regard to period and maximum and minimum values.

13 The lines $y = 2x + 1$ and $y = 2{\cdot}4x + 1$ are drawn in a coordinate system with equal scales along perpendicular axes. Find the angle which each line makes with the x-axis. State the point of intersection of the lines, and the angle between the lines.

Computer Studies

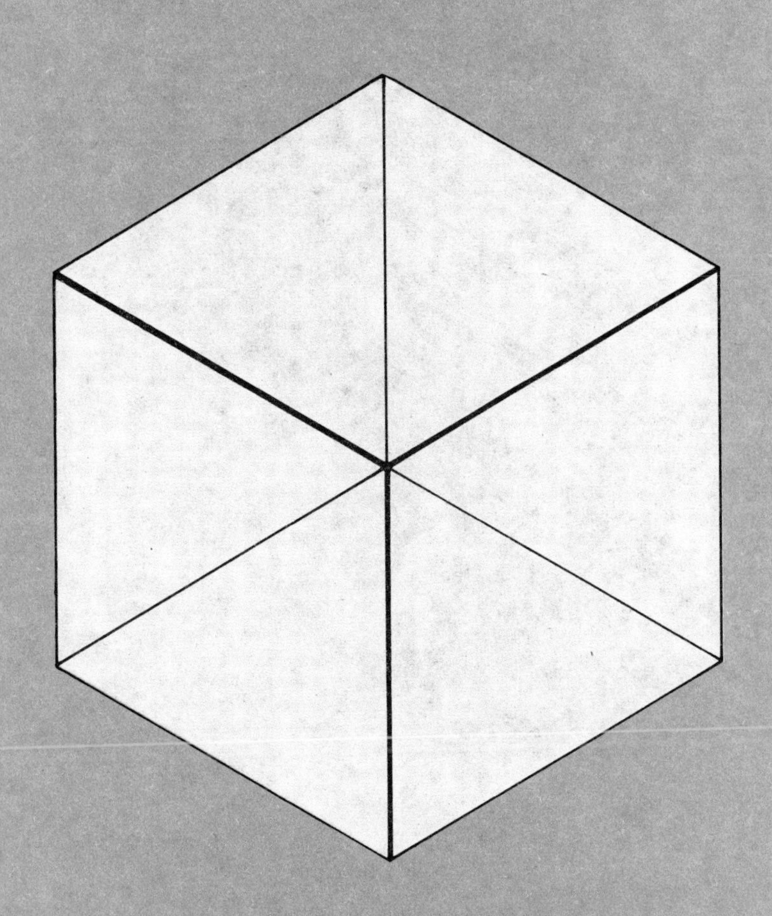

Computer Studies

Communicating with a Computer

1 Programs and data

The computer is a tool which can be used to store and process information, and to solve problems. It has to be supplied with:

(i) the information to be processed, called the *data*

(ii) instructions for carrying out the processing, called the *program*.

In this chapter we will study one method for supplying the computer with the necessary instructions to process the data; that is, we are going to learn how to write programs.

To communicate with a computer we have to write the instructions in a language that it can 'understand', and we have to present the instructions in a form that it can 'read'. At present English is too difficult for computers to understand, but there are several 'languages' available, called *programming languages*, which we can use to write instructions that computers will understand. In this chapter we will use a language called BASIC; some of the conventions adopted may have to be modified for use with certain makes of computer.

It may seem odd that we can write instructions which the computer can understand, but cannot read. Imagine writing a letter to a blind person. We can write the letter in English, a language which the person can understand, but until this is transformed into Braille the person will not be able to read it. Similarly, we can write to a computer in BASIC, a language which it can understand, but until this is coded as a pattern of holes on paper tape or punched cards the computer will not be able to read it. Figure 1 shows a punched card with a simple program in BASIC, and the corresponding code of punched holes.

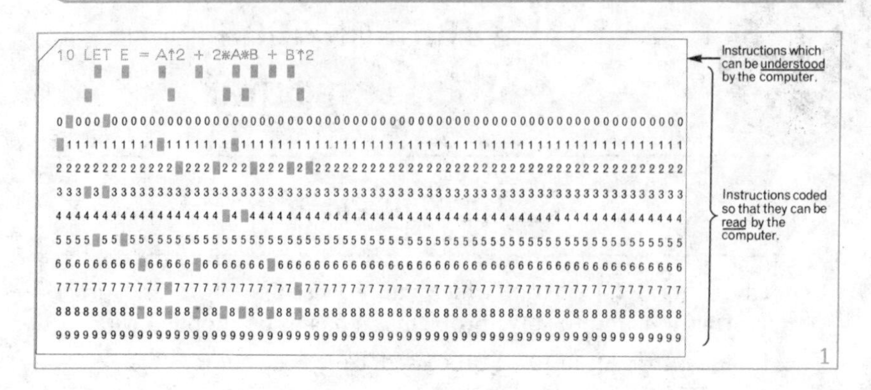

2 A programming language—BASIC

A simple program in BASIC

Example. To add the numbers 7 and 6 we could write the following program.

```
1 LET A = 7
2 LET B = 6
3 LET C = A+B
4 PRINT A, B, C
5 STOP
```

We will use this example to introduce some of the rules that must be obeyed when writing programs in BASIC, and some of the terms that are used.

Statement numbers

The above program consists of five statements, each of which is numbered. When the program is run, the computer will carry out the instructions in ascending order of statement number, i.e. 1 followed by 2, followed by 3, and so on.

All statements in a BASIC program must be numbered.

All writing in a BASIC program must be in the form of capital letters.

Variables

We can imagine part of a computer to consist of a set of stores, like a post office sorting rack, in which we can store numbers. In a computer each of these stores is called a *storage location.* Just as the post office clerk has each space named by a street or district so each storage location is given a name in the form of a capital letter, or a capital letter followed by a single digit, as shown in Figure 2. These names are called *variables*; in the worked example, three variables, A, B and C were used.

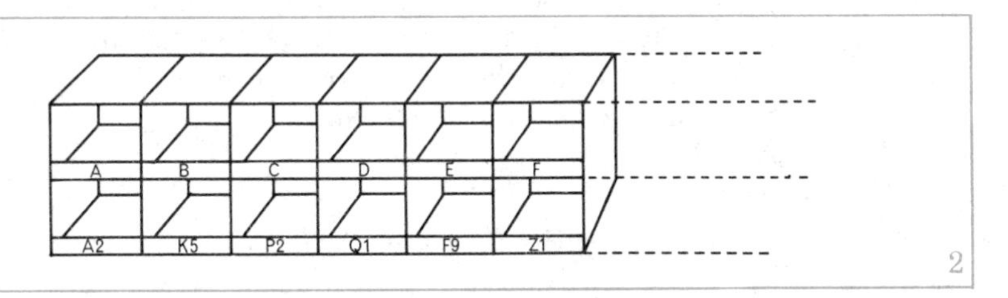

2

A variable, i.e. a storage location name, may be:

(i) a single capital letter, e.g. A, B, C, ...
(ii) a capital letter followed by a single digit, e.g. A2, K5, Z3, ...

Assignment statements

To put numbers in the storage locations we replace the variables by the numbers, that is we assign values to the variables by means of *assignment statements.* In the worked example we used three assignment statements:

1 LET A = 7, which means 'Assign to storage location A the number 7'.
2 LET B = 6, which means 'Assign to storage location B the number 6'.
3 LET C = A+B, which means 'Assign to storage location C the sum of the numbers in A and B'.

Notice that each assignment statement begins 'LET ...', and that the third one above, LET C = A+B, ensures that the operation of addition is carried out.

Print statements

We get the results of a calculation by using a *print statement* such as:
4 PRINT A, B, C, which means 'Print out the numbers in storage locations A, B and C'.

Stop statements

It is necessary to tell the computer when we have reached the end of our instructions. This was achieved in the worked example by the *stop statement*:

5 STOP

Input statements

The program in the worked example is not very useful since it can only be used to add 7 and 6. However, we can rewrite the program in a way that allows us to add any pair of numbers.

We replace the two assignment statements:

1 LET A = 7
2 LET B = 6

by an *input statement* and *data*, as follows:

Original program	*New program*
1 LET A = 7	1 INPUT A, B
2 LET B = 6	2 LET C = A+B
3 LET C = A+B	3 PRINT C
4 PRINT C	4 STOP
5 STOP	Data 7,6

In both cases the computer will print out '13', but in the second program any pair of numbers can be used as data.

Exercise 1

1 Correct the notation for the following BASIC variables where necessary:

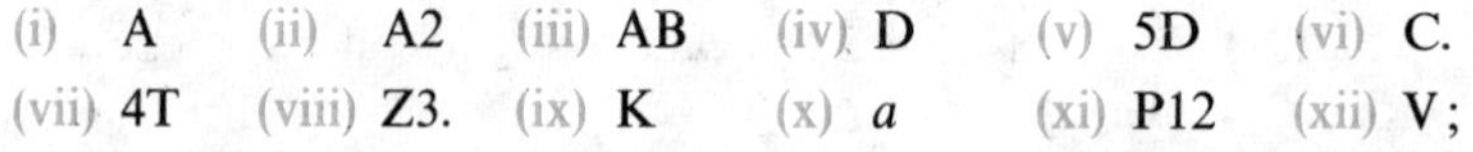

(i) A (ii) A2 (iii) AB (iv) D (v) 5D (vi) C.
(vii) 4T (viii) Z3. (ix) K (x) *a* (xi) P12 (xii) V;

2 There are errors in the following BASIC programs. Write the programs correctly.

(i)	(ii)	(iii)
1 LET A = 12	1 INPUT A, 6	1 Input P, Q
2 LET B = 7	2 C = A+B	2 WRITE K = P−Q
3 LET *c* = A+B	3 PRINT C	3 PRINT R
5 Print A, B, *c*	4 DATA 5, 6	STOP
4 STOP		

3 Write a BASIC program for each of the following. In (i)–(iii) use Assignment, Print and Stop statements as in the Worked Example in Section 1. In (iv)–(vi) use Input statements and Data also.

(i) To add 9 and 8.

(ii) To subtract 25 from 33.

(iii) To add 12, 18 and 29.

(iv) To add two numbers.

(v) To subtract two numbers.

(vi) To add five numbers.

3 *Programs and flow charts*

In the Computer Chapter in Book 4 we saw that the solution of a problem could be broken down into a sequence of simple steps which could then be shown diagrammatically in a flow chart. It is often very useful to draw a flow chart before writing a program, especially with more complicated programs.

Example 1. Draw a flow chart to show how to find the mean of four numbers, and then write a computer program in BASIC.

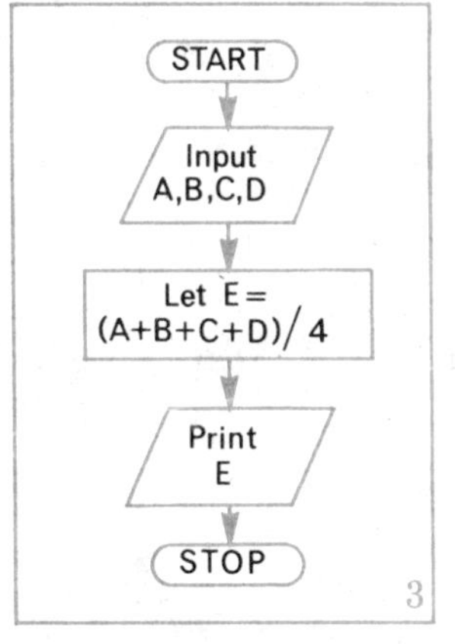


```
10 REM PROGRAM TO FIND
       MEAN OF 4 NUMBERS
20 INPUT A, B, C, D
30 LET E = (A+B+C+D)/4
40 PRINT E
50 STOP
```

Note that the *input* and *output statements* are shown in parallelogram boxes in the flow chart.

We now look at some special features of the above program.

REM statements

Sometimes a program is used, and then set aside for a period before being required again. In order to identify the program quickly, and to remind us of what it can do, we insert a REM (for REMARK) statement at the beginning to describe the purpose of the program. The computer accepts the words following REM simply as comments, and not instructions.

Statement numbers

Notice that while the statement numbers 10, 20, 30, 40, 50 are in ascending order they are not consecutive as in the previous worked example. The advantage of using 10, 20, 30, ... is that we may wish to insert another line or instruction later, and this can be done easily as follows;

```
10 REM PROGRAM TO FIND MEAN OF 4 NUMBERS
20 INPUT A, B, C, D
30 LET E = (A+B+C+D)/4
36 PRINT A, B, C, D                    (an additional line)
40 PRINT E
50 STOP
```

Assignment statement

The assignment statement:

30 LET E = (A+B+C+D)/4

means 'Assign to E the sum of A, B, C, D divided by 4'.

(A+B+C+D)/4 is called an *arithmetic expression.*

Arithmetic expressions

These consist of variables combined by some of the following operations:

+ addition
− subtraction
* multiplication (× could be confused with X, and is not used)
/ division (÷ is not permitted)
↑ raise to a power (A↑3 is the same as A * A * A)

Examples

Arithmetic expression	BASIC *equivalent*
$a+b$	A+B
$c-d$	C−D
xy	X * Y

$2pq$	2 * P * Q
$\frac{ab}{c}$	A * B/C, or A/C * B
x^2	X↑2, or X * X

Example 2. Make a flow chart and a program to find the volume of a cuboid, given its length, breadth and height. The length, breadth and height should be printed out as well as the volume. Illustrate for data 3, 4, 5.

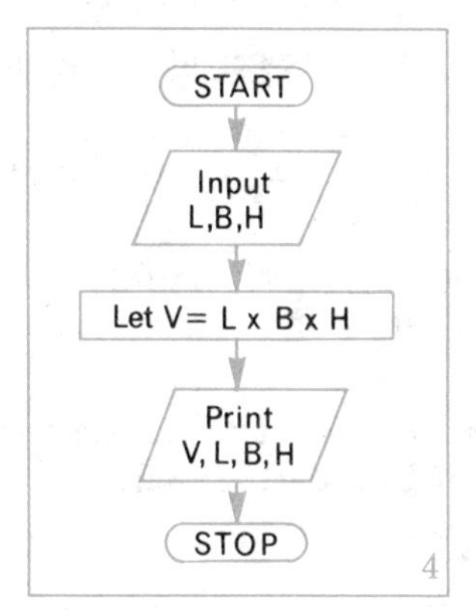


```
10 REM PROGRAM FOR
   VOL OF CUBOID
20 INPUT L, B, H
30 LET V = L * B * H
40 PRINT V, L, B, H
50 STOP
Data 3, 4, 5
```

The *output* for this example will be printed by the computer as:

```
60    3    4    5
```

If we replaced

```
40 PRINT V, L, B, H
```

by

```
40 PRINT V
41 PRINT L, H, B
```

the output would appear as:

```
60
 3    4    5
```

Exercise 2

1 Give the BASIC equivalent for each of the following arithmetic expressions:

(i) $p+q$ (ii) $s-t$ (iii) ab (iv) $5xy$ (v) $\frac{a}{b}$

(vi) a^2 (vii) $3b^2$ (viii) $\frac{pq}{r}$ (ix) $2x+2y$ (x) x^2+y^2

2 Draw a flow chart and write a program containing REM, Input, Assignment, Print and Stop statements for each of the following:

(i) To find the mean of three numbers, with data 9, 7, 2.

(ii) To find the area of a rectangle, given length and breadth as data.

(iii) To find the circumference of a circle, given the radius as data.

(iv) To find the area of a circle; data 5.

(v) To find the volume of a cuboid, given the length, breadth and height.

(vi) To find the volume of a cylinder, given the height and radius of base. ($V = \pi r^2 h$)

(vii) To find the perimeter of a rectangle, given the length and breadth; data 9·8, 7·5.

3 Draw a flow chart and write a program to calculate the rates payable by householders in a town where the rate is 95 pence in the £, being given the rateable values as data.

4 Draw a flow chart and write a program to calculate the sale price of goods in a shop which are reduced by 22% during a sale.

4 *Evaluation of arithmetic expressions*

In an arithmetic expression, operations are given priority as follows:

(i) Content of brackets.

(ii) Raising to a power.

(iii) Multiplication and division, in the order in which they occur.

(iv) Addition and subtraction, in the order in which they occur.

The computer takes the operations from left to right, comparing pairs of operations to decide priority.

Example

```
10 INPUT A, B, C, D
20 LET E = B↑2 + A * C − D
30 PRINT E
40 STOP
Data 2, 3, 4, 5
```

The arithmetic expression B↑2 + A * C − D will be evaluated as follows:

E = 3↑2+2 * 4−5
  = 9+2 * 4−5
  = 9+8−5
  = 17−5
  = 12

Brackets are sometimes necessary to make the meaning of an arithmetic expression clear.

In BASIC, $\frac{a+b}{c-d}$ must be written (A+B)/(C−D).

Without brackets, A+B/C−D would be equivalent to $a+\frac{b}{c}-d$, which is not what was wanted.

Also, 2+3 * 4+5 = 2+12+5 = 19 and (2+3) * (4+5) = 5 * 9 = 45

Exercise 3

1 Spot the errors in the following:

(i) LET AB = M+N
(ii) LET C = 2A+B
(iii) LET J = A^2+B * C
(iv) LET 2M = B×B
(v) LET X = Y÷Z
(vi) LET A = $\frac{BC}{D}$

2 Evaluate the following BASIC arithmetic expressions with A = 2, B = 3, C = 4, D = 5:

(i) A+B * B
(ii) (A+B) * B
(iii) A+B/C
(iv) C/A/B
(v) D↑2−2 * B * C
(vi) D−C/A−B
(vii) A−(B+C)
(viii) (A+B)↑2
(ix) B−C/A * D
(x) A * B/C
(xi) (A+B−C)↑3
(xii) (A+B)/(C+D)

3 Insert brackets in the equivalent BASIC expressions in the table.

	(i)	(ii)	(iii)
Arithmetic expression	$\frac{x}{2a}$	$\frac{a+b}{c}$	$\frac{a+b^2}{cd+f^3}$
BASIC equivalent	X/2 * A	A+B/C	A+B↑2/C * D+F↑3

4 Rewrite the equivalent BASIC expressions in the following correctly:

	(i)	(ii)	(iii)
Arithmetic expression	$a^2-2ab+b^2$	$(x+y)(x-y)$	$\frac{a^2+b^2}{xy}$
BASIC equivalent	A↑2−2AB+B↑2	(X+Y)(X−Y)	$(a^2+b^2)\div$XY

5 What output would the computer print for each of the following?

(i)
```
1 INPUT A, B, C
2 LET P = A↑3+B * C
3 PRINT P
4 STOP
Data 4, 3, 2
```

(ii)
```
1 INPUT X, Y
2 LET Z = 6 * X/Y
3 PRINT X, Y, Z
4 STOP
Data 4, 12
```

(iii)
```
1 INPUT P, Q, R, S
2 LET K = (P+Q)/(R−S)
3 PRINT K
4 STOP
Data −5, −4, −3, −2
```

6 Write a program with REM, Input, Assignment, Print and Stop statements for each of the following:

(i) To read two numbers p and q, and to evaluate $p+q^3$.

(ii) To read two numbers x and y, and to evaluate x^2+y^2 and $(x+y)^2$.

(iii) To read four numbers, to print their sum, product, square of their sum, square of their product, and their mean. Given the data 1, 2, 3, 4, what numbers would the computer print out?

5 'GOTO' and 'IF, THEN' statements

In our work so far the programs were executed in the order in which the statements were numbered. Often, however, we wish to alter the order in which instructions are carried out. We can do this by means of 'GOTO' statements, and conditional 'IF, THEN' statements.

The GOTO *statement*
We write a GOTO statement as follows to alter the order in which statements are carried out:

```
70 GOTO 10
```

This ensures that the computer will now *jump* to statement 10 in the program.

Example 1. Write a program to read 100 numbers one at a time, and to print out each number followed by its square.

A possible flow chart and program are as follows:

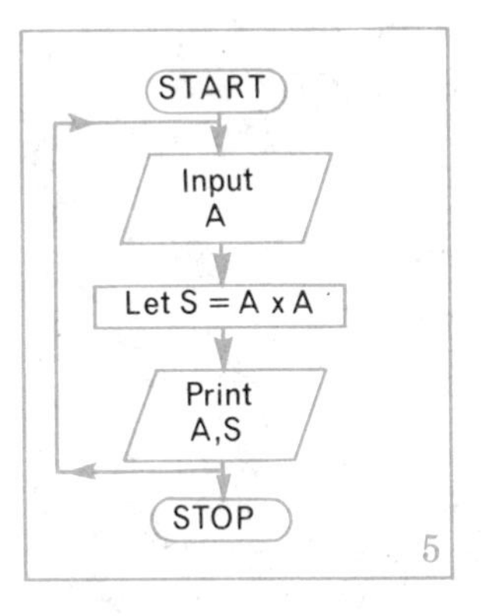


```
10 REM PROGRAM TO
   CALCULATE SQUARES
20 INPUT A
30 LET S = A * A
40 PRINT A, S
50 GOTO 20
60 STOP
```

Note the *loop* in the flow chart corresponding to the GOTO statement. The above program runs into difficulty when the data are exhausted, although in this example the necessary results are obtained. The difficulty is also obvious from the flow chart, as the STOP statement is never reached.

Conditional statements
The last example shows a use for the GOTO statement, and also a limitation of it. To overcome this difficulty we can use a *conditional statement*, including IF . . . , THEN . . . , such as:

```
40 IF C > 100 THEN 90
```

This ensures that the computer will *jump* to statement 90 if the sentence $C > 100$ is true. If the sentence is false (i.e. $C \leqslant 100$) the computer will continue with the next statement in the program.

Using conditional statements there are two ways in which we can avoid the situation that arose in Example 1, that is in which we can get out of the loop.

The first way is by supplying a *counter* to keep a note of how many numbers have been read in. Using this technique we have:

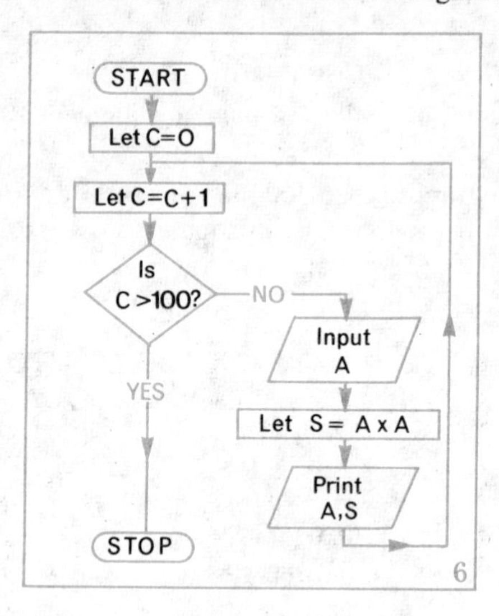


```
10 REM PROGRAM TO
   CALCULATE SQUARES
   OF 100 NUMBERS
20 LET C = 0
30 LET C = C+1
40 IF C > 100 THEN 90
50 INPUT A
60 LET S = A * A
70 PRINT A, S
80 GOTO 30
90 STOP
```

Here we have used storage location C as our counter. In the statement '30 LET C = C+1' the symbol '=' is used as an assignment symbol, and the statement means 'Assign to storage location C the value at present in C, plus 1' or 'Add 1 to the value in storage location C'.

The statement 'LET C = 0' is important since it ensures that we start counting from 1 when we meet '30 LET C = C+1'. The current value of C is then tested, and if this value is greater than 100 the computer jumps to the end of the program.

The second way of getting to the end of a program is to add an extra, easily identified number at the end of the data. For example, if the data are known to consist only of positive numbers or zero, we could add a negative number which could not be confused with the data. Each time a number is read into the computer it is tested; if it is negative the end of the program has been reached.

Example 2. Write a program to read in an unknown number of positive integers, and to print out each integer followed by its square.

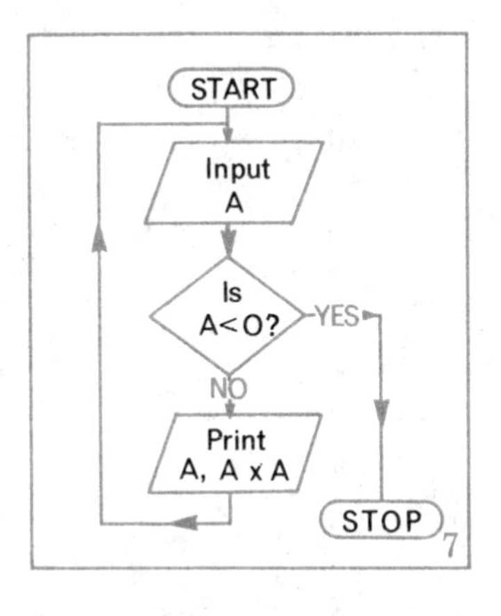


```
10 REM PROGRAM TO
     CALCULATE SQUARES
20 INPUT A
30 IF A < 0 THEN 60
40 PRINT A, A * A
50 GOTO 20
60 STOP
```

Sample data 2, 20, 39, 8, 45, 3, −1

Note. The PRINT statement in this program is different from any met so far:

```
40 PRINT A, A * A
```

In this statement we are asking the computer to print out A, to evaluate A ∗ A, *and* to print this out. In BASIC we can put arithmetic expressions in PRINT statements. Alternatively we could have had:

```
35 LET S = A * A
40 PRINT A, S
```

In the last two examples the conditional statements were:

```
40 IF C > 100 THEN 90
30 IF A < 0 THEN 60
```

In each case, between IF and THEN we have an inequality. The available symbols are:

	Algebraic symbol	BASIC *symbol*
is less than	$<$	<
is greater than	$>$	>
is equal to	$=$	=
is less than or equal to	$\leqslant$	LE
is greater than or equal to	$\geqslant$	GE
is not equal to	$\neq$	NE

Example 3. Data consist of sets of three numbers, each set representing the length, breadth and height of a cuboid. Terminating the data by three negative numbers, write a program which will print out the volume of each cuboid.

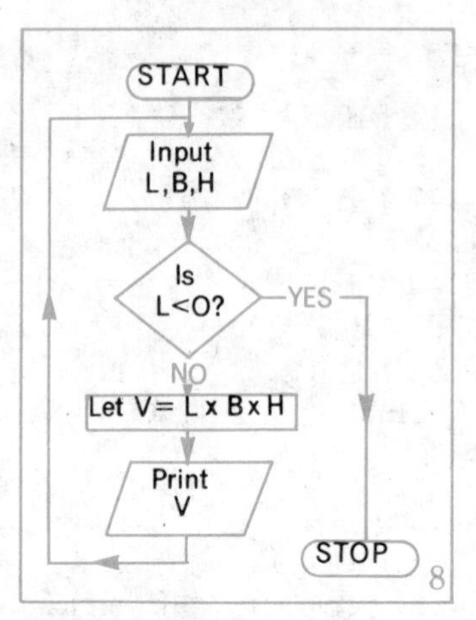


```
10 REM PROGRAM FOR
   VOL OF CUBOID
20 INPUT L, B, H
30 IF L < 0 THEN 70
40 LET V = L * B * H
50 PRINT V
60 GOTO 20
70 STOP
```

Sample data 5, 4, 3; 8, 7, 6; 2, 1, 7; −1, −1, −1

Note. When printing out results it is often useful to label the answers. For example, instead of printing out '12' for a volume it might be more useful to print out:

VOLUME = 12

We can output text in BASIC by enclosing it in inverted commas within a PRINT statement. For example,

```
50 PRINT "VOLUME = ", V
```

Example 4. Write a program to calculate the income tax payable at the rate of 30% for incomes of £800 or more. No tax is paid on incomes or parts of incomes of less than £800.

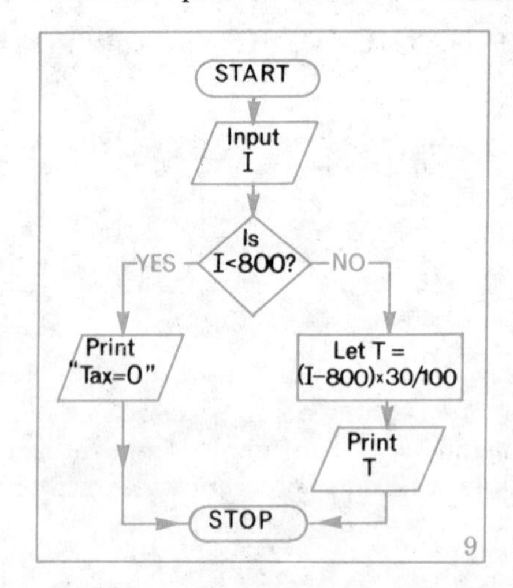


```
10 REM PROGRAM FOR
   INCOME TAX
20 INPUT I
30 IF I < 800 THEN 70
40 LET T = (I−800) * 30/100
50 PRINT T
60 GOTO 80
70 PRINT "TAX = 0"
80 STOP
```

Exercise 4

1 Write a program to read 50 numbers, one at a time, and to print out each number followed by its square.

2 Write a program to read 1000 numbers, one at a time, and to print out each number followed by its cube.

3 Write a program to read an unknown number of positive integers, and to print out each number followed by half its value.

4 Figure 10 shows a flow chart for calculating an agent's commission, based on his weekly sales. Write a program for calculating the commission, being given the weekly sales as data.

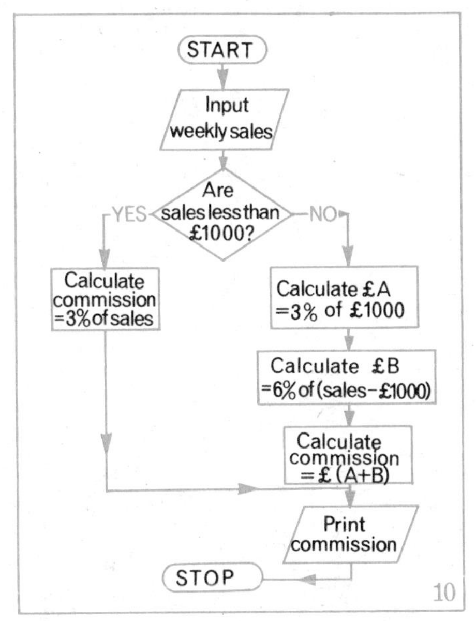

10

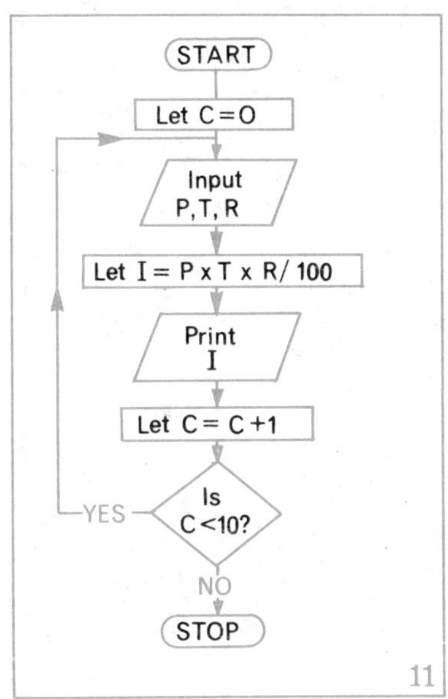

11

5 Figure 11 shows a flow chart for calculating the simple interest (£I) in 10 cases for a sum of money (£P) for T years at rate % per annum R, using the formula $I = \frac{PTR}{100}$. Write a suitable program for the calculations.

6 At a sale goods marked less than £10 have a discount of 5%, and goods marked £10 or more have a discount of 10%. Draw a flowchart and write a BASIC program to calculate the sale price of any item, being given the marked price as data.

7 Write a program for calculating the income tax payable at the rate of $27\frac{1}{2}\%$ on incomes of £1000 or more. No tax is paid on incomes or parts of incomes of less than £1000.

Answers

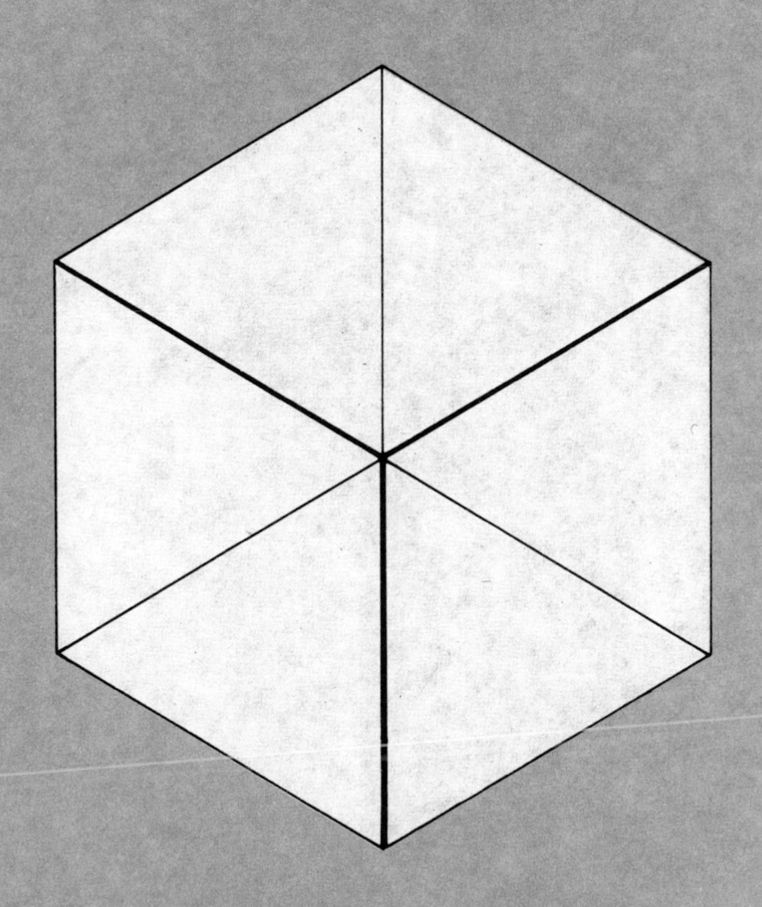

Answers

Algebra—Answers to Chapter 1

Page 3 Exercise 1

1	$10a$	2	$2a$	3	$-2a$	4	$-10a$
5	$9x^2$	6	$3x^2$	7	$6x^2$	8	0
9	ab	10	$11a^3$	11	$2xy$	12	$2pq$
13	$2x+4y$	14	0	15	$7x^2$	16	$-y^2+6y$
17	$3k^3-3k$	18	$2p^3+2p$	19	$4x^2-3xy$	20	$3a^2-5ab$
21	$5x-y$	22	$5a-b$	23	$7x-8$	24	$7y$
25	$6x^2-2$	26	$8y^2+2y$	27	$8x+11$	28	$7a$
29	20	30	$2y^2+3y$	31	$6a$	32	$12x-12y$
33	$2a+1$	34	$3p+2q$	35	$2x^2+3x+4$	36	$3x^2-3x$
37	$-a-2$	38	$-2b^2+8$	39	$2a+b+3c$	40	$6p-5q-2r$

Page 5 Exercise 2

1	$-2x+3$	2	$-3x-1$	3	$-x+1$	4	$-2+3a$
5	$-4-b$	6	$2c-3$	7	$-q+p$	8	$a+b-c$
9	$-2a+3b-4c$	10	$4x+2$	11	$3x-1$	12	$2x+3$
13	$-2x+9$	14	$-3x+2$	15	$-4x$	16	$x+3$
17	$5y-12$	18	k^2+4	19	$-b^2+2$	20	$4x^2+4x$
21	$3a^2-3a$	22	$2a+3b-6$	23	$-2x+2y+6$	24	$-2x$
25	$2a^2-4a+7$	26	$x-9$	27	$-y+14$	28	$2x^2$
29	y^2+2						

Page 5 Exercise 2B

1	$3a^2-7ab$	2	$2a^2$	3	$5p-q$	4	$-4t+1$
5	$5x^3-3x^2-2$	6	$2a^2+a$	7	$-4u-8v$	8	$5i+2k$
9	$-a^2-3a-2$	10	$3x+20y+6$	11	$-2bc$	12	$\frac{2}{3}a^2-4$
13	0	14	$3x^2-9x$				

Page 6 Exercise 3

1	$\{1\}$	2	$\{x:x>3\}$	3	$\{3\}$	4	$\{x:x<2\}$
5	$\{2\}$	6	$\{x:x\geqslant -4\}$	7	$\{\frac{4}{5}\}$	8	$\{x:x>\frac{7}{9}\}$
9	$\{-10\}$	10	$\{x:x\leqslant 5\}$	11	$\{\frac{1}{2}\}$	12	$\{x:x<-2\}$
13	$\{\frac{7}{11}\}$	14	$\{x:x\geqslant -\frac{3}{8}\}$	15	$\{\frac{3}{4}\}$		

Page 8 Exercise 4

1	-15	2	-6	3	6	4	20
5	-100	6	3	7	-28	8	1
9	$-6x$	10	$20y$	11	$12x$	12	$-15m$
13	$-8ab$	14	$10xy$	15	0	16	$-15p^2$
17	$2x+6$	18	$3x-12$	19	$-12x-4$	20	$-14x+35$
21	$2x^2+x$	22	$3x^2-4x$	23	$x-x^2$	24	$-x^2+2x$
25	x^2+4x+3	26	$x^2+7x+12$	27	a^2+2a+1	28	$x^2+7x+10$
29	$y^2+10y+21$	30	z^2+2z+1	31	$2a^2+7a+3$	32	$4b^2+12b+9$
33	$12c^2+23c+10$	34	x^2+x-2	35	x^2-x-12	36	$x^2+3x-10$
37	a^2-5a+6	38	$b^2-8b+16$	39	c^2-6c+5	40	$3d^2-5d-2$
41	$2e^2+3e-20$	42	$8f^2+10f-3$				

Page 9 Exercise 5

1	x^2+4x+3	2	$x^2+8x+15$	3	x^2+6x+8
4	$a^2+10a+25$	5	a^2+7a+6	6	a^2+5a+6
7	$2x^2+5x+3$	8	$3x^2+7x+2$	9	$2x^2+11x+15$
10	$4a^2+12a+9$	11	$9b^2+6b+1$	12	$8c^2+14c+3$
13	x^2-5x+6	14	x^2-3x+2	15	$x^2-7x+12$
16	$a^2-12a+36$	17	a^2-6a+5	18	a^2-2a+1
19	$2x^2-7x+5$	20	$4x^2-9x+2$	21	$2x^2-9x+9$
22	$4a^2-4a+1$	23	$4b^2-8b+3$	24	$6c^2-19c+10$
25	$x^2+3x-10$	26	y^2+3y-4	27	z^2+z-6
28	p^2+p-2	29	q^2+q-6	30	r^2-1
31	$2s^2+3s-2$	32	$4t^2-11t-3$	33	$3u^2+u-4$
34	$4v^2-9$	35	$6w^2-w-1$	36	$20x^2-7x-6$
37	$a^2+8a+12$	38	$b^2-7b+12$	39	$c^2+3c-10$
40	$d^2-9d+14$	41	$2x^2+7x-4$	42	$2y^2+y-6$
43	$6z^2+16z+8$	44	$6k^2-19k+15$	45	$20m^2+2m-6$
46	$x^2+3xy+2y^2$	47	$x^2-4xy+3y^2$	48	$x^2-3xy-10y^2$

Page 10 Exercise 5B

1	$2a^2+5ab+3b^2$	2	$2x^2-3xy+y^2$	3	$2p^2-3pq-2q^2$
4	$9m^2-4n^2$	5	$3s^2+7st-6t^2$	6	$6u^2-16uv+8v^2$
7	$15+11x+2x^2$	8	$1+2y-3y^2$	9	$20-31x+12x^2$
10	$xy+2x+y+2$	11	$ab-a+3b-3$	12	$cd-5c-4d+20$
13	x^3+5x^2+7x+2	14	$6x^3-7x^2-13x+10$	15	$8t^3-18t^2-23t+15$
16	a^3+b^3	17	a^3-b^3	18	p^4-q^4
19	$2x^4-7x^2-4$	20	$3x^4-4x^3-8x^2+4x+5$		

Page 10 Exercise 6

1	$x^2+2xy+y^2$	*2*	$p^2+2pq+q^2$	*3*	$m^2-2mn+n^2$
4	$u^2-2uv+v^2$	*5*	x^2+6x+9	*6*	$x^2+10x+25$
7	x^2+2x+1	*8*	$x^2+20x+100$	*9*	a^2-4a+4
10	a^2-6a+9	*11*	$a^2-8a+16$	*12*	$a^2-16a+64$
13	$4x^2+4x+1$	*14*	$9x^2+12x+4$	*15*	$25x^2+30x+9$
16	$100x^2+20x+1$	*17*	$4a^2-12a+9$	*18*	$9a^2-6a+1$
19	$4a^2-20a+25$	*20*	$25a^2-40a+16$	*21*	$a^2+4ab+4b^2$
22	$9x^2+6xy+y^2$	*23*	$16c^2-8cd+d^2$	*24*	$4m^2-12mn+9n^2$

25 $(x+3)^2 = x^2+6x+9$ *26* $(y-5)^2 = y^2-10y+25$

27 $(a+1)^2 = a^2+2a+1$ *28* $(b-1)^2 = b^2-2b+1$

29 $(2x+3)^2 = 4x^2+12x+9$ *30* $(3y-5)^2 = 9y^2-30y+25$

31	$2x^2+12x+20$	*32*	$2a^2-12a+26$	*33*	$12y$
34	$-20b+20$	*35*	x^4+6x^2+9	*36*	y^4-4y^2+4
37	$a^4+2a^2b^2+b^4$	*38*	$p^4-2p^2q^2+q^4$	*39*	$a^2+2+\frac{1}{a^2}$
40	$a^2-2+\frac{1}{a^2}$	*41*	$4x^2+2+\frac{1}{4x^2}$	*42*	$9x^2-6+\frac{1}{x^2}$

Page 12 Exercise 7

1	2	*2*	-1	*3*	$-\frac{1}{2}$	*4*	1
5	0	*6*	-1	*7*	-2	*8*	$-\frac{1}{5}$
9	0	*10*	$-\frac{4}{3}$	*11*	$\frac{1}{5}$	*12*	$\frac{1}{2}$
13	$\frac{1}{2}$	*14*	-1	*15*	$x = 29$; 20 cm, 29 cm, 21 cm		

16 $x = 7{\cdot}5$; 12·5 m by 4·5 m, 7·5 m by 7·5 m

Page 12 Exercise 7B

1	7·5	*2*	-2	*3*	5	*4*	$\frac{11}{2}$
5	$-\frac{1}{2}$	*6*	$-\frac{2}{7}$	*7*	$x = 4$; 25 cm, 24 cm, 7 cm		

8 16 cm^2; $(4+t)^2$ cm^2, or $(16+8t+t^2)$ cm^2; $(8t+t^2)$ cm^2; 0·81 cm^2

9 $(r+w)$ mm; $A = \pi(r+w)^2-\pi r^2$, etc

10a $(r-12)$ m *b* $r^2 = (r-12)^2+18^2$ *c* $r = 19{\cdot}5$

Algebra—Answers to Chapter 2

Page 18 Exercise 1

1 a *(1)* 3 *(2)* 4 *b* 5, 6, 7, 8 *c* 3, 7, 11 *d* *(1)* 1 *(2)* 11 *(3)* 8

e *(1)* first row, fourth column *(2)* third row, first column

(3) second row, fourth column *(4)* third row, third column

(5) second row, third column *(6)* second row, first column

2 a $\begin{pmatrix} 12 & 4 & 4 & 0 & 0 \\ 8 & 0 & 1 & 2 & 1 \end{pmatrix}$ 2 rows, 5 columns

b $\begin{pmatrix} 20 & 8 & 72 & 4 \\ 19 & 9 & 65 & 4 \\ 12 & 4 & 28 & 2 \end{pmatrix}$ 3 rows, 4 columns

4 (*1*) 2 rows, 2 columns; 2 (*2*) 2 rows, 3 columns; a
(*3*) 1 row, 1 column; 5 (*4*) 3 rows, 1 column; 1
(*5*) 3 rows, 3 columns; -2 (*6*) 1 row, 5 columns; d

5 a 6 rows, 7 columns *b* (*1*) first column (*2*) fifth column
(*3*) first row or column

6 a $\begin{pmatrix}2 & 3\\6 & 4\end{pmatrix}$ *b* $\begin{pmatrix}3 & -1\\1 & 2\end{pmatrix}$ *c* $\begin{pmatrix}-1 & 1\\5 & -4\end{pmatrix}$ *7* $\begin{pmatrix}0 & 1\\1 & 10\end{pmatrix}, \begin{pmatrix}0 & 0\\0 & 1\end{pmatrix}$

Page 20 Exercise 2

1 a 2×3 *b* 3×2 *c* 3×3 *d* 3×4 *e* 1×3 *f* 3×1

2 6, 6, 9, 12, 3, 3 respectively

3 a 9 *b* 12 *c* 1 *d* n *e* mn *f* n^2

4 a e.g. $\begin{pmatrix}2 & 1 & 0 & 4\\1 & -3 & 6 & 8\end{pmatrix}$ *b* e.g. $\begin{pmatrix}a & h & g\\h & b & f\\g & f & c\end{pmatrix}$ *c* e.g. (-4) *d* e.g. $\begin{pmatrix}3\\4\\5\end{pmatrix}$

5 $A = C, E = G, H = L$

6 $1\times3, 1\times3, 1\times3; 2\times1, 2\times1, 2\times1, 2\times1; 2\times2, 2\times2, 2\times2, 2\times2$

7 a 1, 4 *b* 2, 1 *c* 3, -4 *d* 4, -2 *e* 5, -1 *f* 7, 0

8 a $\begin{pmatrix}3 & 2\\1 & 3\\4 & 5\end{pmatrix}, 3\times2$ *b* $\begin{pmatrix}3 & 1 & 4\\2 & 3 & 5\end{pmatrix}, 2\times3$ *c* $\begin{pmatrix}a & h & g\\h & b & f\\g & f & c\end{pmatrix}, 3\times3$

d $\begin{pmatrix}3 & 1 & 2\\2 & 2 & 0\\1 & 3 & 1\\4 & 0 & 3\end{pmatrix}, 4\times3$ *e* $\begin{pmatrix}-1\\-2\\-3\end{pmatrix}, 3\times1$ *f* $(u \quad v \quad w), 1\times3$

9 4, -2

Page 22 Exercise 3

1 $\begin{pmatrix}4\\6\end{pmatrix}$ *2* $\begin{pmatrix}-1\\3\end{pmatrix}$ *3* $\begin{pmatrix}3\\-2\end{pmatrix}$ *4* $\begin{pmatrix}6a\\-2b\end{pmatrix}$ *5* $\begin{pmatrix}m+1\\n+2\end{pmatrix}$

6 $\begin{pmatrix}p+r\\q+s\end{pmatrix}$ *7* $(5 \quad 4)$ *8* $(-2 \quad 3)$ *9* $(3u \quad v)$ *10* $\begin{pmatrix}6\\4\\4\end{pmatrix}$

11 $(6 \quad 4 \quad -2)$ *12* $\begin{pmatrix}3 & 1\\3 & 5\end{pmatrix}$ *13* $\begin{pmatrix}6 & 5\\3 & 6\end{pmatrix}$ *14* $\begin{pmatrix}8 & 2\\3 & 6\end{pmatrix}$ *15* $\begin{pmatrix}1 & 0\\0 & 1\end{pmatrix}$

16 $\begin{pmatrix}\frac{3}{4} & \frac{1}{2}\\\frac{1}{4} & 1\end{pmatrix}$ *17* $\begin{pmatrix}3a & 3b\\-a & 0\end{pmatrix}$ *18* $\begin{pmatrix}3x & 2y\\-2x & 4y\end{pmatrix}$ *19* $\begin{pmatrix}5 & 3\\7 & 9\end{pmatrix}$ twice

20a $\begin{pmatrix}a+f & b+g\\c+h & d+k\end{pmatrix}$ and $\begin{pmatrix}f+a & g+b\\h+c & k+d\end{pmatrix}$ *b* yes; commutative law

21a (*1*) $\begin{pmatrix}6 & 8\\10 & 12\end{pmatrix}$ (*2*) $\begin{pmatrix}8 & 10\\16 & 18\end{pmatrix}$ (*3*) $\begin{pmatrix}9 & 12\\19 & 22\end{pmatrix}$ (*4*) $\begin{pmatrix}9 & 12\\19 & 22\end{pmatrix}$

b yes; associative law

24a $\begin{pmatrix}-4\\-5\end{pmatrix}$ *b* $\begin{pmatrix}-2\\-3\\1\end{pmatrix}$ *c* $\begin{pmatrix}-3 & 1\\0 & 2\end{pmatrix}$ *d* $\begin{pmatrix}3 & -1 & 0\\-4 & 2 & 1\end{pmatrix}$ *e* (m)

Page 25 Exercise 4

1 a $\begin{pmatrix}3\\1\end{pmatrix}$ *b* $\begin{pmatrix}5\\2\end{pmatrix}$ *c* $\begin{pmatrix}1\\-4\end{pmatrix}$ *d* $\begin{pmatrix}-1\\-1\end{pmatrix}$

2 a $\begin{pmatrix}2 & 4\\1 & 2\end{pmatrix}$ *b* $\begin{pmatrix}2 & 0\\5 & 7\end{pmatrix}$ *c* $\begin{pmatrix}-3 & 3\\-1 & -2\end{pmatrix}$ *d* $\begin{pmatrix}2x & 3\\2 & 6y\end{pmatrix}$ *e* $\begin{pmatrix}10 & 3\\7 & 2\end{pmatrix}$

3 a $\begin{pmatrix}-1 & 5\\3 & 5\end{pmatrix}$ *b* $\begin{pmatrix}6 & 4\\2 & 4\end{pmatrix}$ *c* $\begin{pmatrix}5 & 9\\5 & 9\end{pmatrix}$ *d* $\begin{pmatrix}7 & -1\\-1 & -1\end{pmatrix}$ *e* $\begin{pmatrix}3 & -1\\3 & 3\end{pmatrix}$

f $\begin{pmatrix}7 & -1\\-1 & -1\end{pmatrix}$ ***d*** and ***f***

4 a $\begin{pmatrix}2 & -1\\-1 & 2\end{pmatrix}$ *b* $\begin{pmatrix}2 & -4\\-1 & -3\end{pmatrix}$ *c* $\begin{pmatrix}7 & 12\\3 & 1\end{pmatrix}$ *d* $\begin{pmatrix}4 & 1\\-1 & 13\end{pmatrix}$

5 a $x = 3, y = -1, z = -5$ *b* $x = -9, y = 4, z = 1$

6 $p = 5, q = 8, r = 1, s = 3$

Page 27 Exercise 5

1 a $\begin{pmatrix}3\\6\end{pmatrix}$ *b* $\begin{pmatrix}10\\6\\4\end{pmatrix}$ *c* $(15 \quad 5 \quad 10)$ *d* $(3 \quad 4)$

e $\begin{pmatrix}6 & 2\\8 & 4\end{pmatrix}$ *f* $\begin{pmatrix}-3 & 0\\0 & 3\end{pmatrix}$ *g* $\begin{pmatrix}2 & 1\\0 & \frac{1}{2}\end{pmatrix}$ *h* $\begin{pmatrix}5a & 10b\\15a & 20b\end{pmatrix}$

2 a $\begin{pmatrix}6 & 8\\2 & 4\end{pmatrix}$ *b* $\begin{pmatrix}9 & 12\\3 & 6\end{pmatrix}$ *c* $\begin{pmatrix}15 & 20\\5 & 10\end{pmatrix}$ *d* $\begin{pmatrix}-3 & -4\\-1 & -2\end{pmatrix}$

e $\begin{pmatrix}-3 & -4\\-1 & -2\end{pmatrix}$, last two

3 b $\begin{pmatrix}6 & 8\\2 & 4\end{pmatrix} = 2X$ *c* $\begin{pmatrix}9 & 12\\3 & 6\end{pmatrix} = 3X$ *d* $\begin{pmatrix}6 & 8\\2 & 4\end{pmatrix} = 2X$

4 a $\begin{pmatrix}1 & 2\\3 & 4\end{pmatrix}$ *b* $\begin{pmatrix}2 & -1\\-3 & 0\end{pmatrix}$ *c* $\begin{pmatrix}1 & 0\\-3 & 2\end{pmatrix}$ *d* $\begin{pmatrix}4 & 6\\1 & 2\end{pmatrix}$

5 a $\begin{pmatrix}4 & -6\\8 & 0\end{pmatrix}$ *b* $\begin{pmatrix}-6 & 8\\4 & -2\end{pmatrix}$ *c* $\begin{pmatrix}6 & -9\\12 & 0\end{pmatrix}$ *d* $\begin{pmatrix}10 & -15\\20 & 0\end{pmatrix}$

e $\begin{pmatrix}-1 & 1\\6 & -1\end{pmatrix}$ *f* $\begin{pmatrix}-2 & 2\\12 & -2\end{pmatrix}$ *g* $\begin{pmatrix}5 & -7\\2 & 1\end{pmatrix}$ *h* $\begin{pmatrix}10 & -14\\4 & 2\end{pmatrix}$

i $\begin{pmatrix}-2 & 2\\12 & -2\end{pmatrix}$ *j* $\begin{pmatrix}10 & -14\\4 & 2\end{pmatrix}$ *k* $\begin{pmatrix}10 & -15\\20 & 0\end{pmatrix}$ *l* $\begin{pmatrix}0 & -1\\16 & -2\end{pmatrix}$

Equal matrices are ***d, k***; ***i, f***; ***h, j***.

6 a $\begin{pmatrix}8 & -1 & -9\\21 & 2 & 4\end{pmatrix}$ *b* $\begin{pmatrix}7 & 0 & -1\\7 & 3 & 4\end{pmatrix}$

7 a $(13 \quad 6 \quad 7)$ *b* $(18 \quad 11 \quad 7)$ *c* $(2 \quad -1 \quad -2)$
d $(4 \quad -2 \quad -4)$ *e* $(9 \quad 8 \quad 6)$ *f* $(17 \quad 19 \quad 13)$

8 a $\begin{pmatrix}2\\-1\\3\end{pmatrix}$ *b* $\begin{pmatrix}2\\-2\\1\end{pmatrix}$ *c* $\begin{pmatrix}2\\3\\1\end{pmatrix}$ *9 a* $\begin{pmatrix}2 & -1\\3 & 4\end{pmatrix}$ *b* $\begin{pmatrix}3 & 2\\-1 & 3\end{pmatrix}$ *c* $\begin{pmatrix}2 & 1\\1 & 5\end{pmatrix}$

Page 31 Exercise 6

1 *a* (11) *b* (26) *c* (11) *d* (13) *e* (26)
f (4) *g* $(a+4)$ *h* $(8+5c)$ *i* $(3m)$

2 *a* (6) *b* (13) *c* (-2) *d* (6) *e* (17)
f (-17) *g* (-17) *h* (a^2-10) *i* $(4a-3b)$

3 *a* 4 *b* 1 *c* 3 *d* 4

4 *a* (14) *b* (1) *c* (5) *d* $(3x+y+5z)$

5 *a* $\begin{pmatrix}3\\4\end{pmatrix}$ *b* $\begin{pmatrix}3\\-4\end{pmatrix}$ *c* $\begin{pmatrix}-3\\4\end{pmatrix}$ *d* $\begin{pmatrix}-3\\-4\end{pmatrix}$ *e* $\begin{pmatrix}-4\\3\end{pmatrix}$
f $\begin{pmatrix}-1\\7\end{pmatrix}$ *g* $\begin{pmatrix}1\\8\end{pmatrix}$ *h* $\begin{pmatrix}8\\14\end{pmatrix}$ *i* $\begin{pmatrix}-7\\7\end{pmatrix}$

6 *a* $\begin{pmatrix}8\\10\end{pmatrix}$ *d* (37) *e* $\begin{pmatrix}4\\12\end{pmatrix}$ *f* $\begin{pmatrix}3\\7\end{pmatrix}$

7 *a* $x = 5, y = 3$ *b* $x = 3, y = -1$ *c* $x = 2, y = 3$
d $x = -3, y = 4$ *e* $x = 3, y = -2$ *f* $x = 2, y = 3$

8 Each = (50). The results illustrate the commutative and associative laws for the multiplication of real numbers and matrices.

Page 34 Exercise 7

1 *a* $x' = -1.x+0.y$, $y' = 0.x+1.y$ *b* $\begin{pmatrix}x'\\y'\end{pmatrix}=\begin{pmatrix}-1&0\\0&1\end{pmatrix}\begin{pmatrix}x\\y\end{pmatrix}$ *c* P′(−6, 2) *d* $\begin{pmatrix}-1&0\\0&1\end{pmatrix}$

2 *a* $x' = -1.x+0.y$, $y' = 0.x+(-1).y$ *b* $\begin{pmatrix}x'\\y'\end{pmatrix}=\begin{pmatrix}-1&0\\0&-1\end{pmatrix}\begin{pmatrix}x\\y\end{pmatrix}$ *c* P′(−6, −2) *d* $\begin{pmatrix}-1&0\\0&-1\end{pmatrix}$

3 *a* $x' = 0.x+(-1).y$, $y' = 1.x+0.y$ *b* $\begin{pmatrix}x'\\y'\end{pmatrix}=\begin{pmatrix}0&-1\\1&0\end{pmatrix}\begin{pmatrix}x\\y\end{pmatrix}$ *c* P′(−2, 6) *d* $\begin{pmatrix}0&-1\\1&0\end{pmatrix}$

4 *a* $x' = 0.x+1.y$, $y' = -1.x+0.y$ *b* $\begin{pmatrix}x'\\y'\end{pmatrix}=\begin{pmatrix}0&1\\-1&0\end{pmatrix}\begin{pmatrix}x\\y\end{pmatrix}$ *c* P′(2, −6) *d* $\begin{pmatrix}0&1\\-1&0\end{pmatrix}$

5 *a* identity, or 'no change' *b* reflection in x-axis
c enlargement by the factor 2 *d* reflection in the line $y = x$

6 *a* $\begin{pmatrix}x'\\y'\end{pmatrix}=\begin{pmatrix}2&-1\\1&2\end{pmatrix}\begin{pmatrix}x\\y\end{pmatrix}$ *b* (0, 5) *c* (−5, 0), (0, −5), (5, 0)

7 $a = 4, b = 2, c = 2, d = 1$

8 $\begin{pmatrix}1&1\\-2&1\end{pmatrix}$

Page 37 Exercise 8

1 *a* $\begin{pmatrix}4&11\\7&8\end{pmatrix}$ *b* $\begin{pmatrix}5&5\\16&7\end{pmatrix}$ *c* $\begin{pmatrix}5&10\\10&20\end{pmatrix}$ *d* $\begin{pmatrix}7&14\\9&18\end{pmatrix}$ *e* $\begin{pmatrix}2&3\\13&18\end{pmatrix}$ *f* $\begin{pmatrix}8&9\\11&12\end{pmatrix}$

2 *a* $\begin{pmatrix}2&3\\-4&-5\end{pmatrix}$ *b* $\begin{pmatrix}7&-1\\12&-1\end{pmatrix}$ *c* $\begin{pmatrix}-4&6\\10&0\end{pmatrix}$ *d* $\begin{pmatrix}1&4\\0&9\end{pmatrix}$ *e* $\begin{pmatrix}1&0\\0&1\end{pmatrix}$ *f* $\begin{pmatrix}0&10\\7&19\end{pmatrix}$

3 *a* $\begin{pmatrix}5&10\\13&22\end{pmatrix}$ *b* $\begin{pmatrix}9&14\\13&18\end{pmatrix}$ *c* $\begin{pmatrix}12&17\\14&9\end{pmatrix}$ *d* $\begin{pmatrix}11&24\\10&10\end{pmatrix}$ *e* $\begin{pmatrix}8&7\\18&19\end{pmatrix}$ *f* $\begin{pmatrix}17&24\\6&10\end{pmatrix}$

4 a $\begin{pmatrix}1 & 2\\3 & 4\end{pmatrix}$ *b* $\begin{pmatrix}3 & -2\\-4 & 5\end{pmatrix}$ *c* $\begin{pmatrix}a & b\\c & d\end{pmatrix}$ *d* $\begin{pmatrix}1 & 2\\3 & 4\end{pmatrix}$ *e* $\begin{pmatrix}3 & -2\\-4 & 5\end{pmatrix}$ *f* $\begin{pmatrix}a & b\\c & d\end{pmatrix}$

5 Each is $\begin{pmatrix}0 & 0\\0 & 0\end{pmatrix}$ *6 a* $\begin{pmatrix}a+c & b+d\\a-c & b-d\end{pmatrix}$ *b* $\begin{pmatrix}a+kc & b+kd\\c & d\end{pmatrix}$

c $\begin{pmatrix}a & b\\kc & kd\end{pmatrix}$ *d* $\begin{pmatrix}a+b & a-b\\c+d & c-d\end{pmatrix}$ *e* $\begin{pmatrix}a & ak+b\\c & ck+d\end{pmatrix}$ *f* $\begin{pmatrix}a & bk\\c & dk\end{pmatrix}$

Page 38 Exercise 8B

1 a $\begin{pmatrix}6 & -1\\4 & 0\end{pmatrix}$ *b* $\begin{pmatrix}8 & 0\\36 & -4\end{pmatrix}$ *c* $\begin{pmatrix}6 & 4\\13 & 2\end{pmatrix}$ *d* $\begin{pmatrix}2 & -4\\23 & -6\end{pmatrix}$ *e* $\begin{pmatrix}8 & 0\\36 & -4\end{pmatrix}$
the distributive law for matrix multiplication over addition

2 a $\begin{pmatrix}8 & 13\\18 & 31\end{pmatrix}$ *b* $\begin{pmatrix}-3 & 3\\-1 & 1\end{pmatrix}$ *c* $\begin{pmatrix}-5 & 5\\-13 & 13\end{pmatrix}$ *d* $\begin{pmatrix}-5 & 5\\-13 & 13\end{pmatrix}$

the associative law for matrix multiplication

3 a $\begin{pmatrix}4 & 7\\16 & -2\end{pmatrix}$, $\begin{pmatrix}-4 & 8\\12 & 6\end{pmatrix}$, no

4 a $\begin{pmatrix}7 & 10\\15 & 22\end{pmatrix}$ *b* $\begin{pmatrix}-1 & -4\\8 & 7\end{pmatrix}$ *c* $\begin{pmatrix}37 & 54\\81 & 118\end{pmatrix}$ *d* $\begin{pmatrix}-9 & -11\\22 & 13\end{pmatrix}$

6 a $x = -4p+13q,\ y = 7p+q$

Page 40 Exercise 9

1 (2, −2), (4, 0), (6, −5) *2* (0, 1), (−1, 3), (0, 5) *3* (4, 4), (0, 2), (3, 1)
4 (3, −5), (0, −5), (1, 1) *5* (0, 0), (1, −6), (−1, −6) *6* (−2, −2), (−1, 3), (4, 1)
7 a (0, 0), (3, 0), (3, 3), (0, 3) *b* (0, 0), (2, 1), (3, 3), (1, 2) *c* (0, 0), (2, 1), (4, 4), (2, 3)
d (0, 0), (−2, 0), (−2, −2), (0, −2) *8* (0, 0), (6, 2), (7, 14), (−5, 10)

9 $\begin{pmatrix}x'\\y'\end{pmatrix} = \begin{pmatrix}3 & -4\\3 & 3\end{pmatrix}\begin{pmatrix}x\\y\end{pmatrix}$ {(4, 22), (−15, 5), (−21, 22)}

10 $\begin{pmatrix}x'\\y'\end{pmatrix} = \begin{pmatrix}1 & 1\\2 & 2\end{pmatrix}\begin{pmatrix}x\\y\end{pmatrix}$ {(0, 0), (2, 4)}

11a (20, 10), (6, 3), (−2, −1), (−34, −17) *b* $y = \frac{1}{2}x$

Page 42 Exercise 10

7 a $\begin{pmatrix}1 & -1\\-1 & 2\end{pmatrix}$ *b* $\begin{pmatrix}5 & -3\\-3 & 2\end{pmatrix}$ *c* $\begin{pmatrix}7 & -3\\-9 & 4\end{pmatrix}$ *d* $\begin{pmatrix}2 & 3\\1 & 2\end{pmatrix}$

e $\begin{pmatrix}4 & 5\\7 & 9\end{pmatrix}$ *f* $\begin{pmatrix}-4 & 7\\-3 & 5\end{pmatrix}$ *8 b* $\begin{pmatrix}\frac{1}{2} & 0\\0 & \frac{1}{2}\end{pmatrix}$ *9* $\begin{pmatrix}2 & -5\\-1 & 3\end{pmatrix}$; yes

10 Products I. *a* reflection in 'x-axis' twice *b* reflection in the origin twice
c reflection in line $y = x$ twice *d* reflection in line $y = -x$ twice
11 rotations of +90° and −90° about the origin

12a A′(6, 5), B′(4, 1), C′(−3, 2) *b* $\begin{pmatrix}1 & -1\\-1 & 2\end{pmatrix}$
c (1, 4), (3, −2), (−5, 7) respectively
d 'undoes' the mapping with matrix $\begin{pmatrix}2 & 1\\1 & 1\end{pmatrix}$

Page 45 Exercise 11

1 $\frac{1}{2}\begin{pmatrix}2 & -3\\-2 & 4\end{pmatrix}$ *2* $\frac{1}{2}\begin{pmatrix}3 & -1\\-4 & 2\end{pmatrix}$ *3* $\frac{1}{4}\begin{pmatrix}2 & -2\\-1 & 3\end{pmatrix}$ *4* none

5 $\begin{pmatrix}-9 & 4\\16 & -7\end{pmatrix}$ *6* $\frac{1}{5}\begin{pmatrix}4 & -3\\-1 & 2\end{pmatrix}$ *7* none *8* $\frac{1}{3}\begin{pmatrix}9 & -7\\-6 & 5\end{pmatrix}$

9 $\begin{pmatrix}0 & 1\\1 & -1\end{pmatrix}$ *10* $\frac{1}{13}\begin{pmatrix}5 & 3\\-1 & 2\end{pmatrix}$ *11* $-\frac{1}{14}\begin{pmatrix}4 & -6\\-3 & 1\end{pmatrix}$ *12* $\frac{1}{7}\begin{pmatrix}2 & -3\\1 & 2\end{pmatrix}$

13 $\frac{1}{5}\begin{pmatrix}2 & 1\\1 & 3\end{pmatrix}$ *14* $-\frac{1}{2}\begin{pmatrix}-1 & -4\\-1 & -2\end{pmatrix}$ *15* $-\frac{1}{3}\begin{pmatrix}3 & -2\\0 & -1\end{pmatrix}$

16a $\begin{pmatrix}7 & 19\\1 & 3\end{pmatrix}$ *b* $\begin{pmatrix}4 & 11\\2 & 6\end{pmatrix}$ *c* $\frac{1}{2}\begin{pmatrix}1 & -3\\0 & 2\end{pmatrix}$ *d* $\begin{pmatrix}3 & -5\\-1 & 2\end{pmatrix}$

e $\frac{1}{2}\begin{pmatrix}3 & -19\\-1 & 7\end{pmatrix}$ *f* $\frac{1}{2}\begin{pmatrix}6 & -11\\-2 & 4\end{pmatrix}$ *g* $\frac{1}{2}\begin{pmatrix}3 & -19\\-1 & 7\end{pmatrix}$ *h* $\frac{1}{2}\begin{pmatrix}6 & -11\\-2 & 4\end{pmatrix}$

equal ***e***, ***g***; ***f***, ***h***

Page 47 Exercise 12

1 {(8, 3)}	*2* {(4, 4)}	*3* {(1, −2)}	*4* {(3, −2)}	*5* {(4, 7)}
6 {(4, −3)}	*7* {(2, −3)}	*8* {(7, −2)}	*9* {(3, 4)}	*10* {(6, −1)}
11 {(1, 1)}	*12* {(4, −1)}	*13* {(2, 3)}	*14* {(3, 2)}	*15* {($\frac{1}{5}$, −1)}

16 The matrices have no inverses. In ***a***, the straight lines given by the equations are coincident, and in ***b*** they are parallel; in neither case has the system of equations a unique solution.

17a {(1, 2)} *b* {(4, −1)} *c* {(1, $-\frac{1}{2}$)}

Algebra—Answers to Chapter 3

Page 52 Exercise 1

1 (ii), (iv) and (v)

2 a domain {1, 2, 3, 4, 5} and range {2, 4, 8}

b

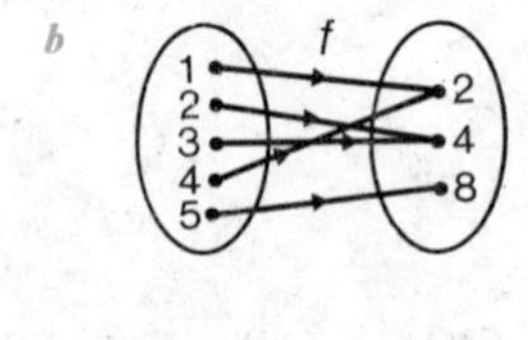

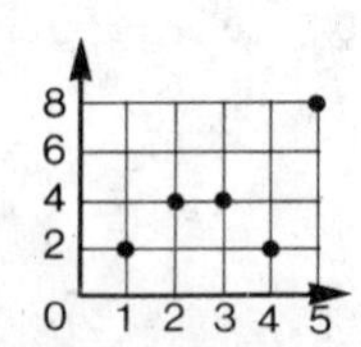

3 a

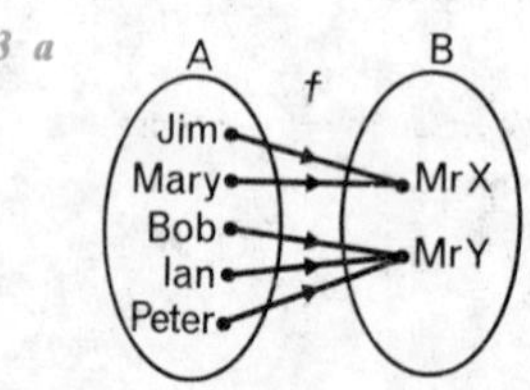

b yes, since f relates each element of A to exactly one element of B

4 *a* $B = \{\text{George, Tom, Bill, Ian, David, Alan}\}$
$S = \{\text{North St., South St., Market St.}\}$

b

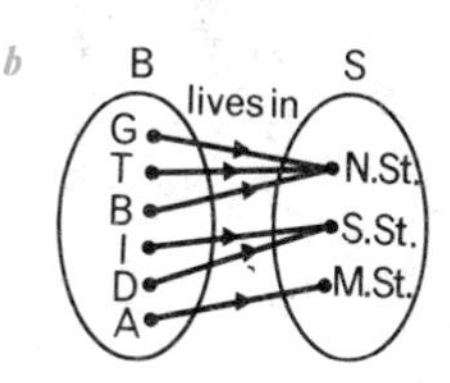

5 *a* 'is the capital of'

b

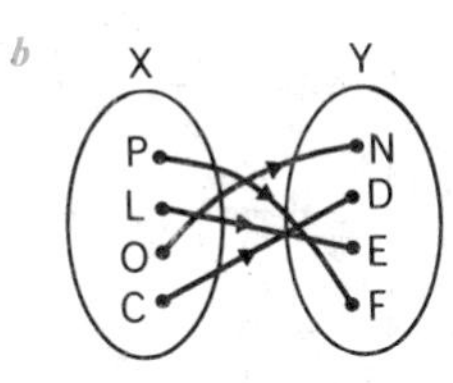

6 *a* range = {1, 2, 3, 4, 5}

b

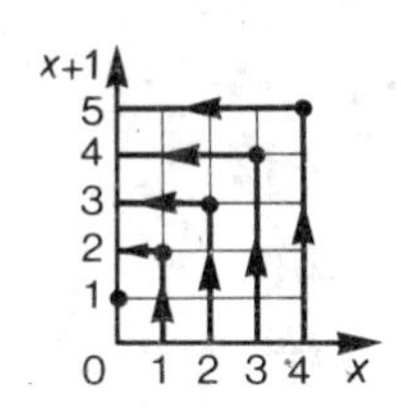

7 *a* domain = {1, 2, 3, 4}
b range = {3, 6, 9, 12}

c

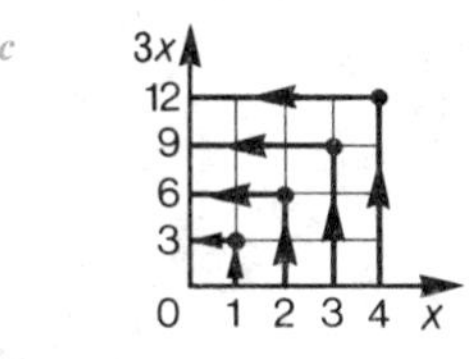

8 range = {0, 1, 4, 9}

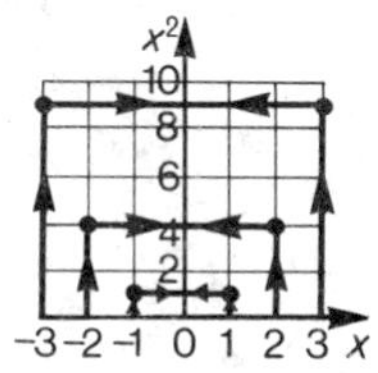

9

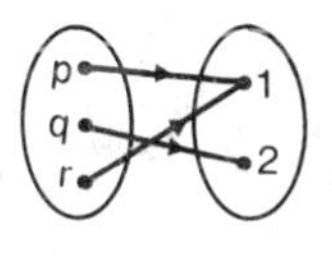

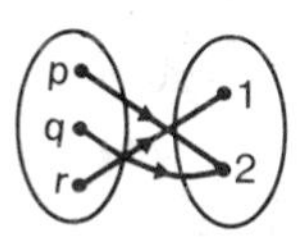

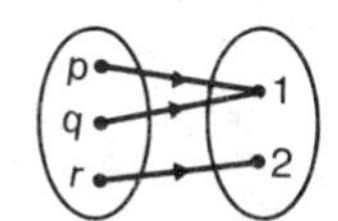

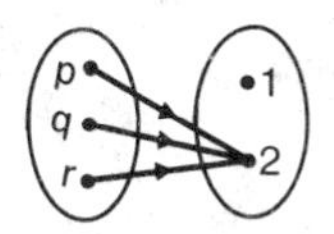

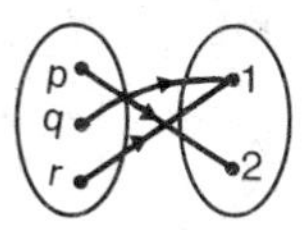

10 range = {4}

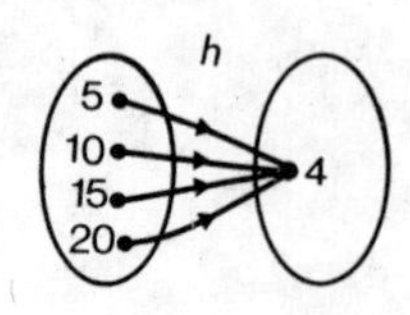

11a range = {27, 8, 1, 0, −1, −8, −27}

b

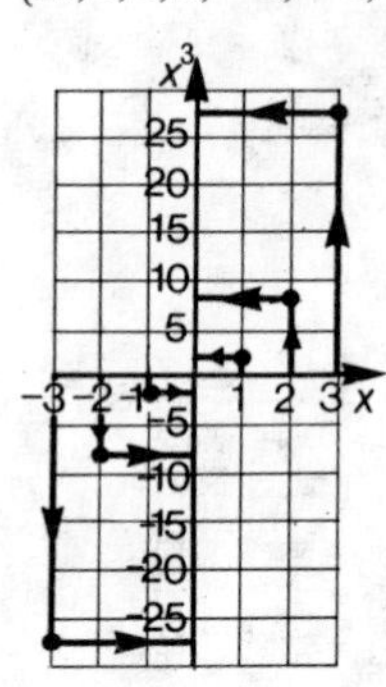

12 8, 3, 0, −1, 0, 3, 8

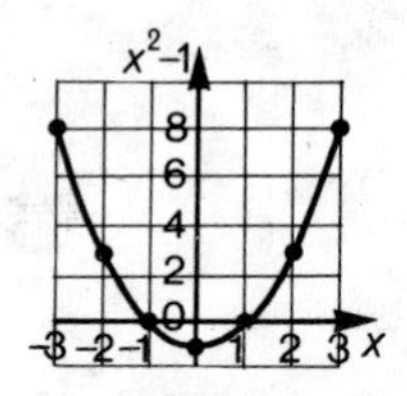

Page 55 Exercise 2

1 a 4 *b* 5 *c* 0 *d* 3 *e* $3\frac{1}{2}$

2 a −1 *b* −9 *c* 11 *d* 1

3 a 5, 2, 17, 10 *b* $a^2+1 = 50$; 7 or −7

4 a 1, −4, −5, −13, 3 *b* $2a-5 = 99$ giving $a = 52$

5 a 4, 2·5, 1 *b* $\frac{1}{2}(x+5) = 0$ giving $x = -5$

6 a $3x-1 = 20$ so $x = 7$ *b* $3x-1 = -16$ so $x = -5$

7 $a = 2, b = 5$ *8* $p = 1, q = 2$ *9* $a = 2, b = -3$; $g(4) = 5$

10a 8, 4, 2, 1, $\frac{1}{2}$, $\frac{1}{4}$ *b* 6

Page 56 Exercise 2B

1 ***a, c, d, f*** are true *2* 3, 1, 2; {1, 2, 3} *3* 2, 4, 4; {2, 4}

4 Image set is {0, 2, 6}

5 a 0·699, 1·398, 1·699 *b* 316

6 a 0·500, 0·866, 1·000 *b* 48·6 *c* $\{s: -1 \leqslant s \leqslant 1, s \in R\}$

7 a 3, 7, 39

8 a 9, 20, 35, 54 *b* 170 *c* $\{4, 5, 6, \ldots\}, \{2, 5, 9, \ldots\}$

9 0, 35, 80, $48\frac{3}{4}$, 0; has returned to its starting level

10a $-\frac{1}{3}, \frac{1}{5}, -\frac{1}{4}, -\frac{4}{15}$ *b* $a = 4$ or $a = -4$

Page 60 Exercise 3

1 a (0, 4) *b* 4 *c* $-2, 2$ *d* $\{y: -5 \leqslant y \leqslant 4\}$ *e* $x = 0$

2 a $(-1, -4)$ *b* -4 *c* $-3, 1$ *d* $\{y: -4 \leqslant y \leqslant 12\}$ *e* $x = -1$

3 a *(1)* 20, 13, 8, 5, 4, 5, 8, 13, 20 *(2)* 12, 5, 0, -3, -4, -3, 0, 5, 12
c $f(x) = x^2 + 4$

4 a (0, 0) *b* 0 *c* 0 *d* $\{y: 0 \leqslant y \leqslant 18\}$ *e* $x = 0$

5 a $(3, -9)$ *b* -9 *c* 0, 6 *d* $\{y: -9 \leqslant y \leqslant 7\}$ *e* $x = 3$

6 a (2, 0) *b* 0 *c* 2 *d* $\{y: 0 \leqslant y \leqslant 16\}$ *e* $x = 2$

7 a (1, 4) *b* 4 *c* $-1, 3$ *d* $\{y: -5 \leqslant y \leqslant 4\}$ *e* $x = 1$

8 a $(-1, 4)$ *b* 4 *c* none *d* $\{y: 4 \leqslant y \leqslant 13\}$ *e* $x = -1$

9 a $(-1, 9)$ *b* 9 *c* $-4, 2$ *d* $\{y: -16 \leqslant y \leqslant 9\}$ *e* $x = -1$

10a values of f: 4, 0, -2, -2, 0, 4, $-1{\cdot}25$, $-2{\cdot}25$, $-1{\cdot}25$
b $x > 4$ or $x < 1$ *c* $-2{\cdot}25$

11a values of f: -9, 0, 5, 6, 3, -4, -15, 5, 6
b *(1)* $f(2{\cdot}5) = -9$ *(2)* $-2 < x < 1{\cdot}5$ *(3)* $x = 1$ or $x = -1{\cdot}5$

12a 25 *b* $k = 21$ *c* $-5, 5$ *d* $h = -6$

Page 65 Exercise 4

1 b *(1)* 6·2 m² *(2)* 3·5 m, 1·5 m *(3)* 6·25 m², 2·5 m by 2·5 m
The rectangle of maximum area is a square of side 2·5 m

2 *(1)* 3 seconds and 45 metres *(2)* 4·5 seconds approximately

3 a 60 cm² *b* at P, R: $\frac{1}{2}x(6-x)$ and at Q, S: $\frac{1}{2}x(10-x)$
c area $= 60 - x(6-x) - x(10-x) = 60 - 16x + 2x^2$ cm² *e* 28 cm²

4 a BC $= (40-2x)$ cm so $A(x) = (40-2x)x = 40x - 2x^2$
c 10 cm, 20 cm

5 2

Algebra—Answers to Revision Exercises

Page 69 Revision Exercises 1A

1 a $2x^2$ *b* $-4y^2$ *c* $-2z^2$ *d* 0
e x *f* $-2p^2 + 3q^2$ *g* $5x + y$ *h* $2m - 8n$

2 a $-4x + 15y$ *b* $6p$ *c* $2c - 9d - 2e$ *d* $4x^2 - 2x + 1$

3 a $-8x + 9y$ *b* $10a - b$ *c* $4p + 6q + 2r$ *d* $-6c + 3d - 7e$

4 a $5a - 6b$ *b* $2x + 4y$

5 a $\{-1\}$ *b* $\{x: x > 2\}$ *c* $\{4\}$ *d* $\{x: x \geqslant 1\}$

6 a $x^2 + 3x + 2$ *b* $2x^2 + 7x + 3$ *c* $6x^2 + 13x + 6$ *d* $y^2 - 6y + 8$

e $2y^2-3y+1$ *f* $8y^2-14y+3$ *g* a^2+a-6 *h* b^2-4b-5
i c^2-36 *j* $6x^2+x-1$ *k* $10y^2-29y+10$ *l* $8z^2+22z-21$

7 a $c^2+2cd+d^2$ *b* $m^2+2mn+n^2$ *c* $u^2-2uv+v^2$ *d* $x^2-2xy+y^2$
e $x^2+12x+36$ *f* y^2-6y+9 *g* $4x^2+20x+25$ *h* $9y^2-12y+4$

8 a $\frac{1}{2}$ *b* 4 *c* -1 *d* 3·4
e $-\frac{5}{3}$ *f* $-\frac{5}{2}$

10 $(a+2)$ m, $(b+2)$ m; $(2a+2b+4)$ m²

Page 70 Revision Exercise 1B

1 a $2x^2-6x$ *b* $-2b^3$ *c* $3x^2-6$ *d* y^2+6y
e $-i+4j-6k$ *f* $-5k$ *g* $4x^2-x$ *h* $-3x^2-2x$

2 a $\{1\}$ *b* $\{x:x<-\frac{2}{3}\}$ *c* $\{-\frac{2}{3}\}$ *d* $\{x:x\geqslant-\frac{1}{5}\}$

3 a a^2-b^2 *b* $2a^2-5ab-3b^2$ *c* $15a^2-16ab+4b^2$ *d* $10+7x+x^2$
e $4-5x+x^2$ *f* $4-9x^2$ *g* $-y^2+y+6$ *h* $-x^2-3x+4$
i $-z^2+5z-6$

4 a $ab+5a+2b+10$ *b* $xy-4x-y+4$ *c* x^4-y^4
d $2x^3+5x^2-6x-9$ *e* $15y^3-26y^2+33y-10$

5 a $4ab$ *b* $-4pq$ *c* $-5x^2+5$ *d* $2x^2+\frac{2}{x^2}$

6 a $3\frac{1}{2}$ *b* 4 *c* 2 *d* -1

9 1·012

10 $15^2-x^2=13^2-(14-x)^2$; $x=9$; $h=12$; area $=84$ cm²

Page 71 Revision Exercise 2A

1 a 2, 2, 2 × 2, 7 *b* 2, 4, 2 × 4, −4
c 4, 1, 4 × 1, 3 *d* 3, 3, 3 × 3, 2

2 a First and last; third *b (1)* 15 *(2)* pq

3 $\begin{pmatrix}1&3\\0&1\\1&4\end{pmatrix}$ *6 a* 6, −2 *b* 0, $\frac{1}{2}$ *c* 4, −5

7 (0 −1 2), (6 −1 6), (0 −1 2), (−6 1 −6), (−3 −2 2),
(24 −6 28)

8 a 2, −3 *b* −4, 3

9 $a=-3, b=2, c=0, d=-1$

10 $\begin{pmatrix}-1&-1\\1&1\end{pmatrix}, \begin{pmatrix}0&0\\0&0\end{pmatrix}$ *11* (5, −3), (−17, 13)

12a 4 *b* 6 *c* 7

13a $\begin{pmatrix}20&-11\\29&-16\end{pmatrix}, \begin{pmatrix}7&3\\17&7\end{pmatrix}, \begin{pmatrix}34&14\\12&5\end{pmatrix}$ *b* $\begin{pmatrix}39&16\\56&23\end{pmatrix}, \begin{pmatrix}-1&1\\-5&4\end{pmatrix}, \begin{pmatrix}18&10\\5&3\end{pmatrix}$

c $\begin{pmatrix}3&-2\\-7&5\end{pmatrix}, \begin{pmatrix}3&-1\\5&-2\end{pmatrix}, \begin{pmatrix}\frac{1}{2}&-1\\-\frac{1}{2}&2\end{pmatrix}$ *d* $\begin{pmatrix}69&29\\100&42\end{pmatrix}, \begin{pmatrix}69&29\\100&42\end{pmatrix}, \begin{pmatrix}42&1\\60&1\end{pmatrix}$

14a (2, −1), (6, −1), (6, −3) *b* (−2, 1), (−6, 1), (−6, 3)
c (−1, −2), (−1, −6), (−3, −6)

15a −2, 3 *b* 2, −1 *c* 3, 1

Page 73 Revision Exercise 2B

1 a A, C *b* A, B *c (1)* $\begin{pmatrix}0 & 0\\0 & 0\end{pmatrix}$ *(2)* $\begin{pmatrix}2 & 2\\-2 & 8\end{pmatrix}$ *(3)* $\begin{pmatrix}0 & -2\\2 & 0\end{pmatrix}$

(4) $\begin{pmatrix}-2 & 0\\0 & -8\end{pmatrix}$ *(5)* $\begin{pmatrix}4 & 8\\12 & 16\end{pmatrix}$ *(6)* $\begin{pmatrix}0 & -8\\-12 & 0\end{pmatrix}$ *(7)* $\begin{pmatrix}4 & 0\\0 & 16\end{pmatrix}$ *(8)* $\begin{pmatrix}4 & 16\\24 & 16\end{pmatrix}$

2 $\begin{pmatrix}3\\4\end{pmatrix}$ and $\begin{pmatrix}2\\-1\end{pmatrix}$

3 $(0\ \ 0\ \ 0), (2a\ \ -2b\ \ 2c), (0\ \ 0\ \ 0),$
$(-2a\ \ 2b\ \ -2c), (-a\ \ -b\ \ -c), (8a\ \ -8b\ \ 8c)$

4 $p = 2, q = 5, r = 5, s = 4$

5 $a = 5, b = 8, c = 0, d = 0, e = -1, f = -1$

6 a $5x+3y = 7, 2x+3y = 1; 2, -1$
b $3x+y = 4, 2x-y = 6; 2, -2$

7 a $\begin{pmatrix}3 & 9\\-6 & -3\end{pmatrix}$ *b* $\begin{pmatrix}9 & 1 & -14\\-1 & -11 & 14\end{pmatrix}$ *c* $\begin{pmatrix}a^2-b^2 & 0\\-2ab & a^2+b^2\end{pmatrix}$

d $\begin{pmatrix}a^2-b^2 & 2ab\\-2ab & a^2-b^2\end{pmatrix}$

8 $(10, -5), (12, -13), (-17, 18), (7, -13)$

9 $\begin{pmatrix}1-k & 2\\2 & 1-k\end{pmatrix}$. $\det(A-kI) = 0 \Leftrightarrow (1-k)^2-4 = 0 \Leftrightarrow k = 3$ or -1

10 $a = 1, b = 2, c = 4, d = -3$ *11* $\begin{pmatrix}1 & -2\\-1 & 3\end{pmatrix}$; $X = \begin{pmatrix}8 & -1\\-10 & 4\end{pmatrix}$

12a $\frac{1}{3}\begin{pmatrix}2 & -1\\1 & 1\end{pmatrix}$ *b* $\frac{1}{8}\begin{pmatrix}2 & -4\\1 & 2\end{pmatrix}$ *c* none *d* none

13 $\begin{pmatrix}1 & 2\\-2 & 8\end{pmatrix}$ *14a* $4, -2$ *b* $-3, 2$ *c* $3, -\frac{1}{2}$

15a 9 *b* (9), 9 routes from A to C *c* (12) 12 routes from A to C

16b 8 from x_1, 7 from x_2; matrix product $\begin{pmatrix}4 & 4 & 0\\0 & 2 & 5\end{pmatrix}$

Page 76 Revision Exercise 3A

1 a

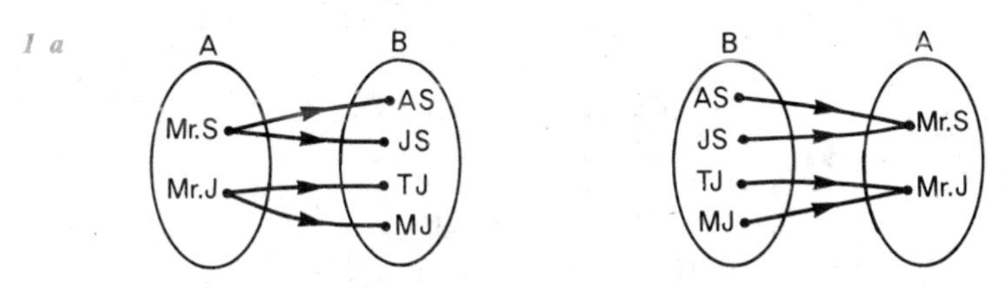

b the second one. Each element of B is related to exactly one element of A.

2 a 1 *b* 1 and 4 *c* $\{1, 2, 4\}$ *d* $\{(1, 2), (2, 1), (3, 4), (4, 2)\}$

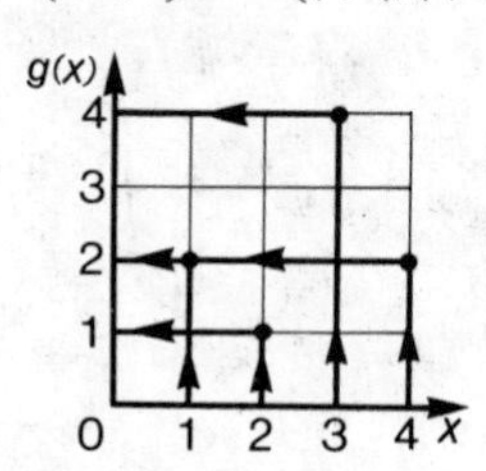

3 a 2 *b* 1, 2, 3 and 4 *c* $\{2\}$ *d* $\{(1, 2), (2, 2), (3, 2), (4, 2)\}$

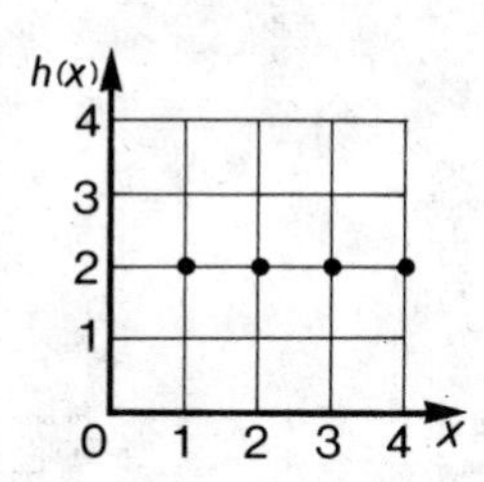

4 $\{0, 1, 16\}$

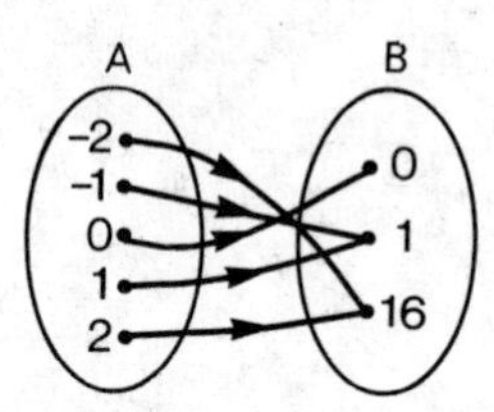

5 a $f(x) = x^2+1$ *b* $g(x) = 2x-1$ *c* $h(x) = \frac{1}{2}x^2$

6 a -5 *b* 7 *c* -11 *d* $-3{\cdot}5$; 15

7 a (*1*) 2 (*2*) 50 (*3*) 18 (*4*) $8t^2$
b 2 or -2 *8* $= 3, b = -2$

9 $p = \frac{1}{2}, q = 5$; 12 *10c* $(-1{\cdot}5, -6{\cdot}3)$; minimum turning value $-6{\cdot}3$
d (*1*) 4·6 (*2*) $-5{\cdot}4$

11 (0, 2) minimum turning point. No zeros of f.

12a $-2, 4$ *b* (1, 9) maximum turning point $x = 1$
d $4 < x \leqslant 6$ or $-4 \leqslant x < -2$

13 100 m^2; length and breadth each 10 m

14 3 and -1 *15* $\{-1, 1, 3, 5\}$

Page 78 Revision Exercise 3B

1 *is the cube of* Each element of P is related to one element of Q.

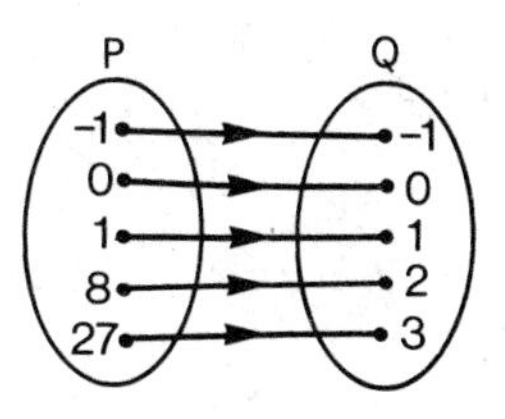

2 a 12 *b* -3 *c* -2 *d* 0 *e* -3

3 a domain = $\{-3, -2, -1, 0, 1, 2, 3\}$; range = $\{-3, -1, 1, 3, 5, 7, 9\}$
c $a = 2, b = 3$

4 a 3, 2, 1, 0, 1, 2, 3 *b* $\{(-3, 3), (-2, 2), (-1, 1), (0, 0), (1, 1), (2, 2), (3, 3)\}$

c

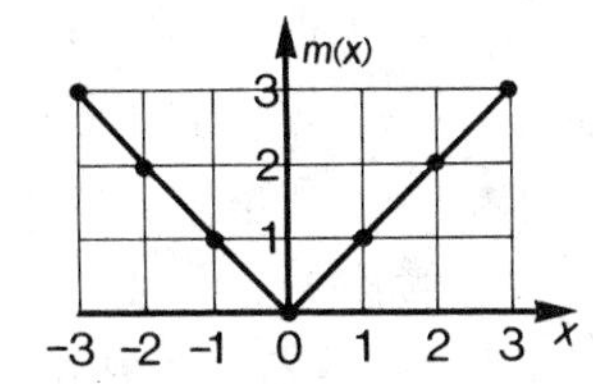

5 a $-3, 0, -1, -2\frac{1}{2}$ *b* $\frac{1}{3}$ *c* The value of the function is not defined.

6 $a = 1, b = -3$ *7 a* $s(t) = 16t + \frac{1}{2}t^2$ *b* *(1)* 210 m *(2)* 264 m
c 27 m/s *8* $-3, 2$ *9 a* T *b* F *c* T *d* T *e* F

10a $-4, -1\frac{1}{2}, 0, \frac{1}{2}, 0, -1\frac{1}{2}, -4$ *b* $\frac{1}{2}$ *c* $-2{\cdot}6$ *d* $0 < x < 2$

11a $-2, 5$ *b* $-10, q = -10$ *c* $p = -3$ *d* $\{x : 1{\cdot}5 \leqslant x \leqslant 6\}$ *e* $-12{\cdot}3$

12c *(1)* $(\frac{1}{2}, -16)$, minimum *(2)* $-1\frac{1}{2}$ or $2\frac{1}{2}$ *(3)* -14 approximately

13c *(1)* $(\frac{1}{2}, 16)$, maximum *(2)* $-1\frac{1}{2}$ or $2\frac{1}{2}$ *(3)* 14 approximately

14 Minimum A occurs when $x = 5$, giving 25 cm^2 as minimum area.

Geometry—Answers to Chapter 1

Page 85 Exercise 1

1 10 m, 8 m *2* 16 m, 12 m *3* 1 cm represents 1 m
4 1 cm represents 33 km *5* 4·5 cm *6* 105 m by 75 m, 18 m

Page 87 Exercise 2

1 10·8 m *2* 8·8 m *3* 9 cm *4 a* 10 cm *b* 20 cm
5 55 cm *6* 68 cm *7* 8 cm, 4 cm *8* 15 cm
9 4·2 m *10* 16 cm, 1:4

Page 88 *Exercise 3*

1 **b** *2* ***a, c, d*** *3* (ii) and (iv) might be; (iii) and (v) cannot be
4 no *5* 6 cm *6 a* yes *b* no
7 ***a, c, d, f***

Page 90 *Exercise 4B*

1 yes *3* (iii) and (iv) are similar
5 remove 2 top squares *a* each pair in (i) and (ii) *b* all *c* as in ***a***
sides, 1:2; perimeters, 1:2; areas, 1:4
6 a 1:2 *b* 1:4 *c* 1:8
7 6 cm by 3 cm by 2 cm *a* 4:1 *b* 16:1 *c* 64:1
m:1, m^2:1, m^3:1

Page 92 *Exercise 5*

1 a corresponding angles of parallels *b* 2:1
2 a corresponding angles of parallels, *b* 2:3 *3* yes *4* no *5* A, B

Page 95 *Exercise 6*

1 $\angle A = \angle P$, $\angle B = \angle Q$, $\angle C = \angle R$; $\frac{a}{p} = \frac{b}{q} = \frac{c}{r}$
2 {ACB, EDF, MNO, PQR}, sides in proportion; {GHI, LJK}, equiangular
3 equiangular, $\frac{AB}{QR} = \frac{BC}{RP} = \frac{CA}{PQ}$
4 sides in proportion, $\angle D = \angle Y$, $\angle E = \angle Z$, $\angle F = \angle X$
5 no, sides not in proportion *6* ***b, c, d, e*** to the first; ***a, f*** to the second
7 19 m *8* 7·5 m

Page 96 *Exercise 7*

1 (i) $\frac{a}{c} = \frac{f}{e} = \frac{b}{d}$ (ii) $\frac{a}{d} = \frac{e}{f} = \frac{b}{c}$ (iii) $\frac{a}{b} = \frac{d}{c} = \frac{f}{e}$
(iv) $\frac{a}{a+b} = \frac{d}{d+c} = \frac{e}{f}$ *2* 42·7 cm
3 EAB, BDX, BFE *a* $\frac{EC}{EA} = \frac{EX}{EB} = \frac{CX}{AB}$ *b* $\frac{BD}{BF} = \frac{BX}{BE} = \frac{DX}{FE}$
4 b $\frac{AB}{AD} = \frac{AC}{AE} = \frac{BC}{DE}$ *c* 6 cm, 2 cm *d* 6 *5* 12 cm, 10 cm
6 PQR, PLN, QLM; 12 m *7* $a = 6, b = 3, c = 5, d = 8\frac{1}{3}$

Page 98 *Exercise 7B*

1 4·5, $5\frac{5}{9}$ *2* $a = 50, b = 20, c = 40, d = 76$ *3* 8 cm, 10 cm
4 b 6 cm *5 b* 9 cm *c* 7 cm, $\sqrt{63}$ cm, $\sqrt{112}$ cm *6 b* 4·8 cm, 7·2 cm

7 a They cut OC in the same ratio. *b* Use corresponding angles.

8 b ay *c* 7·5 cm, 6 cm *9 b* 9 m *10b* Join BD, draw AXY‖BC, draw BLM‖AD.

Page 102 Exercise 8

1 $m_{AB} = \frac{2}{3}$, $m_{CD} = -1$, $m_{EF} = \frac{2}{7}$, $m_{GH} = 0$, $m_{KH} = \frac{5}{2}$, $m_{MN} = -6$

2 a 4 *b* 10 *c* 0 *d* -3 *e* 3 *f* cannot be calculated

3 a upwards from left to right *b* downwards from left to right *c* parallel to the x-axis *4 a* $y = 2x$ *b* 3

5 a $y = -\frac{1}{2}x$ *b* -2 *6 a* 8, 17, 62, -10; $y = 3x+2$ *b* $\frac{1}{2}$

7 a $-5, -3, -1, 1, 5$ *b* -1 *8 a* $y = 4x$ *b* $y = \frac{2}{3}x$ *c* $y = \frac{3}{2}x$ *d* $y = -4x$ *e* $y = -\frac{2}{3}x$ *f* $y = -\frac{3}{2}x$ *g* $y = 0$

9 a $y = 4x+4$ *b* $y = \frac{2}{3}x+4$ *c* $y = \frac{3}{2}x+4$ *d* $y = -4x+4$ *e* $y = -\frac{2}{3}x+4$ *f* $y = -\frac{3}{2}x+4$ *g* $y = 4$

10a $y = 2x+4$ *b* $y = 2x$ *c* $y = 2x+1$ *d* $y = 2x-4$

11a $y = -x+4$ *b* $y = -x$ *c* $y = -x+1$ *d* $y = -x-4$

12a 2, (0, 5) *b* -1, (0, 3) *c* -2, (0, -4) *d* $\frac{3}{4}$, (0, -3) *e* 1, (0, 2) *f* 3, (0, 4) *g* -1, (0, 4) *h* -2, (0, 5) *i* 5, (0, 7) *j* $\frac{1}{2}$, (0, 3) *k* $-\frac{2}{3}$, (0, -2) *l* $\frac{4}{3}$, (0, $-\frac{2}{3}$)

Geometry—Answers to Chapter 2

Page 106 Exercise 1

1 b 55 km, 3 km *c* a circle, centre A, radius 29 km; direction

2 17 km; 250° *3 a* $\overrightarrow{DB}$ *b* $\overrightarrow{CD}$ *c* 0

4 a $\overrightarrow{PQ}$, $\overrightarrow{SR}$ *b* $\overrightarrow{PS}$, $\overrightarrow{QR}$ *c* $\overrightarrow{PR}$ *d* $\overrightarrow{QS}$

5 a $\begin{pmatrix}4\\1\end{pmatrix}$ *b* $\begin{pmatrix}4\\-3\end{pmatrix}$ *c* $\begin{pmatrix}-4\\-2\end{pmatrix}$ *d* $\begin{pmatrix}0\\2\end{pmatrix}$

6

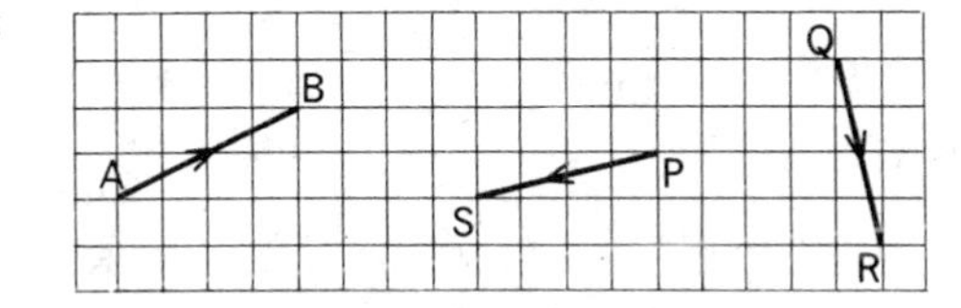

7 b (1) $\begin{pmatrix}6\\-2\end{pmatrix}$ *(2)* $\begin{pmatrix}8\\-2\end{pmatrix}$ *(3)* $\begin{pmatrix}-4\\0\end{pmatrix}$ *8 a* $\begin{pmatrix}4\\4\end{pmatrix}$ *b* $\begin{pmatrix}4\\-3\end{pmatrix}$ *c* $\begin{pmatrix}-a\\b\end{pmatrix}$ *d* $\begin{pmatrix}2\\7\end{pmatrix}+\begin{pmatrix}3\\-3\end{pmatrix}$

9 $\sqrt{13}$, $\sqrt{17}$, 5, $\sqrt{20}$

Page 109 *Exercise 2*

2 Commutative law

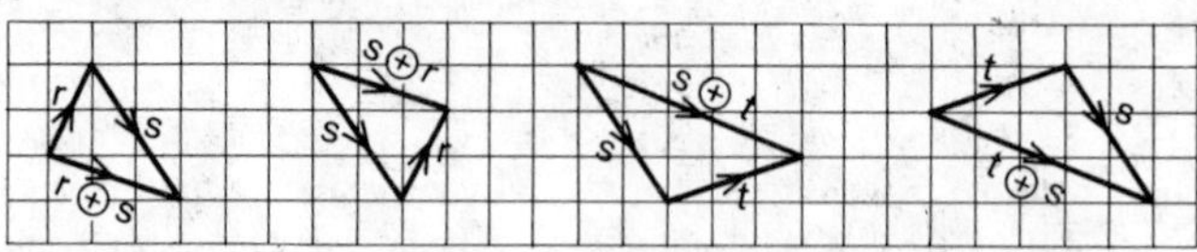

3 Associative law

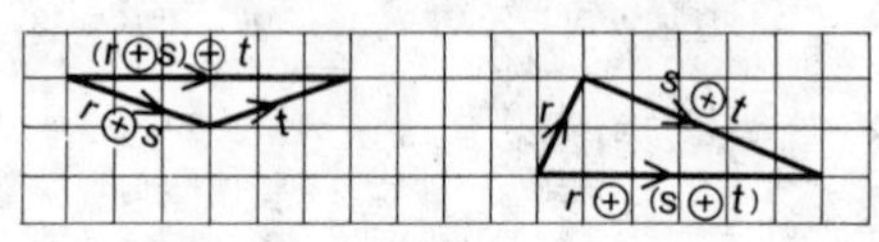

5 *a* $\overrightarrow{AC}$, $\overrightarrow{AD}$ *b* $\overrightarrow{BD}$, $\overrightarrow{AD}$ 6 *b* 260 km/h, 023°

Page 110 *Exercise 3*

1 *c* $\overrightarrow{HL}$ *d* $\overrightarrow{FH}$ 2 *a* no *b* yes 3 *a* yes *b* equal magnitudes
4 *a* $\overrightarrow{AC}$ *b* $\overrightarrow{DE}$ *c* $\overrightarrow{AC}$ *d* $\overrightarrow{CD}$ 5 *a* $\overrightarrow{EB}$ *b* $\overrightarrow{DE}$ *c* $\overrightarrow{EB}$
d $\overrightarrow{BD}$ *e* $\overrightarrow{CD}$

Page 113 *Exercise 4*

1 *a* $\begin{pmatrix}1\\3\end{pmatrix}, \begin{pmatrix}4\\-2\end{pmatrix}, \begin{pmatrix}-4\\2\end{pmatrix}, \begin{pmatrix}-1\\-3\end{pmatrix}, \begin{pmatrix}5\\1\end{pmatrix}, \begin{pmatrix}5\\1\end{pmatrix}, \begin{pmatrix}-5\\-1\end{pmatrix}$ *b* $\begin{pmatrix}1+4\\3-2\end{pmatrix}$ i.e. add the components

2

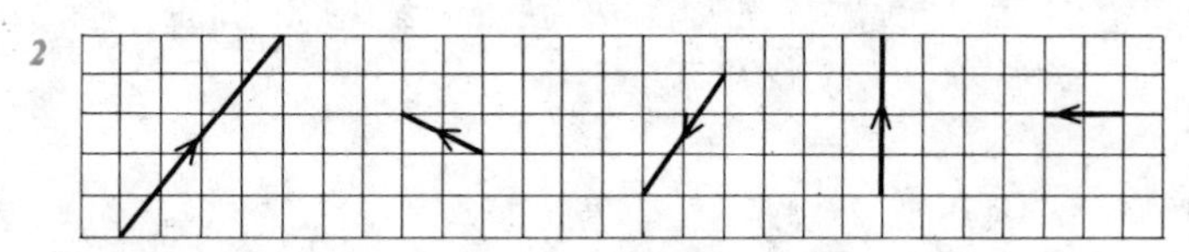

3 *b* $\begin{pmatrix}1\\4\end{pmatrix}$ 4 *b* $\begin{pmatrix}-3\\-1\end{pmatrix}$ 5 *a* $\begin{pmatrix}1\\1\end{pmatrix}$ *b* $\begin{pmatrix}3\\-2\end{pmatrix}$ *c* $\begin{pmatrix}-6\\-5\end{pmatrix}$
commutative law

6 *a* $\begin{pmatrix}2\\7\end{pmatrix}$ *b* $\begin{pmatrix}0\\6\end{pmatrix}$ *c* $\begin{pmatrix}-3\\2\end{pmatrix}$ *d* $\begin{pmatrix}0\\6\end{pmatrix}$

7 *a* $\begin{pmatrix}4\\13\end{pmatrix}$ *b* $\begin{pmatrix}1\\15\end{pmatrix}$ *c* $\begin{pmatrix}2\\6\end{pmatrix}$ *d* $\begin{pmatrix}1\\15\end{pmatrix}$; associative law

8 $a = c, b = d$

Page 114 *Exercise 5*

1 zero length, direction not defined 2 zero line segment

3 zero line segment in each case 4 *a* $\begin{pmatrix}0\\0\end{pmatrix}$ *b* $\begin{pmatrix}0\\0\end{pmatrix}$ *c* $\begin{pmatrix}0\\0\end{pmatrix}$

5 *a* $\overrightarrow{BD}$ *b* $\overrightarrow{CE}$ *c* $\boldsymbol{0}$ *d* $\boldsymbol{0}$ 6 *a* $\boldsymbol{u}$ *b* $\boldsymbol{v}$ *c* $\boldsymbol{u}+\boldsymbol{v}$

7 a $\begin{pmatrix} 2 \\ -3 \end{pmatrix}$ *b* $\begin{pmatrix} -4 \\ 5 \end{pmatrix}$ *c* $\begin{pmatrix} -2 \\ -8 \end{pmatrix}$ *8* $\overrightarrow{AB}$, $\overrightarrow{DC}$, $\overrightarrow{QR}$

9 a $\overrightarrow{DB}$ *b* $\overrightarrow{AB}$ *c* $\overrightarrow{CA}$

10

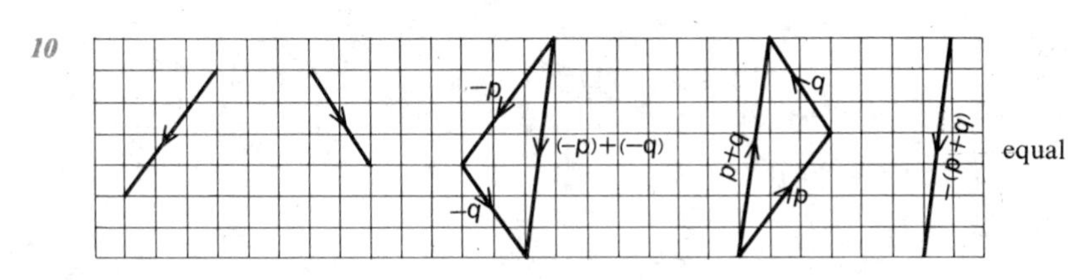

11a $\begin{pmatrix} -5 \\ -4 \end{pmatrix}$ *b* $\begin{pmatrix} 3 \\ -5 \end{pmatrix}$ *c* $\begin{pmatrix} -4 \\ 7 \end{pmatrix}$ *d* $\begin{pmatrix} 0 \\ 0 \end{pmatrix}$ *e* $\begin{pmatrix} -a \\ -b \end{pmatrix}$

12a $\begin{pmatrix} -3 \\ 4 \end{pmatrix}$ *b* $\begin{pmatrix} 4 \\ 7 \end{pmatrix} + \begin{pmatrix} -4 \\ -7 \end{pmatrix}$ *c* $\begin{pmatrix} -a \\ b \end{pmatrix}$

Page 116 Exercise 6

1 a $\overrightarrow{BC}$ *b* $\overrightarrow{CA}$ *c* $\overrightarrow{AC}$ *d* $\overrightarrow{BA}$ *e* $\overrightarrow{AB}$; $\overrightarrow{LS}$

2 a $\overrightarrow{HF}$ *b* $\overrightarrow{FH}$ *c* $\overrightarrow{EG}$ *d* $\overrightarrow{HE}$

3 a $\begin{pmatrix} 1 \\ 4 \end{pmatrix}$ *b* $\begin{pmatrix} -5 \\ -3 \end{pmatrix}$ *c* $\begin{pmatrix} a-c \\ b-d \end{pmatrix}$ *4 a* $\begin{pmatrix} 3 \\ 4 \end{pmatrix}$ *b* $\begin{pmatrix} 7 \\ 10 \end{pmatrix}$

5 a $\boldsymbol{v}$ *b* $\boldsymbol{u}$ *c* $\boldsymbol{0}$ *d* $\boldsymbol{0}$

6 a $\boldsymbol{u}$ *b* $\boldsymbol{a}$ *c* $\boldsymbol{u}$ *d* $\boldsymbol{v}+\boldsymbol{w}$

7 $\overrightarrow{AE}$, where $\overrightarrow{CE} = \overrightarrow{BC}$ *8 a* $\begin{pmatrix} 3 \\ 2 \end{pmatrix}$ *b* $\begin{pmatrix} 6 \\ 8 \end{pmatrix}$ *c* $\begin{pmatrix} 5 \\ -6 \end{pmatrix}$

9 a $\begin{pmatrix} -8 \\ 11 \end{pmatrix}$ *b* $\begin{pmatrix} 16 \\ 1 \end{pmatrix}$ *c* $\begin{pmatrix} 2 \\ -7 \end{pmatrix}$

Page 118 Exercise 7

1 a $2\boldsymbol{p}$ *b* $\frac{3}{2}\boldsymbol{p}$ *c* $\frac{1}{2}\boldsymbol{p}$

2 c $\begin{pmatrix} 4 \\ -2 \end{pmatrix}, \begin{pmatrix} 6 \\ -3 \end{pmatrix}$ *d* $\begin{pmatrix} 2\times 2 \\ 2\times -1 \end{pmatrix}, \begin{pmatrix} 3\times 2 \\ 3\times -1 \end{pmatrix}$

4

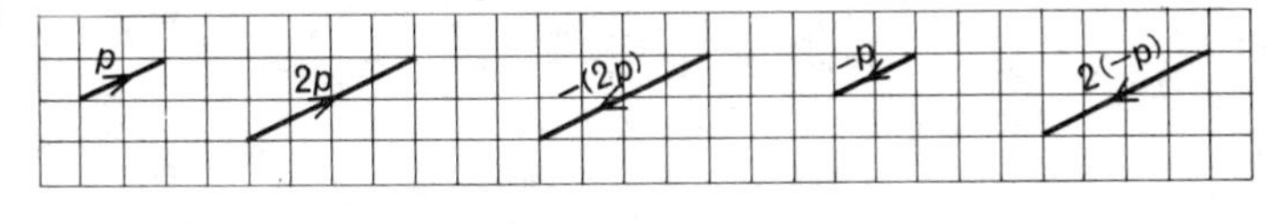

$-(2\boldsymbol{p}) = 2(-\boldsymbol{p})$

Page 120 Exercise 7B

1

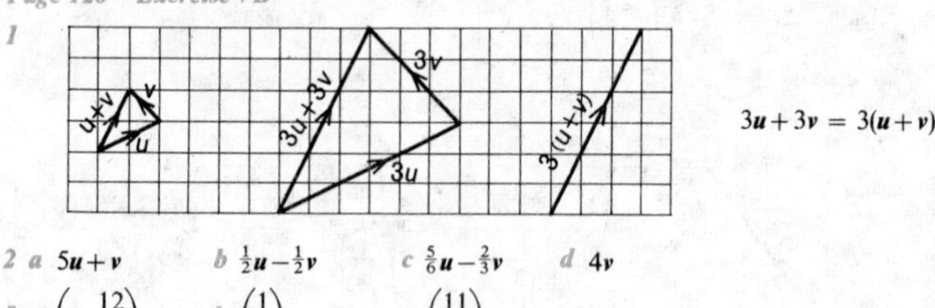

$3\boldsymbol{u}+3\boldsymbol{v} = 3(\boldsymbol{u}+\boldsymbol{v})$

2 a $5\boldsymbol{u}+\boldsymbol{v}$ *b* $\frac{1}{2}\boldsymbol{u}-\frac{1}{2}\boldsymbol{v}$ *c* $\frac{5}{6}\boldsymbol{u}-\frac{2}{3}\boldsymbol{v}$ *d* $4\boldsymbol{v}$

5 a $\begin{pmatrix}12\\-12\end{pmatrix}$ *b* $\begin{pmatrix}1\\3\end{pmatrix}$ *c* $\begin{pmatrix}11\\25\end{pmatrix}$

6 a $\boldsymbol{x} = \dfrac{5\boldsymbol{v}-3\boldsymbol{u}}{3}$ *b* $\boldsymbol{x} = \dfrac{5\boldsymbol{u}-7\boldsymbol{v}}{3}$

7 a $\begin{pmatrix}3\\2\end{pmatrix}$ *b* $\begin{pmatrix}4\\-7\end{pmatrix}$ *8 a* $\begin{pmatrix}3\\0\end{pmatrix}$ *b* $\begin{pmatrix}6\\0\end{pmatrix}$ *c* $\begin{pmatrix}6\\9\end{pmatrix}$ *d* $\begin{pmatrix}3\\1\end{pmatrix}$

9 a k times magnitude and same direction

b $-k$ times magnitude, but opposite direction

10a $\boldsymbol{u} = \boldsymbol{0}$ *b* $\boldsymbol{u}$ and $\boldsymbol{v}$ have same direction *c* $a = b = 0$

Page 122 Exercise 8

1 a $\begin{pmatrix}2\\3\end{pmatrix}, \begin{pmatrix}7\\5\end{pmatrix}$ *b* $\begin{pmatrix}5\\2\end{pmatrix}$ *2 a* $\begin{pmatrix}0\\-1\end{pmatrix}, \begin{pmatrix}4\\2\end{pmatrix}$ *b* $\begin{pmatrix}4\\3\end{pmatrix}$

3 $\begin{pmatrix}4\\0\end{pmatrix}, \begin{pmatrix}6\\2\end{pmatrix}, \begin{pmatrix}-2\\1\end{pmatrix}; \begin{pmatrix}2\\2\end{pmatrix}, \begin{pmatrix}-8\\-1\end{pmatrix}, \begin{pmatrix}6\\-1\end{pmatrix}$

4 a PQ is equal and parallel to SR. *b* (5, 3) *5 b* (7, 3)

6 a $\begin{pmatrix}4\\2\end{pmatrix}, \begin{pmatrix}2\\6\end{pmatrix}$ *b* $\begin{pmatrix}-2\\4\end{pmatrix}, \begin{pmatrix}2\\-4\end{pmatrix}$ *c* $\begin{pmatrix}1\\-2\end{pmatrix}$ *d* $\begin{pmatrix}3\\4\end{pmatrix}$ *f* (3, 4)

7 a $\begin{pmatrix}7\\6\end{pmatrix}, \begin{pmatrix}9\\-4\end{pmatrix}; \begin{pmatrix}2\\-10\end{pmatrix}, \begin{pmatrix}-2\\10\end{pmatrix}; \begin{pmatrix}-1\\5\end{pmatrix}; \begin{pmatrix}8\\1\end{pmatrix}$ (8, 1)

b $\begin{pmatrix}-2\\8\end{pmatrix}, \begin{pmatrix}-4\\-7\end{pmatrix}\begin{pmatrix}-2\\-15\end{pmatrix}, \begin{pmatrix}2\\15\end{pmatrix}; \begin{pmatrix}1\\7\frac{1}{2}\end{pmatrix}; \begin{pmatrix}-3\\\frac{1}{2}\end{pmatrix}$ $(-3, \frac{1}{2})$

9 a (3, −4) *b* They are the same point.

10a (5, 2) *b* (−11, −7) *c* (13, 10)

11a (4, 3) *b* The coordinates of T are the average of those of Q and S.

Page 125 Exercise 8B

2 $\begin{pmatrix}6\\3\end{pmatrix}$, (6, 3) *3 b* (3, 1) *4 b* (7, −2)

5 b *(1)* (6, 4) *(2)* (1, −4) *(3)* (3, 4)

7 a $\begin{pmatrix}4\\3\end{pmatrix}, \begin{pmatrix}5\\1\end{pmatrix}, \begin{pmatrix}-3\\5\end{pmatrix}$ *b* $\begin{pmatrix}1\\3\end{pmatrix}$ *c* (1, 3) *d* $\begin{pmatrix}2\\3\end{pmatrix}$ *e* (2, 3)

8 *a* $\frac{1}{2}(\boldsymbol{a}+\boldsymbol{b}), \frac{1}{2}(\boldsymbol{a}+\boldsymbol{c})$ *b* $\boldsymbol{c}-\boldsymbol{b}, \frac{1}{2}(\boldsymbol{c}-\boldsymbol{b})$ *c* $\overrightarrow{DE} = \frac{1}{2}\overrightarrow{BC}$, DE$\|$BC

9 *a* $\frac{1}{2}(\boldsymbol{a}+\boldsymbol{b}), \frac{1}{2}(\boldsymbol{b}+\boldsymbol{c}), \frac{1}{2}(\boldsymbol{c}+\boldsymbol{d}), \frac{1}{2}(\boldsymbol{d}+\boldsymbol{a})$ *b* $\frac{1}{2}(\boldsymbol{c}-\boldsymbol{a})$ *c* parallelogram
d The mid-points of the sides are the vertices of a parallelogram.

10*a* $\frac{1}{2}(\boldsymbol{a}+\boldsymbol{b}), \frac{1}{2}(\boldsymbol{c}+\boldsymbol{d})$ *b* $\frac{1}{4}(\boldsymbol{a}+\boldsymbol{b}+\boldsymbol{c}+\boldsymbol{d})$ *c* $\frac{1}{4}(\boldsymbol{a}+\boldsymbol{b}+\boldsymbol{c}+\boldsymbol{d})$
d The joins of the midpoints of opposite sides and of the diagonals are concurrent and bisect one another.

Page 128 Exercise 9

1 each 10 units; one; eight 2 *a* $\sqrt{5}$ *b* $\sqrt{65}$ *c* $\sqrt{65}$ *d* 13
3 *a* false *b* true 4 *a* $\sqrt{41}$ *b* 10 *c* $\sqrt{41}$ *d* $\sqrt{58}$
e $5a$ *f* $m\sqrt{53}$ 5 *a* collinear *b* same direction
6 Two sides of a triangle are together greater than the third side.

Geometry—Answers to Revision Exercises

Page 131 Revision Exercise 1

1 8·5 cm 2 2·9 cm, 9 m 3 120 mm 4 90 m by 60 m, 9 m
5 *a* 1 cm represents 5 m *b* largest and smallest similar
6 Δs equiangular, $\frac{AB}{DE} = \frac{BC}{EF} = \frac{CA}{FD}$
7 sides in proportion, $\angle A = \angle P$, $\angle B = \angle R$, $\angle C = \angle Q$
8 DHX, QZE; sides in proportion; $\angle A = \angle D = \angle Q$, $\angle B = \angle H = \angle Z$, $\angle C = \angle X = \angle E$
9 *a* yes *b* no 11 *a* T *b* F *c* F *d* T *e* T *f* T *g* F
12 14 cm 13*b* $\frac{AB}{DE} = \frac{BC}{EF} = \frac{CA}{FD}$ *d* $\frac{AX}{DY} = \frac{AB}{DE} = \frac{BC}{EF}$
14 $\frac{PS}{XW} = \frac{PQ}{XY} = \frac{QR}{YZ}$ 15 $c = 6, d = 12$ 16 $x = 3{\cdot}5, a = 6, b = 9$
17*a* 4:1 *b* 9:1 *c* 1:16 *d* 4:9 *e* 16:9 *f* 4:25
g $m^2:1$ *h* $m^2:n^2$
18*a* 8:1 *b* 27:1 *c* 27:8 *d* 8:27
20*a* 50 cm *b* 1:100 *c* 1:1000 *d* 0·05 litre
21 4·2 m 22 15 23 $x = 6{\cdot}4, y = 4{\cdot}8, p = 10, q = 7{\cdot}5$
24*a* $\frac{1}{2}$, (0, 2) *b* -2, (0, -5) *c* -3, (0, 4) *d* $\frac{1}{3}$, (0, -2)
e $-\frac{1}{4}$, (0, 2) *f* $-\frac{2}{3}$, (0, $\frac{5}{3}$)
25*a* $\frac{1}{7}$ *b* $-\frac{5}{7}$ *c* $\frac{4}{3}$
d cannot be calculated *e* 0 *f* 1
26*a* $y = \frac{1}{2}x+3$ *b* $y = -4x+1$ *c* $y = -\frac{1}{3}x-4$
d $y = 4$ *e* $y = 3x-1$ *f* $y = kx+p$
27 3:5 28*a* equiangular *c* 5·04 m *d* 2·25 m

Page 136 Revision Exercise 2

1 a false *b* true *c* true *d* false

2 a $\overrightarrow{AD}$ *b* $\overrightarrow{AG}$ *c* $\overrightarrow{GD}$

3 a $\overrightarrow{PT}$ *b* $\overrightarrow{SR}$ *c* $\overrightarrow{PT}$

4 a $\begin{pmatrix}6\\-1\end{pmatrix}$ *b* $\begin{pmatrix}12\\-2\end{pmatrix}$ *c* $\begin{pmatrix}2\\5\end{pmatrix}$ *d* $\begin{pmatrix}8\\-4\end{pmatrix}$ *e* $\begin{pmatrix}-6\\1\end{pmatrix}$ *f* $\begin{pmatrix}4\\-2\end{pmatrix}$

5 a $\begin{pmatrix}2\\-4\end{pmatrix}$ *b* $\begin{pmatrix}-6\\1\end{pmatrix}$ *c* $\begin{pmatrix}4\\3\end{pmatrix}$ *d* $\begin{pmatrix}-2\\4\end{pmatrix}$ *e* $\begin{pmatrix}-4\\-3\end{pmatrix}$ *f* $\begin{pmatrix}-2\\4\end{pmatrix}$

6 a $\overrightarrow{GD}$ *b* $\overrightarrow{RF}$ *c* $\overrightarrow{NE}$ *d* $\overrightarrow{HL}$

7 a $2\overrightarrow{AB}$ *b* $3\overrightarrow{AB}$ *c* $3\overrightarrow{AB}$ *d* $2\overrightarrow{AB}$

8 a $\begin{pmatrix}4\\3\end{pmatrix}$ *b* $\begin{pmatrix}8\\6\end{pmatrix}$ *c* $\begin{pmatrix}12\\9\end{pmatrix}$ *d* $\begin{pmatrix}-16\\-12\end{pmatrix}$ *e* $\begin{pmatrix}4\\0\end{pmatrix}$ *f* $\begin{pmatrix}0\\6\end{pmatrix}$

9

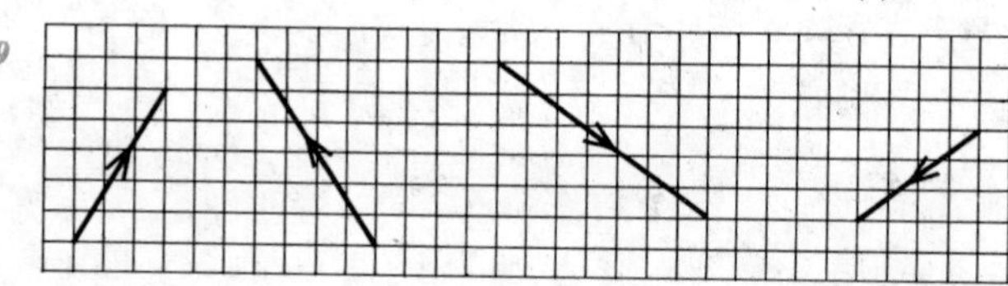

12a $\begin{pmatrix}-2\\-3\end{pmatrix}$ *b* $\begin{pmatrix}4\\-1\end{pmatrix}$ *c* $\begin{pmatrix}0\\0\end{pmatrix}$ *d* $\begin{pmatrix}-h\\k\end{pmatrix}$ *e* $\begin{pmatrix}-p+q\\-p-q\end{pmatrix}$

13a $\begin{pmatrix}-1\\3\end{pmatrix}$ *b* $\begin{pmatrix}3\\2\end{pmatrix}$ *c* $\begin{pmatrix}5\\-7\end{pmatrix}$

14a $\begin{pmatrix}1\\5\end{pmatrix}$ *b* $\begin{pmatrix}-7\\2\end{pmatrix}$ *c* $\begin{pmatrix}3\\3\end{pmatrix}$ *d* $\begin{pmatrix}p\\-q\end{pmatrix}+\begin{pmatrix}-p\\q\end{pmatrix}$

15a (3, 6) *b* (5, 10) *c* (−2, 3) *d* (4, −3); (−5, −3)

16a $\begin{pmatrix}8\\6\end{pmatrix}, \begin{pmatrix}2\\-2\end{pmatrix}, \begin{pmatrix}0\\10\end{pmatrix}$ *b* $\begin{pmatrix}5\\2\end{pmatrix}, \begin{pmatrix}1\\4\end{pmatrix}, \begin{pmatrix}4\\8\end{pmatrix}$, (5, 2), (1, 4), (4, 8)

17 5, $\sqrt{26}$, $\sqrt{29}$

Page 139 Revision Exercise 2B

1 a $\begin{pmatrix}0\\1\end{pmatrix}$ *b* $\begin{pmatrix}8\\-10\end{pmatrix}$ *c* $\boldsymbol{u}+7\boldsymbol{v}$ *d* $\boldsymbol{u}-3\boldsymbol{v}$

2 a $\begin{pmatrix}1\\-1\end{pmatrix}$ *b* $\begin{pmatrix}4\\-5\end{pmatrix}$ *c* $\begin{pmatrix}13\\-22\end{pmatrix}$

3 a $\begin{pmatrix}1\\10\end{pmatrix}$ *b* $\begin{pmatrix}1\\-1\end{pmatrix}$ *c* $\boldsymbol{x}=\boldsymbol{a}$ *d* $\boldsymbol{x}=3\boldsymbol{a}+2\boldsymbol{b}$

4 $|\boldsymbol{u}| = \sqrt{50} = |\boldsymbol{v}|$, but $\boldsymbol{u}$ and $\boldsymbol{v}$ have different directions

5 a $x = 2, y = -3$ *b* $x = 3, y = -2$ *c* $x = 0, y = 8$

6 $\begin{pmatrix}2\\1\end{pmatrix}, \begin{pmatrix}8\\-3\end{pmatrix}$ *a* (5, −1) *b* $(3\frac{1}{2}, 0)$ *c* $(6\frac{1}{2}, -2)$

7 a (1) $\overrightarrow{OA'} = \frac{3}{2}\overrightarrow{OA}$ *(2)* $\overrightarrow{OB'} = \frac{3}{2}\overrightarrow{OB}$ *(3)* $\overrightarrow{A'B'} = \frac{3}{2}\overrightarrow{AB}$
b 6 cm, 2 cm, $3\frac{1}{3}$ cm

8 a F *b* T *c* F *d* F *e* T *f* T

9 b $\frac{1}{2}(\boldsymbol{c}-\boldsymbol{a})$ *c* $\overrightarrow{PQ} = \overrightarrow{SR} = \frac{1}{2}\overrightarrow{AC}$, $\overrightarrow{PQ}\|\overrightarrow{SR}\|\overrightarrow{AC}$ *d* a parallelogram
e none *f* no

10a (4, 5) *b* $\begin{pmatrix}4\\2\end{pmatrix}$ *c* $\begin{pmatrix}0\\3\end{pmatrix}$ *d* $\begin{pmatrix}4\\3\end{pmatrix}$, (4, 3) *e* $\begin{pmatrix}3\\2\end{pmatrix}$

f A, P, C are collinear, and AP = $\frac{1}{3}$AC

Arithmetic—Answers to Chapter 1

Page 146 Exercise 1

1 a 8, 16, 32, 64 *b* $\log_2 4 = 2$, $\log_2 8 = 3$, $\log_2 16 = 4$, $\log_2 64 = 6$

2 a 0 *b* 2·6 *c* 3 *d* 3·3 *e* 3·9

3 a 4 *b* 16 *c* 2·8 *d* 11·2 *e* 6·6

4 $\log_2 1 = 0$, $\log_2 6 = 2{\cdot}6$, $\log_2 8 = 3$, $\log_2 10 = 3{\cdot}3$, $\log_2 15 = 3{\cdot}9$;
$\log_2 4 = 2$, $\log_2 16 = 4$, $\log_2 2{\cdot}8 = 1{\cdot}5$, $\log_2 11{\cdot}2 = 3{\cdot}5$, $\log_2 6{\cdot}6 = 2{\cdot}7$

5 a 3·6 *b* 2·3 *c* 1·6 *d* 3·8

6 a 8 *b* 5·8 *c* 1·7 *d* 3

Page 148 Exercise 2

1 a 0·3 *b* 0·8 *c* 0·9 *d* 1·1

2 a 1·6 *b* 2·5 *c* 4·0 *d* 6·4

3 a 0·491 *b* 0·740 *c* 0·903 *d* 0·914 *e* 0·915
f 0·820 *g* 0·874 *h* 0·318

4 a 1·78 *b* 2·40 *c* 2·56 *d* 6·05 *e* 1·03
f 2·51 *g* 2·82 *h* 2·86

5 a 0·884, 0·881, 0·845, 0·009, 0·522 *b* 2·51, 2·88, 2·94, 7·96, 1·01

Page 150 Exercise 3

1 6·14 *2* 7·13 *3* 8·77 *4* 2·07 *5* 1·45

6 1·00 *7* 1·44 *8* 4·48 *9* 6·40 *10* 5·48

11 9·62 *12* 8·17 mm² *13* 6·15 cm³ *14* 9·75 m²

Page 152 Exercise 4

1 a 2 *b* 1 *c* 3 *d* 0 *e* 6
f 1 *g* 4 *h* 2 *i* 0 *j* 7

2 a 2·393, 3·556, 0·176, 1·624, 3·000 *b* 23·4, 6·31, 115, 13 300, 100 000

3 90·6 *4* 834 *5* 66 500 *6* 158 *7* 1·07

8 114 *9* 267 *10* 1·79 *11* 47·5 *12* 9·62

13 11·9% *14* 60 400 m² *15* 37·6 m *16* 43·0 cm

17a $9{\cdot}31 \times 10^7$ *b* $3{\cdot}10 \times 10^4$

Page 154 Exercise 5

1 45·9	*2* 26·2	*3* 45 200 000	*4* 1·35	*5* 2·33	
6 3·80	*7* 9·73	*8* 10·1	*9 a* 2·81	*b* 4·56	*c* 18·6
10a 252 m^2	*b* 93·3 cm^2	*11* 31·6 m	*12* 3·09 cm	*13* 12·0	
14 1·58	*15* 2·14				

Page 155 Exercise 6

1 a $\bar{2}$ *b* $\bar{3}$ *c* $\bar{1}$ *d* $\bar{4}$ *e* $\bar{1}$

2 a $\bar{1}$·369, $\bar{2}$·230, $\bar{3}$·903, 1·561, 0·768   *b* 0·221, 0·005 09, 0·0104, 104, 47

3 a $\bar{1}$·500   *b* 2·500   *c* 5·500   *d* $\bar{3}$·500

4 a 2·369   *b* $\bar{1}$·369   *c* $\bar{2}$·756   *d* $\bar{1}$·004   *e* $\bar{4}$·845

Page 156 Exercise 7

1 2·6	*2* 5·2	*3* $\bar{4}$·2	*4* $\bar{1}$·3	*5* $\bar{3}$·3
6 $\bar{2}$·2	*7* 4·2	*8* $\bar{3}$·8	*9* $\bar{5}$·8	*10* $\bar{2}$·9
11 1·8	*12* 3·8	*13* 6·5	*14* $\bar{2}$·5	*15* 1·6
16 0·5	*17* $\bar{8}$·5	*18* $\bar{3}$·8	*19* 1·8	*20* $\bar{3}$·6
21 4·58	*22* 0·705	*23* 20·5	*24* 0·0638	*25* 228
26 423	*27* 0·163	*28* 0·0395	*29* 2·00	*30* 0·253
31 4600	*32* 0·000 048 2	*33* 0·149	*34* 32·9	*35* 6·04
36 38·3	*37* 0·005 94	*38* 0·144	*39* 392	*40* 1·37

Page 157 Exercise 8

1 $\bar{3}$·9	*2* $\bar{4}$·1	*3* $\bar{5}$·8	*4* $\bar{2}$·8	*5* $\bar{9}$·0
6 $\bar{1}$·14	*7* $\bar{1}$·4	*8* $\bar{1}$·9	*9* $\bar{1}$·65	*10* $\bar{2}$·3
11 $\bar{1}$·15	*12* $\bar{1}$·4	*13* $\bar{1}$·65	*14* $\bar{1}$·9	*15* $\bar{1}$·6
16 $\bar{1}$·9	*17* $\bar{2}$·9	*18* 0·759	*19* 0·776	*20* 0·000 399
21 1·05 m^2	*22* 0·914	*23* 0·289	*24* 0·685	*25* 0·565
26 0·444	*27* 0·209	*28* 0·755 m	*29* 2·65	*30* 0·575
31 3·10				

Page 158 Exercise 8B

1 1·39	*2* 94·8	*3* 0·849	*4* 0·603	*5* 0·520
6 83·0	*7* 0·402	*8* 0·714	*9* 0·000 003 59	*10* 0·938
11 0·796	*12* 0·021	*13* 0·002 49	*14* 0·719	*15* 0·650
16 0·443	*17* 0·0658	*18* 0·228	*19* 0·721 cm^2	*20 a* 1860 m
b 275 000 m^2	*21 a* 9620	*b* 0·518	*c* 0·685	*22* 0·959 m
23 2·94	*24* 137	*25* 55·7	*26 a* 12·5	*b* 1·61

Arithmetic—Answers to Chapter 2

Page 164 Exercise 1

1 a 96 cm^2 *b* 12·5 m^2 *c* 30 000 m^2 *d* 30 m
2 a 192 mm^2 *b* 210 m^2 *c* 67·5 cm^2 *d* 3 m
3 a 616 m^2, 88 m *b* 38·5 cm^2, 22 cm *c* 78·5 mm^2, 31·4 mm *d* 7 m, 154 m^2 *4 a* 60 m^3 *b* 4 cm^3 *c* 100 cm^3 *d* 1·5 m
5 128 cm *6* 21 m^2 *7* 90 litres *8* 5·2 tonnes *9* 8·5 hectares
10a 25 m *b* 4·5 m^3 *c* 20·25 m^2

Page 165 Exercise 1B

1 a 17·5 m^2 *b* 13·7 cm^2 *c* 13·5 mm^2 *d* 113 cm^2 *e* 2·88 cm^2
2 1090 cm^2 *3* 20 cm^3 *4 a* 2 m *b* 10 m *c* 40 m
5 a 35·2 cm *b* 77·4 cm^2 *6 a* 21 cm *b* 1·8 m
7 625 tonnes *8* £5·28, £9·24

Page 168 Exercise 2

1 a 4·2 m^3 *b* 27·5 cm^3 *c* 2700 cm^3 *d* 2·7 m
2 108 cm^3 *3 a* 144 cm^3 *b* 100 cm^3 *c* 288 cm^3 *d* 180 m^3
4 a 770 cm^3 *b* 1230 cm^3 *c* 12·6 m^3 *d* 283 cm^3 *e* 0·5 m
5 385 cm^3 *6* 136 *7* 11 000 litres *8* $6{\cdot}28 \times 10^4$ m^3
9 a 12·7 cm *b* 11·3 cm *10* 24 cm *11* 35 kg *12* 1·05 cm

Page 170 Exercise 3

1 a 22 cm *b* 22 cm by 10 cm *2 a* 297 cm^2 *b* 258 cm^2
3 31·4 cm, 31·4 cm by 5 cm; 314 cm^2, 236 cm^2
4 a πr^2 *b* $2\pi rh$ *c* $2\pi rh + \pi r^2$ *d* $2\pi rh + 2\pi r^2$
5 a 440 m^2 *b* 126 cm^2 *c* 8·8 cm^2 or 880 mm^2 *d* 1·26 m^2 or 12 600 cm^2 *6* 2990 cm^2 *7* 1·63 m^2 *8* 25·1 m^2

Page 172 Exercise 4B

2 a 50 m^3 *b* 96 cm^3 *c* 60·5 cm^3 *3* 875 m^3
4 22·5 cm *5 a* 360 cm^2 *b* 12 cm *c* 400 cm^3
6 a $21\frac{1}{3}$ m^3 *b* $10\frac{2}{3}$ m^3; OABFE, OBCGF, OADHE

Page 175 Exercise 5B

1 a 264 m^3 *b* 4620 cm^3 *c* 1260 m^3 *d* 80·5 m^3
2 94·3 cm^3 *3* 452 m^3 *4a* 550 cm^2 *b* 154 cm^2 *c* 24 cm *d* 1230 cm^3
5 a 283 cm^2 *b* 314 cm^3 *6* 5·3 cm *7* 55·2 cm *8* 1·75 m, 9·73 m^3

Page 176 Exercise 6B

1 a 154 cm^2 *b* 1260 m^2 *2a* 4·19 m^3 *b* 1440 mm^3
3 1386 cm^2, 4851 cm^3 *4* 113 m^3 *5* 1·34 cm *6* £2310
7 396 m^3 *8* 16·4 g *9* 434 cm^2 *10* 7 kg
11a 1:4 *b* 4:9 *c* 1:4:9 *d* 1:8 *e* 8:27 *f* 1:8:27
12 6·7 cm

Page 177 Exercise 7

1 220 mm, 3850 mm^2 *2* 26·4 cm, 55·4 cm^2 *3* 70 m
4 3·5 cm, 22 cm *5* 468 *6* 59·9 m^2 *7* 14·3 m^2 *8* 285
9 13 cm *10* 20 cm *11* 15·4 cm^2 *12a* 80 m^2 *b* 1200 m^3

Page 178 Exercise 7B

1 25·1 cm, 50·2 cm^2 *2* 5·2 cm, 15·6 cm^2 *3* 43·5 m^2
4 29 *5* 24 cm *6* 28·2 cm *7* 15 square units
8 a 240 cubic units *b* 80 cubic units *9* 1·02 kg
10a 434 cm^2 *b* 520 cm^3 *11* 1600 m^3 *12b* 2:3

Arithmetic—Answers to Chapter 3

Page 182 Exercise 1

1 a All 20s; tens digit is 2 *b* All 70s; tens digit is 7 *c* units digit is 5
2 sixth row and sixth column; tenth row and tenth column
3 a multiples of 11; digits are equal *b* multiples of 9; sum of digits is 9
4 in the first, third, fifth, seventh and ninth columns
5 c multiples of 21

Page 183 Exercise 2

1 b 1, 2, 3, 4, 5, 6; 1, 3, 6, 10, 15, 21 *3 a* 9, 11 *b* 32, 64 *c* 36, 49
d 1, 5 *e* 20, 25 *f* 4×7, 5×8 *4 a* 11 *b* 20 *c* 71 *d* 20
5 a 8 *b* 3 *c* 1 *d* 63
7 a 2, 4, 8 *b* 16, 8, 4, 2 *c* 7; two preliminary games required
8 a 42, 91, 140; 49; 189 *b* yes; 119; 168 *c* yes
9 a $\frac{1}{2}, \frac{1}{4}, \frac{1}{8}$ *b* $\frac{1}{32}$; the eighth
10a 11, 17 *b* 12, 30 *c* 48, 384 *d* 0·0001, 0·000 000 1
e 25, 64 *f* 11, 19 *g* 5×6, 8×9 *h* $\frac{1}{16}, \frac{1}{128}$
11a They are the first four odd numbers. *c* 25, 36, 49
d 10 000, $497\,283^2$ *e* n^2

Page 188 *Exercise 3*

1 a 11 *b* 8 *c* 16 *d* 6 *e* 17 *f* 132

3 21, 34, 55, 89 *a* 4, 7, 11 *b* 5, 9, 14

Page 189 *Exercise 4*

1 a subtract 4 *b* multiply by 2 *c* add 2 *d* divide by 2
e subtract 1·1 *f* multiply by 10 *g* add $\frac{1}{2}$ *h* multiply by 2

2 a 1, 2, 3, 4, 5, ... *b* 4, 7, 10, 13, 16, ...
c 1, 3, 6, 10, 15, ...; triangular numbers

3 a 1, 3, 5, 7, 9, ... *b* 3, 7, 11, 15, 19, ... *c* 1, 5, 11, 19, 29, ...

4 ABEI, ACEI, ACFI; ACFJ; ACFJN, ACFIN, ACEIN, ABEIN
a 1, 1 *b* It is the sum of these numbers
c 1, 5, 10, 10, 5, 1; 1, 6, 15, 20, 15, 6, 1 *d* 1, 2, 4, 8, 16, 32, 64, 128; 2^{19}; 2^{999}
e 1, 1, 1, 1, 1, 1; 1, 2, 3, 4, 5, 6; 1, 3, 6, 10, 15, 21; 1, 4, 10, 20, 35, 56;
1, 5, 15, 35, 70, 126; 1, 6, 21, 56, 126, 252

5 a $\frac{1}{2}, \frac{1}{2}$ *b* $\frac{1}{4}, \frac{2}{4}, \frac{1}{4}$ *c* $\frac{1}{8}, \frac{3}{8}, \frac{3}{8}, \frac{1}{8}$ *d (1)* $\frac{1}{4}$ (2) $\frac{15}{16}$

6 a 1, 4 *b* 1, 4, 10, 20, 35 *c* yes
d To form the *n*th pyramid number add the first *n* triangular numbers.

7 a 1, 5 *b* 1, 5, 14, 30, 55 *c* no
d To form the *n*th pyramid number add the first *n* square numbers.

Page 192 *Exercise 5B*

1 a $n, n+1, n+4, n+10$ *b* $3n, 3n+1, 3n+5, 3n+9$
c (1) $5n+1$ (2) $4n-1$ (3) $6n-6$

2 a n *b* n^2 *c* $n(n+1)$ *d* n^3 *e* 3^n *f* $1+4n$
g $n(n+1)(n+2)$ *h* $\frac{n}{n+1}$

3 a 5, 8, 11, 14 *b* 10, 20, 40, 80 *c* 189, 183, 177, 171
d 2, 6, 12, 20 *e* 3, 5, 9, 17 *f* $1, \frac{3}{2}, \frac{5}{3}, \frac{7}{4}$
g 0, 5, 14, 27 *h* 0, 1, 3, 6

4 10, 625, 81, 10 100, 23, 2184 *5* 2 m, 6 m, 12 m, 20 m, 30 m; $n(n+1)$; 650 m

6 a 3, 12, 48, 192, 768 *b* $3\times4^{n-1}$ *7 a* 3, 12, 48, 192, 768 *b* $3\times4^{n-1}$

Page 194 *Exercise 6B*

1 a $5n$ *b* 2^n *c* $n+2$ *d* $n-1$ *e* $3n-1$ *f* $\frac{1}{n}$

2 a 6, 15 *b* 1, 19 *c* 1, 1000 *d* 4, 121 *e* 0, 90 *f* 90, 0

3 290 *4* yes *5* 100; $(1+2+3+\ldots+n)^2$

6 $1; 1+2; 1+2+3; 1+2+3+4, \ldots; 1+2+3+\ldots+49; 1+2+3+\ldots+(n-1)$;
$\frac{1}{2}n(n-1)$; $\frac{1}{2}n(n+1)$

7 3, 9, 18, 30, 45, 63; 4, 12, 24, 40, 60, 84; yes, 84

8 $\frac{1}{2}$ unit; $\frac{1}{4}$ unit *a* $1\frac{7}{8}$ *b* $1\frac{31}{32}$ *c* $1\frac{127}{128}$; not possible

Arithmetic—Answers to Revision Exercises

Page 198 Revision Exercise 1A

1 373 *2* 5930 *3* 1680 *4* 21·7 *5* 56·6
6 1·26 *7* 4·07 *8* 259 *9* 8·38 *10* 7·16
11 0·697 *12* 0·202 *13* 49·0 *14* 0·000 558 *15* 0·0692
16 0·600 *17* 0·209 *18* 0·452 *19* 4·83 million *20* 0·017%

Page 198 Revision Exercise 1B

1 2·52 *2* 17·7 *3* 0·182 *4* 41·1 *5* 0·337
6 0·832 *7* 1·35 *8* 0·004 18 *9* 0·0577 *10* 63·8
11 5·85 *12* 0·450 *13* $2{\cdot}12 \times 10^6$ *14* $1{\cdot}43 \times 10^{-4}$
15a 434 *b* 776 *c* 1610 *16a* 2·21 *b* 0·971
c 2·59 *d* 0·220 *17a* (*1*) 420 (*2*) 1·63
b (*1*) 1·34 (*2*) 0·891 *18a* 0·383 *b* $1{\cdot}71 \times 10^{-3}$ *c* $2{\cdot}70 \times 10^7$

Page 199 Revision Exercise 2A

1 52·8 cm; 222cm² *2* 10 *3 a* 5·5 cm *b* 9·6 cm²
4 273 *5 a* 150 cm³ *b* 75 cm³ *6 a* 42·8 cm² *b* 1710 cm³
c 149 cm³ *7 a* 22·5 m² *b* 9000 litres *8 a* 3 m *b* 1080
9 a 176 m³ *b* 201 m² *10* 157 hours *11* £140·45 *12* £140·40
13a 224 m² *b* 448 m³ *14a* 50 m² *b* 750 m³

Page 201 Revision Exercise 2B

1 7 *2 a* 14·7 cm *b* 102·7 cm² *3 a* 23·5 m² *b* 10 500 litres
4 a 8 or 6 *b* 29 m *5* 24 hours *6 a* 896 cm² *b* 1568 cm³
7 a 660 m² *b* 264 cm² *c* 5544 mm²
8 a 1190 mm³ *b* 171 m³ *c* 210 cm³
9 98·1 tonnes *10* 2270 cm³ *11* 1680 km/h *12* 6120

Page 202 Revision Exercise 3A

1 a 15, 18, 21; add 3 *b* 17, 19, 23; the next prime number
c 33, 28, 23; subtract 5 *d* 9, 3, 1; divide by 3
e 13, 21, 34; add previous two terms
f 37, 50, 65; add 2 more than the previous difference
2 a $\frac{1}{11}$ *b* 9 × 10 *c* 99 *d* 100
3 a 82 *b* 18 *c* $\frac{3}{5}$ *d* 24
4 a 3, 5, 7, 9 *b* 1, 2, 3, 5 *c* 5, 10, 20, 40
5 23, 34; 47 *6* 27, 9, 3 *7* £256, £128, £64, £32. 7 years
8 a 5 *b* yes *c* 5·5 *d* It is their average. *e* 4950, 10 440

Page 204 Revision Exercise 3B

1 a 0, 2, 6, 12, 20 *b* $\frac{1}{2}, \frac{2}{3}, \frac{3}{4}, \frac{4}{5}, \frac{5}{6}$ *c* 120, 60, 40, 30, 24
d 1, 5, 14, 30, 55 *e* 1, 9, 36, 100, 225

2 (i) $3n+1$ *(ii)* $8n$ *(iii)* $6n+1$

3 a $2n+5$ *b* $\frac{1}{n+1}$ *c* 2^{n-1} *d* $\frac{1}{3^n}$

4 $\frac{1}{2}, \frac{1}{4}, \frac{1}{8}, \ldots; \frac{1}{2^n}$ *5* $\frac{1}{4}, \frac{1}{16}, \frac{1}{64}; \frac{1}{4^n}$

6 a $(21+23+25+27+29)$, $6^3 = 216$, $10^3 = 1000$, 99^3
b (1) $(1+2+3+4+5)+(1+2+3+4)$, $6^2 = 36$, $10^2 = 100$, 99^2
(2) $(1+2+3+4+5)^2-(1+2+3+4)^2$, $6^3 = 216$, $10^3 = 1000$, 99^3

7 a 11 *b* 13 *8* 17

Trigonometry—Answers to Chapter 1

Page 209 Exercise 1

1 a D *b* J *c* G *d* A

2 a increases to 30° *b* increases to 90° *c* increases to 150°
d increases to 180° *e* increases to 270° *f* increases to 360°

4 90°; zero; OA

5 a OM_2P_2, OM_3P_3 *b* $\frac{OM_2}{OP_2}, \frac{OM_3}{OP_3}$
c Draw a line from any point on OA, perpendicular to OX.

6 a OM_2P_2, OM_3P_3 *b* $\frac{OM_2}{OP_2}, \frac{OM_3}{OP_3}$
c Draw a line from any point on OA, perpendicular to OX.

7 $\frac{M_2P_2}{OP_2}, \frac{M_3P_3}{OP_3}$ *8* $\frac{M_2P_2}{OM_2}, \frac{M_3P_3}{OM_3}$

Page 213 Exercise 2

1 $\frac{4}{5}, \frac{3}{5}, \frac{3}{4}; -\frac{3}{5}, \frac{4}{5}, -\frac{4}{3}; -\frac{12}{13}, -\frac{5}{13}, \frac{5}{12}; \frac{4}{5}, -\frac{3}{5}, -\frac{3}{4}$

2 $\frac{7}{25}, \frac{24}{25}, \frac{24}{7}$ *3* (3, 4), 5; $\frac{3}{5}, \frac{4}{5}, \frac{4}{3}$ *4* $(-5, 1)$ $\sqrt{26}$; $-\frac{5}{\sqrt{26}}, \frac{1}{\sqrt{26}}, -\frac{1}{5}$

5 $(-2, -4)$, $\sqrt{20}$; $-\frac{2}{\sqrt{20}}, -\frac{4}{\sqrt{20}}, 2$ *6* $(6, -4)$, $\sqrt{52}$; $\frac{6}{\sqrt{52}}, -\frac{4}{\sqrt{52}}, -\frac{2}{3}$

7 (5, 0), 5; 1, 0, 0 *8 a* $\frac{1}{\sqrt{2}}, \frac{1}{\sqrt{2}}, 1; -\frac{1}{\sqrt{2}}, \frac{1}{\sqrt{2}}, -1; -\frac{1}{\sqrt{2}}, -\frac{1}{\sqrt{2}}, 1;$
$\frac{1}{\sqrt{2}}, -\frac{1}{\sqrt{2}}, -1$
b first—all; second—sine; third—tangent; fourth—cosine

Page 216 Exercise 3

1 a 0·454 *b* 0·688 *c* 0·500 *d* 0·990 *e* 0·548 *f* 0·033
g 1·000 *h* 2·438 *i* 19·08 *j* 0·500 *k* 0·012 *l* 0·433

2 a 25·9 *b* 89·5 *c* 26·0 *d* 30·0 *e* 60·0 *f* 60·0
g 9·3 *h* 89·8 *i* 85·6

3 a T *b* T *c* F *d* T *e* T *f* F
g T *h* T

4 a $\sqrt{2}$ *b* $\frac{1}{\sqrt{2}}, \frac{1}{\sqrt{2}}, 1$ *5 a* $(1, \sqrt{3}); \frac{1}{2}, \frac{\sqrt{3}}{2}, \sqrt{3}$ *b* $(\sqrt{3}, 1); \frac{\sqrt{3}}{2}, \frac{1}{2}, \frac{1}{\sqrt{3}}$

6 see summary at end of chapter

Page 218 Exercise 4

1 (i) $\frac{4}{5}, \frac{3}{5}, \frac{3}{4}$ (ii) $\frac{12}{13}, \frac{5}{13}, \frac{5}{12}$ (iii) $\frac{8}{17}, \frac{15}{17}, \frac{15}{8}$

2 (i) $\frac{4}{5}, \frac{3}{5}, \frac{3}{4}$ (ii) $\frac{24}{25}, \frac{7}{25}, \frac{7}{24}$ (iii) $\frac{3}{5}, \frac{4}{5}, \frac{4}{3}$ (iv) $\frac{3}{5}, \frac{4}{5}, \frac{4}{3}$

3 (i) $\frac{b}{c}, \frac{a}{c}, \frac{a}{b}$ (ii) $\frac{e}{f}, \frac{d}{f}, \frac{d}{e}$ (iii) $\frac{k}{h}, \frac{g}{h}, \frac{g}{k}$ (iv) $\frac{p}{r}, \frac{q}{r}, \frac{q}{p}$

4 (ii) 2·18 (iii) 9·19 (iv) 22·0

5 (i) 7·95 (ii) 10·2 (iii) 4·04 (iv) 31·2

6 (i) 26·6 (ii) 64·2 (iii) 32·0 (iv) 70·6

Page 221 Exercise 5

1 190 m *2* 82·3° *3 a* 11·6 m *b* 3·1 m

4 324 m *5* 86 m *6 a* 141·4 km *b* 141·4 km

7 a 64·1° *b (1)* 4·36 m (2) 4·37 m

8 a 8·7 cm *b* 5 cm *c* 0 cm *9* 35°, 55°, 35°, 55°

10 ∠P = 64°, ∠Q = 116° *11* 67·4°, 100·4°, 96·1°, 96·1°; 10·8 cm, 7·8 cm

12 321·3°, 32 km *13* 1·6° *14* 71m

Page 222 Exercise 5B

1 (i) 7·2 m (ii) 3·4 m (iii) 9·5 m *2* 31°, 118°

3 2·2 cm, 2·6 cm *4* 7·4 cm *5* 37 cm

6 1·22 km, 263 m *7* 67 km, 130 km

8 29 m, 13 m *9* 143 m *10* 25·1°, 18·4°

Page 224 Exercise 6

1 440°, −280° *2* 480°, 840°, −240°, −600°, etc.

3 345°, 705°, 1065°, −375°, etc. *4* −160°, 560°, 920°, −520°, etc.

5 the set of integers, *Z* *6 a* one *b* an infinite number

Page 225 *Exercise 7*

1 a 5 *b*

4	3	5	$\frac{4}{5}$	$\frac{3}{5}$	$\frac{3}{4}$
−4	3	5	$-\frac{4}{5}$	$\frac{3}{5}$	$-\frac{3}{4}$
−4	−3	5	$-\frac{4}{5}$	$-\frac{3}{5}$	$\frac{3}{4}$
4	−3	5	$\frac{4}{5}$	$-\frac{3}{5}$	$-\frac{3}{4}$

c

sine positive	all positive
tangent positive	cosine positive

2 a 13 *b*

5	12	13	$\frac{5}{13}$	$\frac{12}{13}$	$\frac{12}{5}$
−5	12	13	$-\frac{5}{13}$	$\frac{12}{13}$	$-\frac{12}{5}$
−5	−12	13	$-\frac{5}{13}$	$-\frac{12}{13}$	$\frac{12}{5}$
5	−12	13	$\frac{5}{13}$	$-\frac{12}{13}$	$-\frac{12}{5}$

c as for question ***1***

3 a + *b* − *c* − *d* − *e* − *f* + *g* − *h* +

4 a 80° *b* 50° *c* 27°

Page 226 *Exercise 8*

1 a 0·643 *b* 0·643 *c* −0·643 *d* −0·643 *e* −0·866
f −0·940 *g* 0·5 *h* 0·766 *i* −5·671 *j* 2·356
k 0 *l* 0·364 *m* −0·174 *n* 0·940 *o* −0·577
p −0·809

2 a 50, 130 *b* 55·6, 304·4 *c* 78·5, 258·5 *d* 155, 205 *e* 225, 315
f 116, 296 *g* 24·5, 155·5 *h* 10·9, 190·9 *i* 100, 260

3 a $\frac{1}{\sqrt{2}}, \frac{1}{\sqrt{2}}, 1$ *b* $-\frac{1}{\sqrt{2}}, -\frac{1}{\sqrt{2}}, \frac{1}{\sqrt{2}}$ *c* $\frac{1}{\sqrt{2}}, -\frac{1}{\sqrt{2}}, -\frac{1}{\sqrt{2}}$ *d* −1, 1, −1

4 a $\frac{\sqrt{3}}{2}, \frac{1}{2}, \frac{1}{\sqrt{3}}$ *b* $-\frac{\sqrt{3}}{2}, -\frac{\sqrt{3}}{2}, \frac{\sqrt{3}}{2}$ *c* $\frac{1}{2}, -\frac{1}{2}, -\frac{1}{2}$ *d* $-\frac{1}{\sqrt{3}}, \frac{1}{\sqrt{3}}, -\frac{1}{\sqrt{3}}$

5 a $\frac{1}{2}, \frac{\sqrt{3}}{2}, \sqrt{3}$ *b* $-\frac{1}{2}, -\frac{1}{2}, \frac{1}{2}$ *c* $\frac{\sqrt{3}}{2}, -\frac{\sqrt{3}}{2}, -\frac{\sqrt{3}}{2}$ *d* $-\sqrt{3}, \sqrt{3}, -\sqrt{3}$

6 a 2·16 m *b* −1·25 m *c* 1·25 m *d* 2·5 m

7 2 m, 2·5 m, 4 m, 6 m, 8 m, 9·5 m, 10 m, 9·5 m, 8 m, 6 m, 4 m, 2·5 m

Page 228 *Exercise 9*

1 e see summary *f* 1 and −1 *2 e* see summary *f* 1 and −1

3 d see summary *4* 360° *5* 360°

6 a no *b* yes, cosine *c* yes, sine

7 a (*1*) 0 (*2*) 1 (*3*) 0 *b* (*1*) −1 (*2*) 0 (*3*) 1

8 a (*1*) $90+k.360$ (*2*) $0+k.180$ (*3*) $270+k.360$
b (*1*) $90+k.180$ (*2*) $0+k.360$ (*3*) $180+k.360$

9 fits after a translation of 270.

Trigonometry—Answers to Revision Exercises

Page 235 *Revision Exercise 1A*

1 0·47, 0·88, 1·88 *2* −0·71, −0·71, 1·00

3 a T *b* F *c* F *d* T *e* T *f* T

4 (i) 5·08 (ii) 9·56 (iii) 3·26

5 a 60° *b* 30° *c* 60° *d* 1·73 m *e* 2·73 m^2
6 a 6·66 cm *b* 37·8° *7 a* 18·0 *b* 36·0 *c* 92°
8 a 4·24 *b* 2·65 *c* 40·9 cm^2 *9* 1·1°
10a + *b* − *c* + *d* + *e* − *f* − *g* + *h* −
11 a 0·151 *b* −1·483 *c* 0·545 *d* −0·431 *e* −0·355 *f* −0·454
12 a T *b* F *c* T *13* 0, 1, −1 *14* 0, 360, 720
15 360, 0; $\begin{pmatrix}270\\0\end{pmatrix}$ $y = \sin x°$

Page 236 Revision Exercise 1B

1 0·750, 0·661 *2* $\frac{4}{5}, \frac{4}{3}$; yes *3* (i) 3·25 (ii) 3·48 (iii) 1·72
4 a 30° *b* 126 cm, 124 cm *5* 3630 m *6* 40 m *7* 55°
8 127 km east, 59 km south
9 a T *b* F *c* F *d* T *e* F *f* T
10 180° *12* same period, same maximum values, same minimum values
13 63·4°, 67·4°; (0, 1), 4°

Computer Studies—Answers to Chapter 1

Page 244 Exercise 1

1 (iii) A, or A1, etc. (v) D5 (vi) C (vii) T4 (viii) Z3 (x) A
(xi) P1 (xii) V

2 (i)
```
1 LET A = 12
2 LET B = 7
3 LET C = A+B
4 PRINT A, B, C
5 STOP
```
(ii)
```
1 INPUT A, B
2 LET C = A+B
3 PRINT C
4 STOP
  Data 5, 6
```
(iii)
```
1 INPUT P, Q
2 LET K = P-Q
3 PRINT K
4 STOP
```

3 (i)
```
1 LET A = 9
2 LET B = 8
3 LET C = A+B
4 PRINT C
5 STOP
```
(ii)
```
1 LET A = 33
2 LET B = 25
3 LET C = A-B
4 PRINT C
5 STOP
```
(iii)
```
1 LET P = 12
2 LET Q = 18
3 LET R = 29
4 LET Z = P+Q+R
5 PRINT Z
6 STOP
```
(iv)
```
1 INPUT X, Y
2 LET Z = X+Y
3 PRINT Z
4 STOP
```
(v)
```
1 INPUT D, E
2 LET F = D-E
3 PRINT F
4 STOP
```
(vi)
```
1 INPUT A, B, C, D, E
2 LET F = A+B+C
         +D+E
3 PRINT F
4 STOP
```

Page 247 Exercise 2

1 (i) P+Q (ii) S−T (iii) A*B (iv) 5*X*Y (v) A/B
(vi) A*A, or A↑2 (vii) 3*B*B (viii) P*Q/R (ix) 2*X+2*Y
(x) X↑2+Y↑2

2 (i)
```
10 REM MEAN
   OF 3 NUMBERS
20 INPUT X, Y, Z
30 LET A =
   (X+Y+Z)/3
40 PRINT A
50 STOP
   Data 9, 7, 2
```

(ii)
```
10 REM AREA
   OF RECTANGLE
20 INPUT L, B
30 LET A = L*B
40 PRINT A
50 STOP
```

(iii)
```
10 REM CIRCUM-
   FERENCE OF CIRCLE
20 INPUT R
30 LET C = 2*3·14*R
40 PRINT C
50 STOP
```

(iv)
```
10 REM AREA
   OF CIRCLE
20 INPUT R
30 LET A =
   3·14*R*R
40 PRINT A
50 STOP
   Data 5
```

(v)
```
10 REM VOL
   OF CUBOID
20 INPUT L, B, H
30 LET V =
   L*B*H
40 PRINT V
50 STOP
```

(vi)
```
10 REM VOL OF
   CYLINDER
20 INPUT R, H
30 LET V =
   3·14*R*R*H
40 PRINT V
50 STOP
```

(vii)
```
10 REM PERIM
   OF RECTANGLE
20 INPUT L, B
30 LET P =
   2*L+2*B
40 PRINT P
50 STOP
   Data 9·8, 7·5
```

3
```
10 REM HOUSE-
   HOLDERS
   RATES
20 INPUT V
30 LET R =
   V*95/100
40 PRINT R
50 STOP
```

4
```
10 REM SALE PRICES
20 INPUT C
30 LET S = C*78/100
40 PRINT S
50 STOP
```

Page 249 Exercise 3

1 (i) LET A = M+N (ii) LET C = 2*A+B (iii) LET J = A↑2+B*C
(iv) LET M2 = B*B (v) LET X = Y/Z (vi) LET A = B*C/D

2 (i) 11 (ii) 15 (iii) 2·75 (iv) $\frac{2}{3}$ (v) 1 (vi) 0 (vii) −5
(viii) 25 (ix) −7 (x) 1·5 (xi) 1 (xii) $\frac{5}{9}$

3 (i) X/(2*A) (ii) (A+B)/C (iii) (A+B↑2)/(C*D+F↑3)

4 (i) A↑2−2*A*B+B↑2 (ii) (X+Y)*(X−Y) (iii) (A↑2+B↑2)/(X*Y)

5 (i) 70 (ii) 4 12 2 (iii) 9

6 (i)
```
10 REM EVALUATE P+Q↑3
20 INPUT P, Q
30 LET R = P+Q↑3
40 PRINT R
50 STOP
```

(ii)
```
10 REM EVALUATE X↑2+Y↑2
   AND (X+Y)↑2
20 INPUT X, Y
30 LET P = X↑2+Y↑2
40 LET Q = (X+Y)↑2
50 PRINT P, Q
60 STOP
```

(iii)
```
10 REM QUESTION 6 OF EX 3
20 INPUT A, B, C, D
30 LET S = A+B+C+D
40 LET P = A*B*C*D
50 LET S1 = S*S
60 LET P1 = P*P
70 LET M = (A+B+C+D)/4
80 PRINT S, P, S1, P1, M
90 STOP
   Data 1, 2, 3, 4
   10 24 100 576 2·5
```

Page 255 Exercise 4

1

```
10 REM NUMBERS AND
   SQUARES
20 LET C = 0
30 LET C = C+1
40 IF C > 50 THEN 90
50 INPUT A
60 LET S = A*A
70 PRINT A, S
80 GOTO 30
90 STOP
```

2

```
10 REM NUMBERS AND CUBES
20 LET C = 0
30 LET C = C+1
40 IF C > 1000 THEN 90
50 INPUT A
60 LET U = A↑3
70 PRINT A, U
80 GOTO 30
90 STOP
```

3

```
10 REM NUMBERS AND
   HALVES
20 INPUT A
30 IF A < 0 THEN 70
40 LET B = A/2
50 PRINT A, B
60 GOTO 20
70 STOP
```

4

```
 10 REM AGENTS COMMISSION
 20 INPUT S
 30 IF S < 1000 THEN 80
 40 LET A = 1000*3/100
 50 LET B = (S-1000)*6/100
 60 LET C = A+B
 70 GOTO 90
 80 LET C = S*3/100
 90 PRINT C
100 STOP
```

5

```
10 REM SIMPLE INTEREST
20 LET C = 0
30 INPUT P, T, R
40 LET I = P*T*R/100
50 PRINT I
60 LET C = C+1
70 IF C < 10 THEN 30
80 STOP
```

6

```
10 REM SALE PRICES
20 INPUT P
30 IF P < 10 THEN 70
40 LET S = P*90/100
50 PRINT S
60 GOTO 90
70 LET S = P*95/100
80 PRINT S
90 STOP
```

7

```
10 REM INCOME TAX
20 INPUT I
30 IF I < 1000 THEN 70
40 LET T = (I-1000)*55/200
50 PRINT T
60 GOTO 80
70 PRINT 'TAX = 0'
80 STOP
```